Formwork for
Concrete Structures

About the Authors

Robert L. Peurifoy (deceased) taught civil engineering at the University of Texas and Texas A&I College, and construction engineering at Texas A&M University and Oklahoma State University. He served as a highway engineer for the U.S. Bureau of Public Roads and was a contributing editor to *Roads and Streets Magazine*. In addition to authoring the McGraw-Hill publications *Construction Planning, Equipment, and Methods* and *Estimating Construction Costs*, 5th ed., coauthored with Garold D. Oberlender, Mr. Peurifoy wrote over 50 magazine articles dealing with construction. He was a long-time member of the American Society of Civil Engineers, which presents an award that bears his name.

Garold D. Oberlender, Ph.D, P.E. (Stillwater, Oklahoma), is Professor Emeritus of Civil Engineering at Oklahoma State University, where he served as coordinator of the Graduate Program in Construction Engineering and Project Management. He has more than 40 years of experience in teaching, research, and consulting engineering related to the design and construction of projects. He is author of the McGraw-Hill publications *Project Management for Engineering and Construction*, 2nd ed., and *Estimating Construction Costs*, 5th ed., coauthored with Robert L. Peurifoy. Dr. Oberlender is a registered professional engineer in several states, a member of the National Academy of Construction, a fellow in the American Society of Civil Engineers, and a fellow in the National Society of Professional Engineers.

Formwork for Concrete Structures

Robert L. Peurifoy
Late Consulting Engineer
Austin, Texas

Garold D. Oberlender
Professor Emeritus
Oklahoma State University

Fourth Edition

New York   Chicago   San Francisco
Lisbon   London   Madrid   Mexico City
Milan   New Delhi   San Juan
Seoul   Singapore   Sydney   Toronto

The McGraw·Hill Companies

Cataloging-in-Publication Data is on file with the Library of Congress

1 2 3 4 5 6 7 8 9 0   DOC/DOC   1 6 5 4 3 2 1 0

ISBN 978-0-07-163917-0
MHID 0-07-163917-9

Sponsoring Editor
Joy Bramble

Editing Supervisor
Stephen M. Smith

Production Supervisor
Richard C. Ruzycka

Acquisitions Coordinator
Alexis Richard

Project Manager
Ranjit Kaur, Glyph International

Copy Editor
Susan Giniger

Proofreader
Eina Malik

Art Director, Cover
Jeff Weeks

Composition
Glyph International

Printed and bound by RR Donnelley.

McGraw-Hill books are available at special quantity discounts to use as premiums and sales promotions, or for use in corporate training programs. To contact a representative, please e-mail us at bulksales@mcgraw-hill.com.

This book is printed on acid-free paper.

Contents

v

Preface

This book is written for architects, engineers, and constructors who are responsible for designing and/or building formwork and temporary structures during the construction process. It is also designed to serve either as a textbook for a course in timber and formwork design or as a reference for systematic self-study of the subject.

A new chapter on the design of wood members for formwork and temporary structures has been added to this edition. Numerous example problems have been added throughout the text to illustrate practical applications for calculating loads, stresses, and designing members. New summary tables have been added to assist the reader in understanding the concepts and techniques of designing formwork and temporary structures.

This fourth edition has been developed with the latest structural design recommendations by the National Design Specification (NDS 2005), published by the American Forest & Paper Association (AF&PA). In writing this edition, an effort has been made to conform to the intent of this reference document. The material presented is suggested as a guide only, and final responsibility lies with the designer of formwork and temporary structures.

Many patented systems and commercial accessories are available to increase the speed and safety of erecting formwork. Numerous figures and photographs are presented to introduce the reader to the available forming systems for walls, columns, beams, and slabs.

Garold D. Oberlender

Acknowledgments

The author would like to thank the many manufacturers for permission to use the contents of their publications and technical information, and the many suppliers of formwork materials and accessories for providing illustrative material that is contained in this book. Many individuals, agencies, and manufacturers have assisted the author in obtaining and presenting the information contained in this book. The author expresses his sincere thanks for this assistance.

The author would like to thank Carisa Ramming for her careful review, helpful comments, and advice in the development of this fourth edition, in particular the new chapter on design of wood members for formwork. The author also wishes to recognize the late Robert L. Peurifoy for his pioneering work as an author and teacher of construction education. Throughout the author's career, Mr. Peurifoy was an inspiration as a role model, mentor, and colleague.

Finally, the author greatly appreciates the patience and tolerance of his wife, Jana, and her understanding and support during the writing and editing phases of the fourth edition of this book.

Abbreviations
and Symbols

A	area
ACI	American Concrete Institute
b	width of a beam, in.
c	distance from neutral axis of beam to extreme fiber in bending, in.
cu ft	cubic feet
cu yd	cubic yard
d	depth of a beam, in.
Δ	deflection of a member, in.
E	modulus of elasticity, lb per sq in.
F	allowable unit stress, lb per sq in.
f	applied unit stress, lb per sq in.
fbm	feet board measure, of lumber
ft	feet
h	height of form, ft
I	moment of inertia of a beam about its neutral axis, in.^{4}
Ib/Q	rolling shear constant, in.^{2}
in.	inches
L	length of a beam or column, ft
l	length of a beam or column, in.
lb	pounds
M	bending moment of a beam, in.-lb
NDS	National Design Specification, for wood
P	total concentrated load, lb
p	unit pressure produced by concrete on forms, lb per sq ft

PCA	Portland Cement Association
psf	pressure or weight, lb per sq ft
psi	stress, lb per sq in.
R	rate of filling forms, ft per hour
S	section modulus, in.^{3}
S4S	lumber that is surfaced on all four sides
sq ft	square feet
sq in.	square inches
T	temperature, degrees Fahrenheit
V	total external shear force on a beam, lb
v	velocity, ft per sec
W	total load uniformly distributed along a beam, lb
w	uniformly distributed load, lb per lin ft
$\perp$	denotes a perpendicular direction
$//$	denotes a parallel direction

Formwork for Concrete Structures

CHAPTER 1

Introduction

Purpose of This Book

This book presents the principles and techniques for analysis and design of formwork for concrete structures. Because each structure is unique, the formwork must be designed and fabricated based on the specific requirements of each job. The level of effort required to produce a good formwork system is as important as the level of effort required to produce the right combination of steel and concrete for the structural system of the structure. Formwork for concrete structures has a significant impact on the cost, time, and quality of the completed project.

Formwork is important because it is a major cost of the concrete structure. Too often the designers of concrete structures devote considerable time in selecting the minimum amount of concrete and steel for a structure without devoting adequate attention to the impact of the formwork that must be constructed to form the concrete. For most structures, more time and cost are required to make, erect, and remove formwork than the time and cost to place the concrete or reinforcing steel. For some structures, the cost of formwork exceeds the cost of the concrete and steel combined.

This book presents the methods of analyses of various components of formwork, to assist the designer in developing a formwork system for his or her project. The purpose of formwork is to safely support the reinforced concrete until it has reached adequate strength. Thus, formwork is a temporary support for the permanent steel and concrete. The designer is responsible for producing a forming system that is safe, economical, and easily constructible at the jobsite. The overall quality of the completed project is highly dependent on the formwork.

Many articles and papers have been written related to the design, fabrication, erection, and failure of formwork. At the end of each chapter of this book, references of other publications are provided to assist the reader in better understanding the work that others have produced related to formwork.

1

Safety of Formwork

The failure of formwork is a major concern of all parties involved in a construction project; including the owner, the designer, and the contractor. Although the principles, concepts, and methods that are contained in this book provide the basics for the analysis and design of formwork, it is the responsibility of each designer of formwork to ensure that the forms are designed adequately. This requires a careful analysis of the job conditions that exist at each jobsite, a determination of the loads that will be applied to the formwork, and the selection and arrangement of suitable forming materials that have adequate strength to sustain the loads.

It is the responsibility of the workers at the jobsite to fabricate and erect the formwork in accordance with the design. A careful check of the design and inspection of the work during construction are necessary to ensure the safety and reliability of the formwork. Safety is everyone's responsibility, and all parties must work together as a team with safety as a major consideration.

Economy of Formwork

Economy should be considered when planning the formwork for a concrete structure. Economy involves many factors, including the cost of materials; the cost of labor in making, erecting, and removing the forms, and the cost of equipment required to handle the forms. Economy also includes the number of reuses of the form materials, the possible salvage value of the forms for use elsewhere, and the cost of finishing concrete surfaces after the forms are removed. A high initial cost for materials, such as steel forms, may be good economy because of the greater number of uses that can be obtained with steel.

An analysis of the proposed formwork for a given project usually will enable the job planner to determine, in advance of construction, what materials and methods will be the most economical.

Allowable Unit Stresses in Formwork Material

In order to attain the maximum possible economy in formwork, it is desirable to use the highest practical unit stresses in designing forms. It is necessary to know the behavior of the pressures and loads that act on forms in determining the allowable unit stresses.

When concrete is first placed, it exerts its maximum pressure or weight on the restraining or supporting forms. However, within a short time, sometimes less than 2 hours, the pressure on wall and column forms will reach a maximum value, and then it will decrease to zero. Thus, the forms are subjected to maximum stresses for relatively short periods of time.

Within a few hours after concrete is placed for girders, beams, and slabs, it begins to set and to bond with the reinforcing steel, thereby developing strength to support itself. Although the forms are usually left in place for several days, magnitudes of the unit stresses in the forms will gradually decrease as the concrete gains strength. Thus, the maximum unit stresses in the formwork are temporary and of shorter duration than the time the forms are left in place.

The allowable unit stresses specified for lumber are generally based on a full design load that is applied for a normal load duration of approximately 10 years. If the duration of the load is only a few hours or days, such as with formwork, the allowable unit stress may be adjusted to a higher value. For loads that are applied for a short duration, less than 7 days, the allowable unit stresses may be increased by 25%. The examples and tables contained in this book are based on using increased allowable unit stresses, assuming loads are applied for a short duration.

Care of Forms

Forms are made of materials that are subject to considerable damage through misuse and mishandling. Wood forms should be removed carefully, then cleaned, oiled, and stored under conditions that will prevent distortion and damage. At periodic intervals, all forms should be checked to determine whether renailing, strengthening, or replacing parts is necessary.

Patented Products

There are numerous patented products for concrete structures that have been produced by companies in the construction industry. Many of these products are contained in this book. However, it is not practical to include all of the products that are currently available. Inclusion of the products of some manufacturers and the exclusion of similar products of others should not be interpreted as implying that the products included in this book are superior to those not mentioned. The products described in this book are intended to illustrate only the types of products available for use in concrete formwork.

For most of the products that are included in this book, the manufacturers' specifications, properties, dimensions, and other useful information are given in tables.

Arrangement of This Book

There are 17 chapters in this book. The following paragraphs briefly describe each one.

Chapter 1, Introduction, provides an introduction to this book, including its purpose, the importance of safety, and general information related to allowable stresses for form materials and patented products that are available for forming concrete structures.

Chapter 2, Economy of Formwork, provides information related to the importance of economy in formwork. Because formwork is a major cost of concrete structures, planning and designing the formwork system is an integral part of the process of designing and constructing concrete structures. There are decisions that must be made during the design process that will have major impacts on the construction process and the cost of the structure.

Chapter 3, Pressure of Concrete on Formwork, presents information related to the pressure that concrete exerts on the formwork. When concrete is placed in the forms, it applies vertical loads due to its weight as well as horizontal loads because it is in a liquid state and has not gained sufficient strength to support itself. In addition to the loads on the formwork from concrete and reinforcing steel, the designer must consider the live loads that are applied to the forms due to workers and equipment that are used to place the concrete.

Chapter 4, Properties of Form Material, provides information related to the properties of form materials. The principal materials used for forms include wood, steel, plywood, fiberglass, plastics, aluminum, and other materials. The designer must know the physical properties and the behavior of the materials that are used in building forming systems for concrete structures. Accessories used to attach the components of form materials are also an important part of formwork. The accessories used to fasten the form materials include nails, screws, bolts, form ties, column clamps, and other parts too numerous to mention.

Chapter 5, Design of Wood Members for Formwork, presents the fundamental concepts and equations that are used to design formwork and temporary structures during construction. The design of formwork involves determining the pressures and loads from the concrete placement during construction, analysis of the loads to determine the distribution of the loads through the formwork system, and selecting the sizes of members to sustain the loads adequately. The formwork must be designed with sufficient strength to resist loads that are applied and to restrict the deflection of the forms within an allowable tolerance. Safety, economy, and quality must be major considerations in designing formwork.

Chapter 6, Shores and Scaffolding, provides information related to shores and scaffolding for formwork. Patented shores are often used to support formwork. If patented shores are used, it is important that placement and spacing of the shores be in accordance with the manufacturer's recommendations. In some situations, shores are fabricated by workers at the jobsite. If job-built shores are used, it is important that a qualified person be involved in ensuring the safety of the shoring system because failure of shores is a common cause of formwork failure. Similarly, scaffolding is important for the safety of workers and their efficiency.

Chapter 7, Failures of Formwork, addresses the important issue of the safety of formwork systems. Formwork failure is costly, in terms of both the physical losses at the jobsite and injuries to workers. Physical losses include the loss of materials that are destroyed in the failure and the time and expenses that must be incurred to clean up and reinstall the forms. Injuries and loss of life of workers create suffering of people and can lead to costly legal actions.

Chapter 8, Forms for Footings, provides information related to the design and construction of forms for footings and the fundamental equations that can be used in the design process. Information is also included for placing anchor bolts in concrete foundations.

Chapter 9, Forms for Walls, addresses the design of forms for concrete walls. Equations and tables are presented to facilitate the design of continuous walls and for walls with corbels. Due to the height of walls, the pressure at the bottom of the forms is significant. Therefore, the designer must carefully evaluate the loads that are applied to wall forms to ensure that the forms have sufficient strength to resist the applied load. Accessories for walls including snap ties, coil ties, and form clamps are also presented.

Chapter 10, Forms for Columns, addresses the design of forms for concrete columns. Included in this chapter are square, rectangular, round, and L-shaped columns. Column forms may be made of wood, steel, or fiberglass. Because columns are generally long in height, the pressure of the concrete at the bottom of the forms is an important consideration in the design of forms for concrete columns.

Chapter 11, Forms for Beams and Floor Slabs, presents relevant information on that subject. The size, length, and spacing of joists are addressed considering the strength and deflection criteria. Spacing of shores under beam bottoms and details for framing beams into girders are also presented.

Chapter 12, Patented Forms for Concrete Floor Systems, is devoted to such patented forms. Patented forms are commonly used for floor systems because considerable savings in labor cost can be derived by simply erecting and removing standard forms, rather than fabricating forms at the jobsite.

Chapter 13, Forms for Thin-Shell Roof Slabs, addresses thin-shell roof slabs. Roofing systems that consist of thin-shell reinforced concrete provide large clear spans below the roof with efficient use of concrete. These types of roofs also produce aesthetically pleasing appearances for the exterior of the structures.

Chapter 14, Forms for Architectural Concrete, considers architectural concrete. There are numerous techniques that can be applied to forms to produce a variety of finishes to the concrete surface after the forms are removed. For concrete buildings, the appearance of the completed structure is often a major consideration in the design of the structure. Forms for architectural concrete can apply to both the interior and the exterior of the building.

Chapter 15, Slipforms, addresses the slipform techniques that have been used successfully to form a variety of concrete structures. Slipforms can be applied to horizontal construction, such as highway pavements and curb-and-gutter construction, as well as to vertical construction of walls, columns, elevator shafts, and so on.

Chapter 16, Forms for Concrete Bridge Decks, discusses the decking of bridges, which are continuously exposed to adverse weather conditions and direct contact with wheel loads from traffic. The deck portion of bridges generally deteriorates and requires repair or replacement before the substructure or foundation portions of the bridges. Thus, there is significant time and cost devoted to formwork for bridge decking.

Chapter 17, Flying Deck Forms, describes the use of flying forms for concrete structures. Flying forms is the descriptive name of a forming system that is removed and reused repetitively to construct multiple levels of a concrete structure. This system of formwork has been applied successfully to many structures.

Appendix A indicates dimensional tolerances for concrete structures that can be used by the workers at the jobsite to fabricate and erect forms that are acceptable.

Appendix B provides recommended guidelines for shoring concrete formwork from the Scaffolding, Shoring, and Forming Institute.

Appendix C presents information related to safety regulations that have been established by the United States Occupational Safety and Health Act (OSHA) of 2009.

Appendix D provides a table of multipliers for converting from the U.S. customary system to metric units of measure.

Appendix E contains a directory of organizations and companies related to formwork. This directory contains addresses, phone numbers, fax numbers, and websites to assist the reader in seeking formwork-related information.

References

1. APA—The Engineered Wood Association, *Concrete Forming*, Tacoma, WA, 2004.
2. ACI Committee 347, American Concrete Institute, *Guide to Formwork for Concrete*, Detroit, MI, 2004.
3. ANSI/AF&PA NDS-2005, American Forest & Paper Association, *National Design Specification for Wood Construction*, Washington, DC, 2005.
4. *Design Values for Wood Construction*, Supplement to the National Design Specification, National Forest Products Association, Washington, DC, 2005.
5. U.S. Department of Labor, *Occupational Safety and Health Standards for the Construction Industry, Part 1926, Subpart Q: Concrete and Masonry Construction*, Washington, DC, 2010.
6. American Institute of Timber Construction, *Timber Construction Manual*, 5th ed., John Wiley & Sons, New York, 2005.

CHAPTER 2

Economy of Formwork

Background Information

Formwork is the single largest cost component of a concrete building's structural frame. The cost of formwork exceeds the cost of the concrete or steel, and, in some situations, the formwork costs more than the concrete and steel combined.

For some structures, placing priority on the formwork design for a project can reduce the total frame costs by as much as 25%. This saving includes both direct and indirect costs. Formwork efficiencies accelerate the construction schedule, which can result in reduced interest costs during construction and early occupancy for the structure. Other benefits of formwork efficiency include increased jobsite productivity, improved safety, and reduced potential for errors.

Impact of Structural Design on Formwork Costs

In the design of concrete structures, the common approach is to select the minimum size of structural members and the least amount of steel to sustain the design loads. The perception is "the least amount of permanent materials in the structure will result in the least cost." To achieve the most economical design, the designer typically will analyze each individual member to make certain that it is not heavier, wider, or deeper than its load requires. This is done under the pretense that the minimum size and least weight result in the best design. However, this approach to design neglects the impact of the cost of formwork, the temporary support structure that must be fabricated and installed to support the permanent materials. Focusing only on ways to economize on permanent materials, with little or no consideration of the temporary formwork, can actually increase, rather than decrease the total cost of a structure.

To concentrate solely on permanent material reduction does not consider the significant cost of the formwork, which often ranges

7

from one-third to one-half of the total installed cost of concrete structures. The most economical design must consider the total process,
including material, time, labor, and equipment required to fabricate,
erect, and remove formwork as well as the permanent materials of
concrete and steel.

Table 2-1 illustrates the impact of structural design on the total
cost for a hypothetical building in which the priority was permanent
material economy. The information contained in this illustration is an
excerpt from *Concrete Buildings, New Formwork Perspectives* [1]. For
Design A, permanent materials are considered to be concrete and
reinforcing steel. The total concrete structural frame cost is $10.35 per
square ft. For Design B, the same project is redesigned to accelerate
the entire construction process by sizing structural members that are
compatible with the standard size dimensions of lumber, which
allows for easier fabrication of forms. The emphasis is shifted to constructability, rather than permanent materials savings. The time has
been reduced, with a resultant reduction in the labor cost required to
fabricate, erect, and remove the forms. Note that for Design B the cost
of permanent materials has actually increased, compared to the cost
of permanent materials required for Design A. However, the increase
in permanent materials has been more than offset by the impact of
constructability, that is, how easy it is to build the structure. The result
is lowering the cost from $10.35 per square ft to $9.00 per square ft, a
13% savings in cost.

Cost Item	Emphasis on Permanent Material, Design A		Emphasis on Constructability, Design B		Percent Increase (Decrease)
Formwork Temporary material, labor, and equipment to make, erect, and remove forms	$5.25/ft²	51%	$3.50/ft²	39%	(33)
Concrete Permanent material and labor for placing and finishing concrete	$2.85/ft²	27%	$3.00/ft²	33%	5
Reinforcing steel Materials, accessories, and labor for installation of reinforcing steel	$2.25/ft²	22%	$2.50/ft²	28%	11
Total cost	$10.35/ft²	100%	$9.00/ft²	100%	(13)

TABLE 2-1 Concrete Structural Frame Cost

Suggestions for Design

Economy of concrete structures begins in the design development stage with designers who have a good understanding of formwork logic. Often, two or more structural alternatives will meet the design objective equally well. However, one alternative may be significantly less expensive to build. Constructability, that is, making structural frames faster, simpler, and less costly to build, must begin in the earliest phase of the design effort.

Economy in formwork begins with the design of a structure and continues through the selection of form materials, erection, stripping, care of forms between reuses, and reuse of forms, if any. When a building is designed, consideration should be given to each of the following methods of reducing the cost of formwork:

1. Prepare the structural and architectural designs simultaneously. If this is done, the maximum possible economy in formwork can be ensured without sacrificing the structural and architectural needs of the building.

2. At the time a structure is designed, consider the materials and methods that will be required to make, erect, and remove the forms. A person or computer-aided drafting and design (CADD) operator can easily draw complicated surfaces, connections between structural members, and other details; however, making, erecting, and removing the formwork may be expensive.

3. If patented forms are to be used, design the structural members to comply with the standard dimensions of the forms that will be supplied by the particular form supplier who will furnish the forms for the job.

4. Use the same size of columns from the foundation to the roof, or, if this is impracticable, retain the same size for several floors. Adopting this practice will permit the use of beam and column forms without alteration.

5. Space columns uniformly throughout the building as much as possible or practicable. If this is not practicable, retaining the same position from floor to floor will result in economy.

6. Where possible, locate the columns so that the distances between adjacent faces will be multiples of 4 ft plus 1 in., to permit the unaltered use of 4-ft-wide sheets of plywood for slab decking.

7. Specify the same widths for columns and column-supported girders to reduce or eliminate the cutting and fitting of girder forms into column forms.

8. Specify beams of the same depth and spacing on each floor by choosing a depth that will permit the use of standard sizes

of lumber, without ripping, for beam sides and bottoms, and for other structural members.

It is obvious that a concrete structure is designed to serve specific purposes, that is, to resist loads and deformations that will be applied to the structure, and to provide an appearance that is aesthetically pleasing. However, for such a structure, it frequently is possible to modify the design slightly to achieve economy without impairing the usability of the structure. The designer can integrate constructability into the project by allowing three basic concepts: design repetition, dimensional standards, and dimensional consistency. Examples of these concepts, excerpted from ref. [1], are presented in this chapter to illustrate how economy in formwork may be affected.

Design Repetition

Any type of work is more efficient if it is performed on a repetitive basis. Assembly line work in the automobile manufacturing industry is a good example of achieving efficiency and economy by repetition. This same concept can be applied to the structural design of concrete structures. Repeating the same layout from bay to bay on each floor provides repetition for the workers. Similarly, repeating the same layout from floor to floor from the lower floor levels to the roof provides repetition that can result in savings in form materials and in efficiency of the labor needed to erect and remove forms.

Dimensional Standards

Materials used for formwork, especially lumber and related wood products such as plywood, are available in standard sizes and lengths. Significant cost savings can be achieved during design if the designer selects the dimensions of concrete members that match the standard nominal dimensions of the lumber that will be used to form the concrete. Designs that depart from standard lumber dimensions require costly carpentry time to saw, cut, and piece the lumber together.

During the design, a careful selection of the dimensions of members permits the use of standard sizes of lumber without ripping or cutting, which can greatly reduce the cost of forms. For example, specifying a beam 11.25 in. wide, instead of 12.0 in. wide, permits the use of a 2- by 12-in. S4S board, laid flat, for the soffit. Similarly, specifying a beam 14.5 in. wide, instead of 14 in. wide, permits the use of two 2- by 8-in. boards, each of which is actually 7.25 in. wide. Any necessary compensation in the strength of the beam resulting from a change in the dimensions may be made by modifying the quantity of the reinforcing steel, or possibly by modifying the depth of the beam.

Dimensional Consistency

For concrete structures, consistency and simplicity yield savings, whereas complexity increases cost. Specific examples of opportunities to simplify include maintaining constant depth of horizontal construction, maintaining constant spacing of beams and joists, maintaining constant column dimensions from floor to floor, and maintaining constant story heights.

Repetitive depth of horizontal construction is a major cost consideration. By standardizing joist size and varying the width, not depth, of beams, most requirements can be met at lower cost because forms can be reused for all floors, including roofs. Similarly, it is usually more cost efficient to increase the concrete strength or the amount of reinforcing material to accommodate differing loads than to vary the size of the structural member.

Roofs are a good example of this principle. Although roof loads are typically lighter than floor loads, it is usually more cost effective to use the same joist sizes for the roof as on the floors below. Changing joist depths or beam and column sizes might achieve minor savings in materials, but it is likely that these will be more than offset by higher labor costs of providing a different set of forms for the roof than required for the slab. Specifying a uniform depth will achieve major savings in forming costs, therefore reducing the total building costs. This will also allow for future expansion at minimal cost. Additional levels can be built after completion if the roof has the same structural capabilities as the floor below.

This approach does not require the designer to assume the role of a formwork planner nor restricts the structural design to formwork considerations. Its basic premise is merely that a practical awareness of formwork costs may help the designer to take advantage of less expensive structural alternatives that are equally appropriate in terms of the aesthetics, structural integrity, quality, and function of the building. In essence, the designer needs only to visualize the forms and the field labor required to form various structural members and to be aware of the direct proportion between complexity and cost.

Of all structure costs, floor framing is usually the largest component. Similarly, the majority of a structure's formwork cost is usually associated with horizontal elements. Consequently, the first priority in designing for economy is selecting the structural system that offers the lowest overall cost while meeting load requirements.

Economy of Formwork and Sizes of Concrete Columns

Architects and engineers sometimes follow a practice of reducing the dimensions of columns every two floors for multistory buildings, as the total loads will permit. Although this practice permits reduction in the quantity of concrete required for columns, it may not reduce the cost of a structure; actually, it may increase the cost. Often, the large column size from the lower floors can be used for the upper floors with a reduction in the amount of the reinforcing steel in the upper floor columns, provided code requirements for strength are maintained. Significant savings in labor and form materials can be achieved by reusing column forms from lower to upper floors. If a change in the column size is necessary, increasing one dimension at a time is more efficient.

The column strategy of the structural engineer has a significant impact on formwork efficiency and column cost. By selecting fewer changes in column size, significant savings in the cost of column formwork can be achieved. Fewer changes in sizes can be accomplished by adjusting the strength of the concrete or the reinforcing steel, or both. For example, to accommodate an increase in load, increasing concrete strength or the reinforcing steel is preferable to increasing column size.

Columns that are placed in an orientation that departs from an established orientation cause major formwork disruptions at their intersections with the horizontal framing. For example, a column that is skewed 30° in orientation from other structural members in a building will greatly increase the labor required to form the skewed column into adjacent members. A uniform, symmetrical column pattern facilitates the use of high-productivity systems, such as gang or flying forms for the floor structural system. Scattered and irregular positioning of columns may eliminate the possibility of using these cost-effective systems. Even with conventional hand-set forming systems, a uniform column layout accelerates construction.

The option to use modern, highly productive floor forming systems, such as flying forms or panelization, may not be feasible for certain column designs. The designer should consider adjacent structural members as a part of column layout and sizing. Column capitals, especially if tapered, require additional labor and materials. The best approach is to avoid column capitals altogether by increasing reinforcement within the floor slab above the column. If this is not feasible, rectangular drop panels, with drops equivalent to the lumber dimensions located above columns, serve the same structural purpose as capitals, but at far lower total costs.

Beam and Column Intersections

The intersections of beams and columns require consideration of both horizontal and vertical elements simultaneously. When the widths of beams and columns are the same, maximum cost efficiency is attained because beam framing can proceed along a continuous line. When beams are wider than columns, beam bottom forms must be notched to fit around column tops. Wide columns with narrow beams are the most expensive intersections to form by far because beam forms must be widened to column width at each intersection.

Economy in Formwork and Sizes of Concrete Beams

Cost savings can be accomplished by selecting beam widths that are compatible with the standard sizes of dimension lumber. Consider a concrete beam 18 ft long with a stem size below the concrete slab that is 16 in. deep and 14 in. wide. If 2-in.-thick lumber is used for the soffit or beam bottom, it will be necessary to rip one of the boards in order to provide a soffit that has the necessary 14.0 in. width. However, if the width of the beam is increased to 14.5 in., two pieces of lumber, each having a net width of 7.25 in., can be used without ripping. Thus, two 2- by 8-in. boards will provide the exact 14.5 in. width required for the soffit. The increase in beam width from 14.0 to 14.5 in.— an additional 0.5 in.—will require a small increase in the volume of concrete as shown in the following equation:

$$\text{Additional concrete} = [(16 \text{ in.} \times 0.5 \text{ in.})/(144 \text{ in.}^2/\text{ft}^2)] \times [18 \text{ ft}]$$

$$= 1.0 \text{ cu ft}$$

Because there are 27 cu ft per cu yard, dividing the 1.0 cu ft by 27 reveals that 0.037 cu yards of additional concrete are required if the beam width is increased by 0.5 in., from 14.0 to 14.5 in. If the cost of concrete is $95.00 per cu yard, the increased concrete cost will be only $3.52. The cost for a carpenter to rip a board 18 ft long will likely be significantly higher than the additional cost of the concrete. Also, when the project is finished, and the form lumber is salvaged, a board having its original or standard width will probably be more valuable than one that has been reduced in width by ripping.

There are numerous other examples of economy of formwork based on sizes of form material. For example, a 15.75-in. rip on a 4-ft-wide by 8-ft-long plywood panel gives three usable pieces that are 8 ft long with less than 1 in. of waste. A 14-in. rip leaves a piece 6 in. wide by 8 ft long, which has little value for other uses. With 6 in. of waste for each plywood panel, essentially every ninth sheet of plywood is thrown away.

This is an area in which architects and engineers can improve the economy in designing concrete structures. Designs that are made primarily to reduce the quantity of concrete, without considering the effect on other costs, may produce an increase rather than a decrease in the ultimate cost of a structure. Additional savings, similar to the preceding example, can be achieved by carefully evaluating the dimension lumber required to form beam and column details.

Economy in Making, Erecting, and Stripping Forms

The cost of forms includes three items: materials, labor, and the use of equipment required to fabricate and handle the forms. Any practice that will reduce the combined cost of all these items will save money. With the cost of concrete fairly well fixed through the purchase of ready-mixed concrete, little, if any, saving can be affected here. It is in the formwork that real economy can be achieved.

Because forms frequently involve complicated forces, they should be designed by using the methods required for other engineering structures. Guessing can be dangerous and expensive. If forms are over-designed, they will be unnecessarily expensive, whereas if they are under-designed, they may fail, which also can be very expensive.

Methods of effecting economy in formwork include the following:

1. Design the forms to provide the required strength with the smallest amount of materials and the most number of reuses.

2. Do not specify or require a high-quality finish on concrete surfaces that will not be exposed to view by the public, such as the inside face of parapet, walls or walls and beams in service stairs.

3. When planning forms, consider the sequence and methods of stripping them.

4. Use prefabricated panels where it is possible to do so.

5. Use the largest practical prefabricated panels that can be handled by the workers or equipment on the job.

6. Prefabricate form members (not limited to panels) where possible. This will require planning, drawings, and detailing, but it will save money.

7. Consider using patented form panels and other patented members, which frequently are less expensive than forms built entirely on the job.

8. Develop standardized methods of making, erecting, and stripping forms to the maximum possible extent. Once carpenters learn these methods, they can work faster.

9. When prefabricated panels and other members, such as those for foundations, columns, walls, and decking, are to be reused several times, mark or number them clearly for identification purposes.

10. Use double-headed nails for temporary connections to facilitate their removal.

11. Clean, oil, and renail form panels, if necessary, between reuses. Store them carefully to prevent distortion and damage.

12. Use long lengths of lumber without cutting for walls, braces, stringers, and other purposes where their extending beyond the work is not objectionable. For example, there usually is no objection to letting studs extend above the sheathing on wall forms.

13. Strip forms as soon as it is safe and possible to do so if they are to be reused on the structure, in order to provide the maximum number of reuses.

14. Create a cost-of-materials consciousness among the carpenters who make forms. At least one contractor displayed short boards around his project on which the cost was prominently displayed.

15. Conduct jobsite analyses and studies to evaluate the fabrication, erection, and removal of formwork. Such studies may reveal methods of increasing productivity rates and reducing costs.

Removal of Forms

Forms should be removed as soon as possible to provide the greatest number of uses but not until the concrete has attained sufficient strength to ensure structural stability and to carry both the dead load and any construction loads that may be imposed on it. The engineer-architect should specify the minimum strength required of the concrete before removal of forms or supports because the strength required for the removal of forms can vary widely with job conditions.

The minimum time for stripping forms and removal of supporting shores is a function of concrete strength, which should be specified by the engineering/architect. The preferred method of determining stripping time is using tests of job-cured cylinders or tests on concrete in place. The American Concrete Institute ACI Committee 347 [2] provides recommendations for removing forms and shores.

The length of time that forms should remain in place before removal should be in compliance with local codes and the engineer who has approved the shore and form removal based on strength and other considerations unique to the job. The Occupational Health and Safety Administration (OSHA) has published standard 1926.703(e) for the construction industry, which recommends that forms and shores not be removed until the employer determines that the concrete has gained sufficient strength to support its weight and superimposed loads. Such determination is based on compliance with one of the following: (1) the plans and specifications stipulate conditions for removal of forms and shores and such conditions have been followed, or (2) the concrete has been properly tested with an appropriate American Society for Testing and Materials (ASTM) standard test method designed to indicate the compression strength of the concrete, and the test results indicate that the concrete has gained sufficient strength to support its weight and superimposed loads.

In general, forms for vertical members, such as columns and piers, may be removed earlier than horizontal forms, such as beams and slabs. ACI Committee 347 suggests the following minimum times forms and supports should remain in place under ordinary conditions. Forms for columns, walls, and the sides of beams often may be removed in 12 hours. Removal of forms for joists, beams, or girder soffits depends on the clear spans between structural supports. For example, spans under 10 ft usually require 4 to 7 days, spans of 10 to 20 ft require 7 to 14 days, and spans over 20 ft generally require 14 to 21 days. Removing forms for one-way floor slabs also will depend on clear spans between structural supports. Spans under 10 ft usually require 3 to 4 days, spans 10 to 20 ft usually require 4 to 7 days, and spans over 20 ft require 7 to 10 days.

Building Construction and Economy

Careful planning in scheduling the construction operations for a building and in providing the forms can assure the maximum economy in formwork and also the highest efficiency by labor, both of which will reduce the cost of formwork.

Consider the six-story building in Figure 2-1, to be constructed with concrete columns, girders, beams, and slabs. The floor area is large enough to justify dividing the floor into two equal or approximately equal areas for forms and concreting. A construction joint through the building is specified or will be permitted. If the structure is symmetrical about the construction joint, the builder will be fortunate. If the building is not symmetrical about the construction joint, some modifications will have to be made in the form procedures presented hereafter.

Each floor will be divided into equal units for construction purposes. Thus, there will be 12 units in the building. One unit will be completely constructed each week, weather permitting, which will include making and erecting the forms; placing the reinforcing steel, electrical conduit, plumbing items, etc., and pouring the concrete. The carpenters should complete the formwork for unit 1 by the end of the third day, after which time some of them will begin the formwork for unit 2 while others install braces on the shores and other braces, if they are required, and check; if necessary, the carpenters will adjust the elevations of girder, beam, and deck forms. One or two carpenters should remain on unit 1 while the concrete is being placed. This will consume one week.

```
                                              Floor ───┐
         Construction                                  │
         joint ────────────►       Roof ─┐
                                         │  ↘
       ┌──────────────────┬──────────────────┐
       │     Unit 11      │     Unit 12       │ ▲
       │    (week #11)    │    (week #12)     │ │
       ├──────────────────┼──────────────────┤   6
       │     Unit 9       │     Unit 10       │
       │    (week #9)     │    (week #10)     │
       ├──────────────────┼──────────────────┤   5
       │     Unit 7       │     Unit 8        │
       │    (week #7)     │    (week #8)      │
       ├──────────────────┼──────────────────┤   4
       │     Unit 5       │     Unit 6        │
       │    (week #5)     │    (week #6)      │
       ├──────────────────┼──────────────────┤   3
       │     Unit 3       │     Unit 4        │
       │    (week #3)     │    (week #4)      │
       ├──────────────────┼──────────────────┤   2
       │     Unit 1       │     Unit 2        │
       │    (week #1)     │    (week #2)      │
       └──────────────────┴──────────────────┘   1
```

FIGURE 2-1   Construction schedule for concrete frame of building.

During the second week, and each week thereafter, a unit will be completed. Delays owing to weather may alter the timing but not the schedule or sequence of operations. Figure 2-1 shows a simplified section through this building with the units and elapsed time indicated but with no provision for lost time owing to weather.

Forms for columns and beam and girder sides must be left in place for at least 48 hours, whereas forms for the beam and girder bottoms, floor slab, and vertical shores must be left in place for at least 18 days. However, concrete test cylinders may be broken to determine the possibility of a shorter removal time of shores. Formwork will be transferred from one unit to another as quickly as time requirements and similarity of structural members will permit.

Table 2-2 will assist in determining the number of reuses of form units and total form materials required to construct the building illustrated in Figure 2-1. Although the extent to which given form sections can be reused will vary for different buildings, the method of analyzing reusage presented in this table can be applied to any building and to many concrete structures.

If the schedule shown in Table 2-2 will apply, it will be necessary to provide the following numbers of sets of forms: for columns and beam and girder sides, two sets and for beam bottoms, slab decking, and shores, three sets.

Units	Total Elapsed Time at Start of Unit, Week	Forms for	Source of Forms
1	0	Columns	New material
		Beam sides	New material
		Beam bottoms	New material
		Slab decking	New material
		Shores	New material
2	1	Columns	New material
		Beam sides	New material
		Beam bottoms	New material
		Slab decking	New material
		Shores	New material
3	2	Columns	Unit 1
		Beam sides	Unit 1
		Beam bottoms	New material
		Slab decking	New material
		Shores	New material
4	3	Columns	Unit 2
		Beam sides	Unit 2
		Beam bottoms	Unit 1
		Slab decking	Unit 1
		Shores	Unit 1
5	4	Columns	Unit 3
		Beam sides	Unit 3
		Beam bottoms	Unit 2
		Slab decking	Unit 2
		Shores	Unit 2
6	5	Columns	Unit 4
		Beam sides	Unit 4
		Beam bottoms	Unit 3
		Slab decking	Unit 3
		Shores	Unit 3
7	6	Columns	Unit 5
		Beam sides	Unit 5
		Beam bottoms	Unit 4
		Slab decking	Unit 4
		Shores	Unit 4
8	7	Columns	Unit 6
		Beam sides	Unit 6
		Beam bottoms	Unit 5
		Slab decking	Unit 5
		Shores	Unit 5

TABLE 2-2   Schedule of Use and Reuse of Formwork for a Building

Units	Total Elapsed Time at Start of Unit, Week	Forms for	Source of Forms
9	8	Columns	Unit 7
		Beam sides	Unit 7
		Beam bottoms	Unit 6
		Slab decking	Unit 6
		Shores	Unit 6
10	9	Columns	Unit 8
		Beam sides	Unit 8
		Beam bottoms	Unit 7
		Slab decking	Unit 7
		Shores	Unit 7
11	10	Columns	Unit 9
		Beam sides	Unit 9
		Beam bottoms	Unit 8
		Slab decking	Unit 8
		Shores	Unit 8
12	11	Columns	Unit 10
		Beam sides	Unit 10
		Beam bottoms	Unit 9
		Slab decking	Unit 9
		Shores	Unit 9

TABLE 2-2   Schedule of Use and Reuse of Formwork for a Building (*Continued*)

If structural sections such as columns, girders, beams, and floor panels in odd-numbered units 1 through 11 are similar, and those in the even-numbered units 2 through 12 are similar, but those in units 1 through 11 are not similar to those in units 2 through 12, it will be necessary to move form sections to higher floors above given units. For example, forms for unit 1 cannot be used in unit 2, or those from unit 3 in unit 4, and so on. Under this condition, it will be necessary to provide one set of columns and beam and girder sides for unit 1 and another set for unit 2, which will be sufficient for the entire building. It will be necessary to provide a set of forms for the beam and girder bottoms, slab decking, and shores for unit 1 and another one for unit 3, and similarly for units 2 and 4.

Economy in Formwork and Overall Economy

The specifications for some projects require smooth concrete surfaces. For such projects, it may be good economy to use form liners, such as thin plywood, tempered hardboard, or sheet steel. Although the cost of the forms will be increased, reduction in the cost of finishing the surfaces will be reduced or eliminated. The small fins that sometimes appear on concrete surfaces opposite the joints in the sheets of lining material can be reduced or eliminated by sealing the joints with putty or some other suitable compound prior to placing the concrete.

Numerous papers have described materials and methods of construction for forming economical concrete buildings. The American Concrete Institute, the Portland Cement Association, and other organizations involved in concrete structures have sponsored international conferences on forming economical concrete buildings. The proceedings of these conferences have been published and are available from these institutions as shown in the references at the end of this chapter.

References

1. Ceco Concrete Construction Co., *Concrete Buildings, New Formwork Perspectives*, Kansas City, MO, 1985.
2. ACI Committee 347, American Concrete Institute, *Guide to Formwork for Concrete*, Detroit, MI, 2004.
3. "Forming Economical Concrete Buildings," *Proceedings of the First International Conference*, Portland Cement Association, Skokie, IL, 1984.
4. "Forming Economical Concrete Buildings," *Proceedings of the Second International Conference*, Publication SP-90, American Concrete Institute, Detroit, MI, 1986.
5. "Forming Economical Concrete Buildings," *Proceedings of the Third International Conference*, Publication SP-107, American Concrete Institute, Detroit, MI, 1988.
6. US Department of Labor, *Occupational Safety and Health Standards for the Construction Industry, Part 1926, Subpart Q: Concrete and Masonry Construction*, Washington, DC, 2010.

CHAPTER 3

Pressure of Concrete on Formwork

Behavior of Concrete

Concrete is a mixture of sand and aggregate that is bonded together by a paste of cement and water. The five basic types of cement used in concrete mixtures are

Type I—Ordinary portland cement

Type II—Modified low heat, modified sulfate resistance

Type III—Early high strength, rapid hardening

Type IV—Low heat of hydration

Type V—Sulfate resisting

Admixtures are commonly used in concrete mixes. Additives include liquids, solids, powders, or chemicals that are added to a concrete mix to change properties of the basic concrete mixture of water, cement, sand, and aggregate. They can accelerate or retard setting times, decrease water permeability, or increase strength, air content, and workability. Admixtures include pozzolans such as silica flume, blast-furnace slag, and fly ash. The pressure of concrete on formwork depends on the type of cement and admixtures in the concrete mix.

When concrete is first mixed, it has properties lying between a liquid and a solid substance. It is best described as a semiliquid and is usually defined as a plastic material. With the passage of time, concrete loses its plasticity and changes into a solid. This property of changing from a plastic to a solid makes concrete a valuable building material because it can be easily shaped by forms before attaining its final state.

The ability to change from a semiliquid (or plastic) to a solid state appears to be the result of two actions within the concrete. The former

21

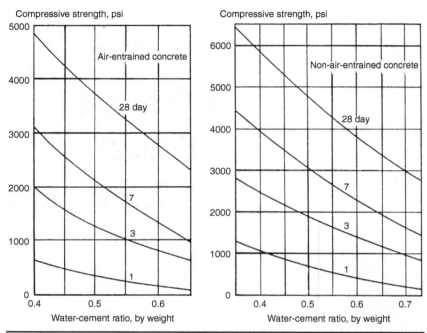

FIGURE 3-1 Relationship between age and compressive strength of concrete for Type I portland cement. Notes: (1) Courtesy, Portland Cement Association. (2) Data based on compressive tests of 6- by 12-in. cylinders using Type I portland cement and moist-curing at 70°F.

action is the result of the setting of the cement, which may begin within 30 min after the concrete, is mixed under favorable conditions, namely, a warm temperature. This action may continue for several hours, especially if the temperature is low. The latter action is the development of internal friction between the particles of aggregate in the concrete that restrains them from moving freely past each other. The magnitude of the internal friction is higher in a dry concrete than in a wet one, and it increases with the loss of water from a concrete.

Figure 3-1 gives illustrative relationships for age-compressive strength of laboratory cured air-entrained and non–air-entrained concrete with different water-cement ratios using Type I portland cement, when the concrete is moist cured at 70°F.

Lateral Pressure of Concrete on Formwork

The pressure exerted by concrete on formwork is determined primarily by several or all of the following factors:

1. Rate of placing concrete in forms
2. Temperature of concrete

3. Weight or density of concrete

4. Cement type or blend used in the concrete

5. Method of consolidating the concrete

6. Method of placement of the concrete

7. Depth of placement

8. Height of form

The American Concrete Institute [1] has devoted considerable time and study to form design and construction practices. ACI Committee 347 identifies the maximum pressure on formwork as the full hydrostatic lateral pressure, as given by the following equation:

$$P_m = wh \qquad (3\text{-}1)$$

where P_m = maximum lateral pressure, lb per sq ft
w = unit weight of newly placed concrete, lb per cu ft
h = depth of the plastic concrete, ft

For concrete that is placed rapidly, such as columns, h should be taken as the full height of the form. There are no minimum values given for the pressures calculated from Eq. (3-1).

Lateral Pressure of Concrete on Wall Forms

For determining pressure of concrete on formwork ACI 347 defines a wall as a vertical structural member with at least one plan dimension greater than 6.5 ft. Two equations are provided for wall form pressure. Equation (3-2) applies to walls with a rate of placement less than 7 ft per hr and a placement height of 14 ft or less. Equation (3-3) applies to all walls with a placement rate of 7 to 15 ft per hr, and to walls placed at less than 7 ft per hr but having a placement height greater than 14 ft. Both equations apply to concrete with a 7 in. maximum slump, and vibration limited to normal internal vibration to a depth of 4 ft or less. For walls with a rate of placement greater than 15 ft per hr, or when forms will be filled rapidly, before stiffening of the concrete takes place, then the pressure should be taken as the full height of the form, $P_m = wh$.

For wall forms with a concrete placement rate of less than 7 ft per hr and a placement height not exceeding 14 ft:

$$P_m = C_w C_c [150 + 9{,}000R/T] \qquad (3\text{-}2)$$

where P_m = maximum lateral pressure, lb per sq ft
C_w = unit weight coefficient as shown in Table 3-1
C_c = chemistry coefficient as shown in Table 3-2
R = rate of fill of concrete in form, ft per hr

T = temperature of concrete in form, degrees Fahrenheit

Minimum value of P_m is $600C_w$, but in no case greater than wh.

Applies to concrete with a slump of 7 in. or less

Applies to normal internal vibration to a depth of 4 ft or less

For all wall forms with concrete placement rate from 7 to 15 ft per hr, and for walls where the placement rate is less than 7 ft per hr and the placement height exceeds 14 ft.

$$P_m = C_w C_c [150 + 43,400/T + 2,800\ R/T] \tag{3-3}$$

where P_m = maximum lateral pressure, lb per sq ft

C_w = unit weight coefficient

C_c = chemistry coefficient

R = rate of fill of concrete in form, ft per hr

T = temperature of concrete in form, °F

Minimum value of P_m is $600C_w$, but in no case greater than wh.

Applies to concrete with a slump of 7 in. or less

Applies to normal internal vibration to a depth of 4 ft or less

Values for the unit weight coefficient C_w in Eqs. (3-2) and (3-3) are shown in Table 3-1 and the values for the chemistry coefficient C_c are shown in Table 3-2. For concrete placed in wall forms at rates of pour greater than 15 ft per hr, the lateral pressure should be wh, where h is the full height of the form. ACI Committee 347 recommends that the form be designed for a full hydrostatic head of concrete wh plus a minimum allowance of 25% for pump surge pressure if concrete is pumped from the base of the form.

Example 3-1

A wall form 12 ft high is filled with 150 lb per cu ft concrete at a temperature of 70°F. The concrete is Type I without a retarder. Concrete will be placed with normal internal vibration to a depth of less than 4 ft. The rate of placement is 5 ft per hr.

From Table 3-1, the value of C_w is 1.0 and from Table 3-2 the value of C_c is 1.0. The rate of placement is less than 7 ft per hr and the placement height does not exceed 14 ft, therefore Eq. (3-2) can be used to calculate the lateral pressure as follows.

$$P_m = C_w C_c [150 + 9,000R/T]$$

$$P_m = C_w C_c [150 + 9,000R/T]$$

$$= (1.0)(1.0)[150 + 9,000(5/70)]$$

$$= 793 \text{ lb per sq ft}$$

Weight of Concrete	Value of C_w
Less than 140 lb per cu ft	0.5 [1 + (w/145 lb per cu ft)], but not less than 0.8
140 to 150 lb per cu ft	1.0
More than 150 lb per cu ft	w/145 lb per cu ft

TABLE 3-1 Values of Unit Weight Coefficient, C_w

Cement Type or Blend	Value of C_c
Types I, II, and III without retarders*	1.0
Types I, II, and III with a retarder	1.2
Other types or blends containing less than 70% slag or 40% fly ash without retarders*	1.2
Other types or blends containing less than 70% slag or 40% fly ash with a retarder*	1.4
Blends containing more than 70% slag or 40% fly ash	1.4

*Retarders include any admixture, such as a retarder, retarding water reducer, retarding mid-range water-reducing admixtures, or high-range water-reducing admixture (superplasticizers), that delay setting of concrete.

TABLE 3-2 Values of Chemistry Coefficient, C_c

Checks on limitations on pressures calculated from Eq. (3-2):

Limited to greater than $600C_w = 600(1.0) = 600$ lb per sq ft
Limited to less than $P_m = wh = 150(12) = 1{,}800$ lb per sq ft

The calculated value from Eq. (3-2) is 793, which is above the minimum of 600 and below the maximum 1,800. Therefore, use 793 lb per sq ft lateral pressure on the forms. The 793 lb per sq ft maximum pressure will occur at a depth of 793/150 = 5.3 ft below the top of the form as shown in Figure 3-2.

Example 3-2

A wall form 8 ft high is filled with 150 lb per cu ft concrete at a temperature of 60°F. The concrete is Type I with a retarder. Concrete will be placed with normal internal vibration to a depth of less than 4 ft. The concrete rate of placement will be 10 ft per hr.

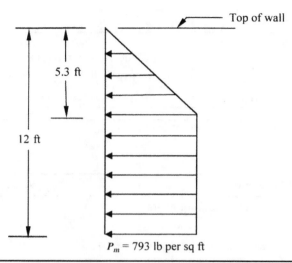

FIGURE 3-2   Distribution of concrete pressure for Example 3-1.

From Table 3-1 the value of C_w is 1.0 and from Table 3-2 the value of C_c is 1.2. The rate of placement is between 7 and 15 ft per hr and the placement height does not exceed 14 ft. Using Eq. (3-3) to calculate the lateral pressure.

$$P_m = C_w C_c[150 + 43{,}400/T + 2{,}800R/T]$$
$$= (1.0)(1.2)[150 + 43{,}400/60 + 2{,}800(10/60)]$$
$$= 1{,}608 \text{ lb per sq ft}$$

Checks on limitations on pressures calculated from Eq. (3-3):

Limited to greater than $600C_w = 600(1.0) = 600$ lb per sq ft
Limited to less than $P_m = wh = 150(8) = 1{,}200$ lb per sq ft

The calculated value from Eq. (3-3) is 1,608 lb per sq ft, which is above the limit of $600C_w$. However, the calculated value 1,608 is greater than the limit of $P_m = wh = 1{,}200$ for an 8-ft-high wall. Therefore, the maximum design concrete lateral pressure is 1,200 lb per sq ft. Figure 3-3 shows the lineal distribution of pressure.

Example 3-3

A concrete wall is 9 ft high, 15 in. thick, and 60 ft long. The concrete will be placed by a pump with a capacity of 18 cu yd per hr at a temperature of 80°F. The concrete density is 150 lb per cu yd with Type I cement without additives or blends; therefore, C_w and $C_c = 1.0$.

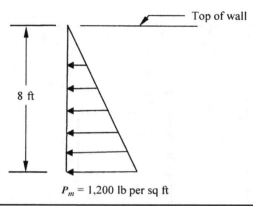

Top of wall

8 ft

$P_m = 1,200$ lb per sq ft

FIGURE 3-3 Distribution of Concrete Pressure for Example 3-2.

The rate of placement R in ft per hr can be calculated as follows:

R = [volume pumped per hour]/[volume pumped in 1 ft height of wall]

Volume pumped per hour = 18 cu yd per hr

Volume of 1 ft height of wall = (1 ft)(60 ft)(15/12 ft)

$\qquad$ = 75 cu ft

$\qquad$ = 2.78 cu yd of concrete in 1 ft height of wall

Therefore, the rate of placement R = [18 cu yd per hr]/[2.8 cu yd per ft]

$\qquad$ = 6.5 ft/hr

The rate of placement is less than 7 ft per hr and wall height does not exceed 14 ft; therefore, Eq. (3-2) can be used to calculate the concrete pressure.

$P_m = C_w C_c [150 + 9,000 R/T]$

$\quad$ = (1.0)(1.0)[150 + 9,000 (6.5/80)]

$\quad$ = 882 lb per sq ft

Checks on limitations on pressures calculated from Eq. (3-2):

Limited to greater than $600 C_w$ = 600(1.0) = 600 lb per sq ft

Limited to less than $P_m = wh$ = 150(9) = 1,350 lb per sq ft

The calculated value from Eq. (3-2) is 882, which is above the minimum of 600 and below the maximum of wh = 1,350. Therefore, use an 882 lb per sq ft lateral pressure on the forms. The 882-lb per sq ft maximum pressure will occur at a depth of 882/150 = 5.9 ft as shown in Figure 3-4.

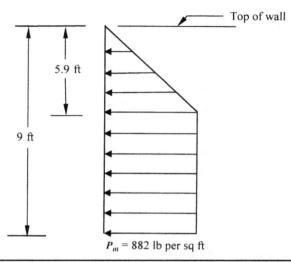

FIGURE 3-4   Distribution of Concrete Pressure for Example 3-3.

Relationship between Rate of Fill, Temperature, and Pressure for Wall Forms

Table 3-3 gives the relationship between the rate of filling wall forms, lateral pressure, and temperature for placement heights up to 14 ft. The pressures are based on 150 lb per cu ft density concrete with no additives, a maximum slump of 7 in., and vibration to a depth of 4 ft or less. For other concrete densities and blends, the pressures should be adjusted by C_w and C_c. For rates of pour greater than 15 ft per hr, the wall pressure should be calculated by $P_m = wh$.

In Table 3-3, the pressures for rates of replacement less than 7 ft per hr are calculated by the equation $P_m = C_w C_c [150 + 9{,}000R/T]$. For rates of placement from 7 to 15 ft per hr, the pressures are calculated by $P_m = C_w C_c [150 + 43{,}400/T + 2{,}800R/T]$. However, these wall pressure equations are limited to a maximum pressure of $P_m = wh$. For example, if a wall form 14 ft deep is filled at a rate of 15 ft per hr at a temperature of 40°F, Table 3-3 indicates a maximum pressure of 2,285 lb per sq ft. However, the maximum pressure is limited to $P_m = wh =$ (150 lb per cu ft) x (14 ft) = 2,100 lb per sq ft.

The wall pressure equations may be used to determine the maximum pressures produced on wall forms, provided the forms are deep enough to permit the calculated pressures to develop. For example, if a wall form 6 ft deep is filled at a rate of 7 ft per hr at a temperature of 50°F, Table 3-3 indicates a maximum pressure of 1,410 lb per sq ft. However, if the concrete weighs 150 lb per cu ft, the maximum

Rate of Filling Forms ft per hr	Lateral Pressure, lb per sq ft for the Temperature Indicated							
	40°F	50°F	60°F	70°F	80°F	90°F	100°F	
1	600	600	600	600	600	600	600	
2	600	600	600	600	600	600	600	
3	825	690	600	600	600	600	600	
4	1,050	870	750	664	600	600	600	$P_m = C_w C_c[150 + 9,000R/T]$
5	1,275	1,050	900	793	713	650	600	↑
6	1,500	1,230	1,050	921	825	750	690	
7	1,725	1,410	1,200	1,050	938	850	780	
8	1,795	1,466	1,247	1,090	973	882	808	↓
9	1,865	1,522	1,293	1,130	1,008	913	836	$P_m = C_w C_c[150 + 43,400/T + 2,800R/T]$
10	1,935	1,578	1,340	1,170	1,043	944	864	
11	2,005	1,634	1,387	1,210	1,078	975	892	
12	2,075	1,690	1,434	1,250	1,113	1,006	920	
13	2,145	1,746	1,450	1,290	1,148	1,037	948	
14	2,215	1,802	1,522	1,330	1,183	1,068	976	
15	2,285	1,858	1,574	1,370	1,218	1,099	1004	

Notes:
1. Do not use design pressure greater than wh.
2. Concrete placement with normal internal vibration to a depth of 4 ft or less.
3. Values are based on concrete with $C_w = 1$ and $C_c = 1$.
4. Concrete without additives with a maximum slump of 7 in.
5. Minimum pressure is $600C_w$ lb per sq ft, but in no case greater than wh.
6. For pour rates greater than 15 ft per hr, use pressure $P_m = wh$.
7. Reference ACI Committee 347 for additional information on concrete form pressures [1].

TABLE 3-3 Relation between Rate of Filling Wall Forms, Lateral Pressure, and Temperature for Placement Heights Up to 14 Feet

pressure will not exceed the pressure of $wh = 150(6) = 900$ lb per sq ft. It should be noted that the left-hand column in Table 3-3 is the rate of filling wall forms and not the height of wall.

Table 3-4 gives the relationship between the rate of filling wall forms, lateral pressure, and temperature for placement heights greater than 14 ft. The pressures are based on 150 lb per cu ft density concrete with no additives, a maximum slump of 7 in., and vibration to a depth

Rate of Filling Forms ft per hr	Lateral Pressure, lb per sq ft for the Temperature Indicated						
	40°F	50°F	60°F	70°F	80°F	90°F	100°F
1	1,305	1,074	920	810	728	664	612
2	1,375	1,130	967	850	763	695	640
3	1,445	1,186	1,014	890	798	726	668
4	1,515	1,242	1,060	930	833	757	696
5	1,585	1,298	1,107	970	868	788	724
6	1,655	1,354	1,154	1,010	903	819	752
7	1,725	1,410	1,200	1,050	938	850	780
8	1,795	1,466	1,247	1,090	973	882	808
9	1,865	1,522	1,293	1,130	1,008	913	836
10	1,935	1,578	1,340	1,170	1,043	944	864
11	2,005	1,634	1,387	1,210	1,078	975	892
12	2,075	1,690	1,434	1,250	1,113	1,006	920
13	2,145	1,746	1,450	1,290	1,148	1,037	948
14	2,215	1,802	1,522	1,330	1,183	1,068	976
15	2,285	1,858	1,574	1,370	1,218	1,099	1004

Notes:
1. Do not use design pressure greater than wh.
2. Concrete placement with normal internal vibration to a depth of 4 ft or less.
3. Values are based on concrete with $C_w = 1$ and $C_c = 1$.
4. Concrete without additives with a maximum slump of 7 in.
5. Minimum pressure is $600C_w$ lb per sq ft, but in no case greater than wh.
6. For pour rates greater than 15 ft per hr, use pressure $P_m = wh$.
7. Reference ACI Committee 347 for additional information on concrete form pressures [1].

TABLE 3-4 Relation between Rate of Filling Wall Forms, Lateral Pressure, and Temperature for Placement Heights Greater Than 14 Feet, Using Equation $P_m = C_w C_c[150 + 43,400/T + 2,800R/T]$

of 4 ft or less. For other concrete densities and blends, the pressures should be adjusted by C_w and C_c. The values in Table 3-4 are based on the equation $P_m = C_w C_c[150 + 43,400/T + 2,800R/T]$. For rates of pour greater than 15 ft per hr, the concrete pressure should be calculated by $P_m = wh$. It should be noted the left-hand column in Tables 3-4 is the rate of filling wall forms and not the height of wall.

Lateral Pressure of Concrete on Column Forms

For determining pressure of concrete on formwork ACI 347 defines a column as a vertical structural member with no plan dimensions greater than 6.5 ft. As previously presented, the American Concrete Institute recommends that formwork be designed for it full hydrostatic lateral pressure as given by Eq. (3-1), $P_m = wh$, where P_m is the lateral pressure (lb per sq ft), w is the unit weight (lb per cu ft) of the newly placed concrete, and h is the depth (ft) of the plastic concrete. Concrete is often placed rapidly in columns with intensive vibration or with self-consolidating concrete. Therefore, h should be taken as the full height of the column form. There are no maximum or minimum values given for the pressure calculated from Eq. (3-1).

For concrete with a slump 7 in. or less and placement by normal internal vibration to a depth of 4 ft or less, formwork for columns can be designed for the following lateral pressure.

$$P_m = C_w C_c [150 + 9,000R/T] \tag{3-4}$$

where P_m = calculated lateral pressure, lb per sq ft
$\quad\quad C_w$ = unit weight coefficient
$\quad\quad C_c$ = chemistry coefficient
$\quad\quad R$ = rate of fill of concrete in form, ft per hr
$\quad\quad T$ = temperature of concrete in form, °F
$\quad\quad\quad$ Minimum value of P_m is $600C_w$, but in no case greater than wh.
$\quad\quad\quad$ Applies to concrete with a slump of 7 in. or less
$\quad\quad\quad$ Applies to normal internal vibration to a depth of 4 ft or less

Values for the unit weight coefficient C_w in Eq. (3-4) are shown in Table 3-1 and the values for the chemistry coefficient C_c are shown in Table 3-2. The minimum pressure in Eq. (3-4) is $600C_w$ lb per sq ft, but in no case greater than wh.

Example 3-4

A column form 14 ft high is filled with 150 lb per cu ft concrete at a temperature of 50°F. The concrete is Type I without a retarder. Concrete will be placed with normal internal vibration to a depth of less than 4 ft. The rate of placement is 7 ft per hr.

From Table 3-1 the value of C_w is 1.0 and from Table 3-2 the value of C_c is 1.0. Using Eq. (3-4), the lateral pressure can be calculated.

$$P_m = C_w C_c [150 + 9,000R/T]$$
$$= (1.0)(1.0)[150 + 9,000(7/50)]$$
$$= 1,410 \text{ lb per sq ft}$$

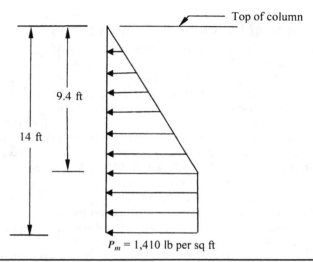

FIGURE 3-5   Distribution of concrete pressure for Example 3-4.

Checks on limitations on pressures calculated from Eq. (3-3):

Limited to greater than $600C_w = 600(1.0) = 600$ lb per sq ft
Limited to less than $P_m = wh = 150(14) = 2,100$ lb per sq ft

The calculated value from Eq. (3-4) is 1,410, which is above the minimum of 600 and below the maximum of $wh = 2,100$. Therefore, use 1,410 lb per sq ft lateral pressure on the forms. The 1,410-lb per sq ft maximum pressure will occur a depth of $1,410/150 = 9.4$ ft below the top of the form.

Example 3-5

A column form 12 ft high is filled with 150 lb per cu ft concrete at a temperature of 60°F. The concrete is Type I with a retarder. Concrete will be placed with normal internal vibration to a depth of less than 4 ft. The concrete will be in 1 hr.

Rate of placement = 12 ft/1 hr = 12 ft per hr

From Table 3-1, the value of C_w is 1.0 and from Table 3-2 the value of C_c is 1.2. Using Eq. (3-4), the lateral pressure can be calculated.

$$P_m = C_w C_c[150 + 9{,}000R/T]$$
$$= (1.0)(1.2)[150 + 9{,}000(12/60)]$$
$$= 2{,}340 \text{ lb per sq ft}$$

Checks on limitations on pressures calculated from Eq. (3-2):

Limited to greater than $600C_w = 600(1.0) = 600$ lb per sq ft
Limited to less than $P_m = wh = 150(12) = 1,800$ lb per sq ft

The calculated value from Eq. (3-4) is 2,340, which is above the maximum of $wh = 150(12 \text{ ft}) = 1,800$. Therefore, use 1,800 lb per sq ft lateral pressure on the forms.

Example 3-6

Concrete with a density of 150 lb per cu yd will be pumped from the base of a column form 10 ft high. Because the concrete is pumped from the base of the form, the maximum lateral concrete pressure on the form is the full hydrostatic pressure of wh plus a minimum allowance of 25% for pump surge pressure. The pressure on the column form can be calculated as follows:

$$P_m = 1.25 \, wh$$
$$= (1.25)(150 \text{ lb per cu ft})(10 \text{ ft})$$
$$= 1,875 \text{ lb per sq ft}$$

Relationship between Rate of Fill, Temperature, and Pressure for Column Forms

Table 3-5 gives the relationship between the rate of filling forms, lateral pressure, and the temperature of the concrete for column forms, using Eq. (3-3) with C_w and $C_c = 1$. As previously presented, values calculated using Eq. (3-4) should not exceed wh. For example, for concrete with a density of 150 lb per cu ft the pressure should not exceed $150h$, where h is the depth, in feet, below the upper surface of freshly placed concrete. Thus, the maximum pressure at the bottom of a form 6 ft high will be $150(6 \text{ ft}) = 900$ lb per sq ft, regardless of the rate of filling the form. Using Eq. (3-4) for a rate of placement of 6 ft per hr at a temperature of 90°F, the calculated value is 750 lb per sq ft. However, for a rate of placement of 6 ft per hr and a temperature of 60°F, using Eq. (3-4), the calculated value is 1,050 lb per sq ft, which is greater than the limiting pressure of $wh = 900$ lb per sq ft. As previously stated, the American Concrete Institute recommends Eq. (3-1) for the design of concrete formwork.

Graphical Illustration of Pressure Equations for Walls and Columns

Figure 3-6 provides a graphical illustration of the relationship between the rate of filling wall and column forms, maximum pressure, and the temperature of concrete. The graphs are based on concrete made with

Rate of Filling Forms, ft per hr	Lateral Pressure, lb per sq ft for the Temperature Indicated						
	40°F	**50°F**	**60°F**	**70°F**	**80°F**	**90°F**	**100°F**
1	600	600	600	600	600	600	600
2	600	600	600	600	600	600	600
3	825	690	600	600	600	600	600
4	1,050	870	750	664	600	600	600
5	1,275	1,050	900	793	713	650	600
6	1,500	1,230	1,050	921	825	750	690
7	1,725	1,410	1,200	1,050	938	850	780
8	1,950	1,590	1,350	1,179	1,050	950	870
9	2,175	1,770	1,500	1,308	1,163	1,050	960
10	2,400	1,950	1,650	1,436	1,275	1,150	1,050
11	2,625	2,130	1,800	1,565	1,388	1,250	1,140
12	2,850	2,310	1,950	1,693	1,500	1,350	1,230
13	3,075	2,490	2,100	1,822	1,613	1,450	1,320
14	3,300	2,670	2,250	1,950	1,725	1,550	1,410
15	3,525	2,850	2,400	2,079	1,838	1,650	1,500
16	3,750	3,030	2,550	2,207	1,950	1,750	1,590
17	3,975	3,210	2,700	2,336	2,062	1,850	1,680
18	4,200	3,390	2,850	2,464	2,175	1,950	1,770
19	4,425	3,570	3,000	2,592	2,288	2,050	1,860
20	4,650	3,750	3,150	2,721	2,400	2,150	1,950
21	4,875	3,930	3,300	2,850	2,513	2,250	2,040
22	5,100	4,110	3,450	2,979	2,730	2,350	2,130
23	5,325	4,290	3,600	3,107	3,525	2,450	2,220
24	5,550	4,470	3,750	3,236	2,850	2,550	2,310
25	5,775	4,650	3,900	3,364	2,963	2,650	2,400

Notes:
1. Note, do not use a design pressure greater than wh.
2. Concrete placement with normal internal vibration to a depth of 4 ft or less.
3. Values based on concrete with $C_w = 1$ and $C_c = 1$.
4. Concrete without additives with a maximum slump of 7 in.
5. Minimum pressure is $600C_w$ lb per sq ft, but in no case greater than wh.
6. Reference ACI Committee 347 for additional information on concrete form pressures [1].

TABLE 3-5   Relation between Rate of Filling Column Forms, Lateral Pressure, and Temperature, $P_m = C_w C_c[150 + 9,000R/T]$

LATERAL CONCRETE PRESSURES FOR VARIOUS TEMPERATURES

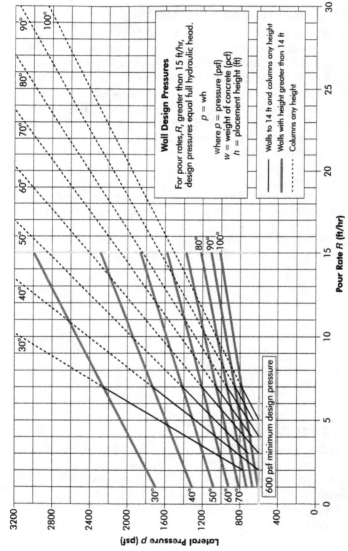

Wall Design Pressures

For pour rates, R, greater than 15 ft/hr, design pressures equal full hydraulic head.

$$p = wh$$

where p = pressure (psf)
w = weight of concrete (pcf)
h = placement height (ft)

——— Walls to 14 ft and columns any height
——— Walls with height greater than 14 ft
- - - - - Columns any height

600 psf minimum design pressure

Pour Rate R (ft/hr)

Lateral Pressure P (psf)

Figure 3-6 Lateral pressures for various rates of pour and temperatures. Notes: (1) Courtesy of APA—The Engineered Wood Association. (2) Source: Concrete Forming Design/Construction Guide, 2004. (3) Concrete made with Type I or Type III cement, weighing 150 lb per cu yd, containing no pozzolans or admixtures, having a slump of 7 in. or less and internal vibration to a depth of 4 ft or less.

Type I or Type III cement, weighting 150 lb per cu ft, containing no prozzolans or admixtures, having a slump of 7 in. or less, and internal vibration to a depth of 4 ft or less.

Effect of Weight of Concrete on Pressure

The unit weight of concrete for most structures is 145 to 150 lb per cu ft. It is common to refer to normal-weight concrete as having a unit weight of 150 lb per cu ft. However, the density of concrete may vary from 100 lb per cu ft for lightweight concrete to 200 lb per cu ft for high-density concrete. For concrete with densities other than 150 lb per cu ft, the pressure can be found by multiplying the normal-weight concrete (150) by the ratios of the densities. For example, if concrete weighing 150 lb per cu ft will produce a pressure of 1,200 lb per sq ft under certain conditions, a low-density concrete weighing 100 lb per cu ft should produce a pressure equal to $(100/150) \times 1,200 = 800$ lb per sq ft under the same conditions. Similarly, if a high-density concrete weighing 200 lb per cu ft will be used, the pressure produced will be equal to $(200/150) \times 1,200 = 1,600$ lb per sq ft under the same conditions.

Equation (3-5) may be used to determine the maximum pressure produced by concrete having a density other than 150 lb per cu ft.

$$P'_m = [(D')/(150)] \times P_m \qquad (3-5)$$

where P'_m = modified pressure, lb per sq ft
D' = density of concrete, lb per cu ft
P_m = maximum pressure for concrete whose density is 150 lb per cu ft when placed under the same conditions

Vertical Loads on Forms

In addition to lateral pressure, vertical loads are also imposed on formwork. These vertical loads are due to the weight of the newly placed concrete and reinforcing steel, the weight of the form materials and hardware that fasten the forms, the weight of tools and equipment, and the weight of workers. For multistory structures, the design of forms must consider the loads that are transmitted from all of the higher floors.

In addition to the weight of concrete, the weight of the forming material, hardware, and reinforcing steel must also be included in the vertical loads that are applied to forms. The weight of the form material and hardware will not be known until the formwork is designed. The designer may estimate a weight of 10 lb per sq ft and then check this value after the design is completed and all member sizes and weights are known. The combined weight of the concrete, reinforcing steel, and form material is often referred to as the dead load on the forms.

Superimposed on the dead loads are the live loads that may be imposed on the forms. Live loads consist of the weights of workers, tools, equipment, and any storage material. ACI Committee 347 recommends a minimum live load of 50 lb per sq ft to provide for workers, tools, and equipment. For conditions where motorized buggies are used, the minimum live load is increased to 75 lb per sq ft, reference Table 3-6. A minimum combined dead and live load of 100 lb per sq ft is recommended, up to 125 lb per sq ft if motorized buggies are used.

Thus, forms for concrete slabs must support workers and equipment (live loads) in addition to the weight of freshly placed concrete and form materials (dead load). For normal weight concrete, 150 lb per cu ft, the weight of the freshly placed concrete will place a load on the forms of $(150 \times 1/12) = 12.5$ lb per sq ft for each inch of slab thickness.

The weight of the concrete on forms may be considered as a concentrated vertical load (lb), a vertical uniform load (lb per ft), or vertical pressure (lb per sq ft), as illustrated in the Examples 3-7 to 3-9.

Thickness of Slab, in.	Design Load, lb per sq ft	
	Nonmotorized Buggies	Motorized Buggies
4	100	125
5	113	138
6	125	150
7	138	163
8	150	175
9	163	188
10	175	200
12	200	225

Notes:
1. All values are based on 150 lb per cu ft concrete.
2. Weight of form materials and reinforcing steel are not included.
3. Values for nonmotorized buggies include 50 lb per sq ft live load.
4. Values for motorized buggies include 75 lb per sq ft live load.

TABLE 3-6 Design Vertical Pressures for Slab Forms

Example 3-7

A floor system has supports at 20 ft in each direction that must support an 8-in.-thick slab of concrete whose weight is 150 lb per cu ft. Motorized buggies will be used to transport and place the concrete. The vertical concentrated load acting on the support can be calculated as follows:

Dead load of concrete = (150 lb per cu ft)[(20 ft × 20 ft) × (8/12 ft)]
$$= 40,000 \text{ lb}$$

Estimated dead load of forms and hardware
$$= (10 \text{ lb per sq ft}) (20 \text{ ft} \times 20 \text{ ft})$$
$$= 4,000 \text{ lb}$$

Live load of workers using buggies = (75 lb per sq ft)(20 ft × 20 ft)
$$= 30,000 \text{ lb}$$

Total design load = 44,000 lb dead load + 30,000 lb live load
$$= 74,000 \text{ lb}$$

Example 3-8

A continuous concrete beam is 16 in. wide and 24 in. deep, with 150 lb per cu ft concrete. The uniform vertical load per foot acting on the bottom of the beam form can be calculated as:

Dead load of concrete = (150 lb per cu ft)(16/12 ft)(24/12 ft)
$$= 400 \text{ lb per lin ft}$$

Estimated dead load of form material = (5 lb per sq ft)(16/12)
$$= 3 \text{ lb per lin ft}$$

Live load of workers and tools = (50 lb per sq ft)(16/12)
$$= 67 \text{ lb per lin ft}$$

Total design load = 403 dead load + 67 live load
$$= 470 \text{ lb per lin ft}$$

Example 3-9

Concrete that weighs 150 lb per cu ft is placed for an elevated 6-in.-thick slab. Motorized buggies will not be used. The vertical pressure on the slab forms can be calculated as:

Dead load of concrete = (150 lb per cu ft)(6/12 ft)
$$= 75 \text{ lb per sq ft}$$

Estimated dead load of form material = 10 lb per sq ft

Live load of workers = 50 lb per sq ft

Total design load = 85 dead load + 50 live load
$$= 135 \text{ lb per sq ft}$$

Placement and Consolidation of Freshly Placed Concrete

Concrete may be placed in the forms directly from a concrete delivery truck, dropped from crane buckets, or pumped from a concrete pumping truck or trailer. If concrete is pumped from the base of the form, ACI Committee 347 recommends that the form be designed for a full hydrostatic head of concrete wh plus a minimum allowance of 25% for pump surge pressure if the concrete is pumped from the base of the form. In certain instances, pressures may be as high as the face pressure of the pump piston.

Cautions must be taken when using external vibration of concrete with shrinkage-compensating or expansive cements because the pressure on the forms may be in excess of the equivalent hydrostatic pressure.

Caution must also be taken with internal vibration of freshly placed concrete. Consolidation techniques may be responsible for formwork failures, either by forcing the concrete to remain semi-liquid longer than anticipated, or by excessive shaking of the forms.

Wind Loads on Formwork Systems

The design of formwork systems must include horizontal wind forces in addition to the vertical weight of concrete and live loads that are placed on the formwork. Horizontal wind applies loads against the side of shoring and the formwork on top of the shoring. The wind can also apply uplift forces against the underside of the formwork.

The impact of wind is influenced by the location, width, length, and height of the formwork system. The impact of wind increases with height. For example, the wind force on the formwork and shoring on the 20th floor of a building will be higher than on the 1st floor of the building. It is necessary for a qualified engineer to properly design the formwork system to adequately resist wind forces.

References

1. ACI Committee 347, "Guide to Formwork for Concrete," American Concrete Institute, Detroit, MI, 2004.
2. APA-The Engineered Wood Association, "Concrete Forming," Tacoma, WA, 2004.
3. J. M. Barnes and D. W. Johnston, "Modification Factors for Improved Prediction of Fresh Concrete Lateral Pressure on Formwork," Institute of Construction, Department of Civil Engineering, 1999.
4. N. J. Gardner, "Pressure of Concrete against Formwork," *ACI Journal Proceedings*, Vol. 77, No. 4, 1980.
5. N. J. Gardner and P. T Ho, "Lateral Pressure of Fresh Concrete," *ACI Journal Proceedings*, Vol. 76, No. 7, 1979.
6. N. J. Gardner, "Lateral Pressure of Fresh Concrete—A Review," *ACI Journal Proceedings*, Vol. 82, No. 5, 1985.
7. R. L. Peurifoy, "Lateral Pressure of Concrete on Formwork" *Civil Engineering*, Vol. 35, 1965.

CHAPTER 4

Properties of Form Material

General Information

Materials used for forms for concrete structures include lumber, plywood, hardboard, fiberglass, plastics, fiber forms, corrugated boxes, steel, aluminum, magnesium, and plaster of paris. Additional materials include nails, bolts, screws, form ties, anchors, and other accessories. Forms frequently involve the use of two or more materials, such as plywood facing attached to steel frames, for wall panels.

Among the properties that form materials should possess are the following:

1. Adequate strength

2. Adequate rigidity

3. Surface smoothness, where required

4. Economy, considering initial cost and number of reuses

Properties of Lumber

Lumber used for formwork that is finished on all sides is designated as Surfaced-4-Sides (S4S) lumber. The cross-sectional dimensions of lumber are designated by nominal sizes, but the actual dimensions are less than the nominal dimensions. For example, a board designated as a nominal size of 2 in. by 12 in. has an actual size of 1½ in. by 11¼ in. after it is surfaced on all sides and edges. Commercial lumber is available in lengths that are multiples of 2 ft, with 18 ft as the common maximum length of an individual board. Table 4-1 provides the nominal and actual sizes for lumber that is commonly used for formwork.

The dressed sizes and properties of lumber appearing in Table 4-1 conform to the dimensions of lumber set forth in the U.S. Department

Nominal Size, in.	Actual Size, Thickness × Width, in.	Net Area, in.^{2}	$x-x$ X-X axis I_x, in.^{4}	S_x, in.^{3}	$y-y$ Y-Y axis I_y, in.^{4}	S_y, in.^{3}
1 × 4	¾ × 3½	2.62	2.68	1.53	0.123	0.328
1 × 6	¾ × 5½	4.12	10.40	3.78	0.193	0.516
1 × 8	¾ × 7¼	5.43	23.82	6.57	0.255	0.680
1 × 10	¾ × 9¼	6.93	49.47	10.70	0.325	0.867
1 × 12	¾ × 11¼	8.43	88.99	15.82	0.396	1.055
2 × 4	1½ × 3½	5.25	5.36	3.06	0.984	1.313
2 × 6	1½ × 5½	8.25	20.80	7.56	1.547	2.063
2 × 8	1½ × 7¼	10.87	47.63	13.14	2.039	2.719
2 × 10	1½ × 9¼	13.87	98.93	21.39	2.603	3.469
2 × 12	1½ × 11¼	16.87	177.97	31.64	3.164	4.219
3 × 4	2½ × 3½	8.75	8.93	5.10	4.56	3.65
3 × 6	2½ × 5½	13.75	34.66	12.60	7.16	5.73
3 × 8	2½ × 7¼	18.12	79.39	21.90	9.44	7.55
3 × 10	2½ × 9¼	23.12	164.88	35.65	12.04	9.64
3 × 12	2½ × 11¼	28.12	296.63	52.73	14.65	11.72
4 × 4	3½ × 3½	12.25	12.50	7.15	12.51	7.15
4 × 6	3½ × 5½	19.25	48.53	17.65	19.65	11.23
4 × 8	3½ × 7¼	25.38	111.14	30.66	25.90	14.80
4 × 10	3½ × 9¼	32.37	230.84	49.91	33.05	18.89
4 × 12	3½ × 11¼	39.37	415.28	73.83	40.20	22.97
5 × 5	4½ × 4½	20.25	34.17	15.19	34.17	15.19
6 × 6	5½ × 5½	30.25	76.25	27.73	76.26	27.73
6 × 8	5½ × 7½	41.25	193.35	51.56	103.98	37.81
6 × 10	5½ × 9½	52.25	393.96	82.73	131.71	47.90
6 × 12	5½ × 11½	63.25	697.06	121.2	159.44	57.98
8 × 8	7½ × 7½	56.25	263.7	70.31	263.67	70.31
8 × 10	7½ × 9½	71.25	535.9	112.8	333.98	89.06
8 × 12	7½ × 11½	86.25	950.5	165.3	404.29	107.81
10 × 10	9½ × 9½	90.25	678.8	142.9	678.75	142.90
10 × 12	9½ × 11½	109.25	1204.0	209.4	821.65	173.98
12 × 12	11½ × 11½	132.25	1457.5	253.5	1,457.50	253.48

TABLE 4-1 Properties of S4S Dry Lumber, Moisture Less Than 19%

of Commerce Voluntary Product Standard PS 20-70 (American Soft-wood Lumber Standard). All information and calculations appearing in this book are based on using lumber whose actual sizes are given in Table 4-1 for dry lumber. Dry lumber is defined as lumber that has been seasoned to a moisture content of 19% or less. Green lumber is defined as lumber having moisture content in excess of 19%. All calculations involving the dimensions of lumber should be based on using the actual sizes of lumber.

When calculations are made to determine certain properties of lumber, definitions, terms, and symbols are used as follows:

Cross Section: A section taken through a member perpendicular to its length, or longitudinal axis.

Neutral Axis: A line through a member (such as a beam or a column) under flexural stress on which there is neither tension nor compression. In Table 4-1, the neutral axis, designated as *X-X* or *Y-Y*, is at the mid-depth of the member and perpendicular to the loading of the member.

Moment of Inertia (I): The moment of inertia of the cross section of a beam is the sum of the products of each of its elementary areas times the square of the distance from the neutral axis of the section to the areas, multiplied by the square of their distance from the neutral axis of the section.

Section Modulus (S): The section modulus of the cross section of a beam is the moment of inertia of the section divided by the distance from the neutral axis to the most distant, or extreme, fiber of the section.

Figure 4-1 shows the cross section of a rectangular beam. The width and depth of the beam are denoted as *b* and *d*, respectively.

The following symbols and equations are generally used when making calculations related to the cross sections of rectangular beams:

A = cross-sectional area of a section
  = bd, in.^{2}
b = width of beam face on which load or force is applied, in.

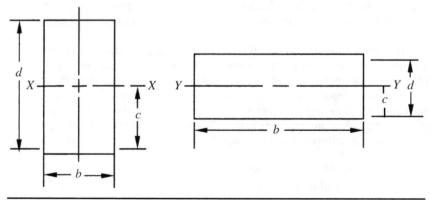

Figure 4-1   Symbols for cross section of rectangular beam.

d = depth or height of beam face parallel to the direction in which the load or force is applied, in.

I = area moment of inertia of the cross section of a beam

  = $bd^3/12$, in.^{4} for a rectangular beam

c = distance from neutral axis to most distant, or extreme, fiber of the beam, in.

S = section modulus of the cross section of a beam

  = $bd^2/6$, in.^{3} for a rectangular beam

r = radius of gyration of the cross section of a beam

  = $\sqrt{I/A}$, in.

E = modulus of elasticity, lb/in.^{2}

Q = static moment of area, a measure of resistance to shear, in.^{3}

X-X = the location of the neutral axis of the cross section of a beam for edgewise bending (load applied to narrow face). The X-X axis is often referred to as the strong axis of the beam

Y-Y = the location of the neutral axis of the cross section of a beam for flatwise bending (load applied to wide face); the Y-Y axis is often referred to as the weak axis of the beam

Allowable Stresses of Lumber

The loads applied to wood members create applied bending moments, shear forces, and compression forces in the member. The lumber used must have sufficient allowable stresses to resist the applied stresses. The magnitudes of the allowable stresses of lumber depend on the grade and species of wood, size of lumber, duration of load, moisture content, and other factors.

Organizations involved in the lumber industry have published design values for various grades and species of lumber based on normal loading conditions. Because the strength of wood varies with the conditions under which it is used, these design values should only be applied in conjunction with appropriate design and service recommendations from the 2005 National Design Specification (NDS) for Wood Construction published by the American Forest & Paper Association [1, 2].

The Supplement to the 2005 NDS provides reference design values for various grades and species of wood. It should be noted that reference design values for dimension lumber are different than for post and timber members. Allowable design stresses are determined by multiplying the reference design values from the NDS Supplement by adjustment factors that are presented in the NDS National Design Specification. The adjustment factors for sawn lumber are

1. Size adjustment factor, C_F

2. Load-duration factor, C_D

3. Wet service factor, C_M

4. Beam stability factor, C_L
5. Column stability adjustment factor, C_P
6. Flat use adjustment factor, C_{fu}
7. Bearing area adjustment factor, C_b
8. Buckling stiffness adjustment factor, C_T
9. Repetitive member adjustment factor, C_r
10. Incising adjustment factor, C_i
11. Temperature factor, C_t

The allowable design stress is obtained by multiplying the reference design value by appropriate adjustment factors as follows:

Allowable Stress for Bending,
$$F'_b = \text{(reference design value for bending)} \bullet [C_F \bullet C_D \bullet C_M \bullet C_L \bullet C_{fu} \bullet C_r \bullet C_i \bullet C_t]$$

Allowable Stress for Shear,
$$F'_v = \text{(reference design value for shear)} \bullet [C_D \bullet C_M \bullet C_i \bullet C_t$$

Allowable Compression Stress Perpendicular to Grain,
$$F'_{c\perp} = \text{(reference design value for compression} \perp \text{to grain)} \bullet [C_M \bullet C_b \bullet C_t \bullet C_i]$$

Allowable Compression Stress Parallel to Grain,
$$F_c = \text{(reference design value for compression // to grain)} \bullet [C_D \bullet C_M \bullet C_F \bullet C_P \bullet C_i \bullet C_t]$$

Allowable Modulus of Elasticity for Beam Deflection Calculations,
$$E' = \text{(reference design value of modulus of elasticity, } E) \bullet [C_M \bullet C_i \bullet C_t]$$

Allowable Modulus of Elasticity for Beam and Column Stability Calculations,
$$E'_{min} = \text{(reference design value of modulus of elasticity, } E_{min}) \bullet [C_M \bullet C_T \bullet C_i \bullet C_t]$$

The designer should become familiar with the specification in ref. [4] in order to properly apply the adjustment factors for the design of wood structures. The most common adjustment factors for construction of temporary structures and formwork are size, load-duration, moisture, beam and column stability, bearing, and flat use. These adjustment factors are presented in subsequent sections of this chapter.

The adjustment factors for temperature, incising, and repetitive member are less common in constructing temporary structures and formwork. Temperature adjustment factors apply when the wood is subjected to sustained elevated temperatures (100 to 150°F). For dimension lumber incised to facilitate preservative treatment, the incising factor C_i should be used. Incising adjustment factors apply to dimension lumber that is incised parallel to grain a maximum depth of 0.4 in., maximum length of ⅜ in., and density of incisions up to 1,100 per sq ft. The repetitive member adjustment factor permits a 15% increase in bending stress under specific conditions for joists, truss chords, rafters, planks, or similar members that are spaced less than 24 in. on centers. The adjustment factors for temperature, incising, and repetitive-member can be found in ref. [4].

Tables 4-2 and 4-3 give basic reference design values for lumber that is commonly used in construction of temporary structures and formwork. Values in these tables must be adjusted based on the particular condition in which the lumber will be used. Table 4-2 gives reference design values for Southern Pine lumber. The values in the table for Southern Pine are adjusted for size; therefore, $C_F = 1.0$. Table 4-3 gives the reference design values for several other species of lumber that are commonly used in construction. The values in Table 4-3 are not adjusted for size. Table 4-3a gives size adjustment factors C_F for the species of lumber shown in Table 4-3. The adjustment factors provided in Tables 4-4 through 4-8 apply to all of the grades and species of lumber given in Tables 4-2 and 4-3.

Adjustment Factor C_D for Load-Duration

The reference design values for lumber tabulated in Tables 4-2 and 4-3 are based on normal load-duration of 10 years. For loads that are applied for durations other than normal load-duration, the values should be multiplied by the appropriate adjustment factor shown in Table 4-4.

For many construction operations, particularly for formwork, the duration of the loads is usually less than 7 days. Thus, the tabulated design values shown in Tables 4-2 and 4-3 may be increased by 25%, provided the load-duration will not be more than 7 days. However, it should be noted that the modulus of elasticity and compression stress perpendicular to grain are not adjusted for load-duration.

Adjustment Factors C_M for Moisture Content

The reference design values shown in Tables 4-2 and 4-3 are based on lumber that with a moisture content of 19% or less. When dimension lumber is used where the moisture will exceed 19%, the values in Tables 4-2 and 4-3 should be reduced by the appropriate adjustment factors given in Table 4-5.

Reference Design Values for Southern Pine, lb per sq in

Nominal Size	Grade	Extreme Fiber in Bending F_b	Shear Parallel to Grain F_v	Compression Perpendicular to Grain $F_{c\perp}$	Compression Parallel to Grain $F_{c//}$	Modulus of Elasticity E	E_{min}
2"-4" thick and 2"-4" wide	No. 1	1,850	175	565	1,850	1,700,000	620,000
	No. 2	1,500	175	565	1,650	1,600,000	580,000
2"-4" thick and 6" wide	No. 1	1,650	175	565	1,750	1,700,000	620,000
	No. 2	1,250	175	565	1,600	1,600,000	580,000
2"-4" thick and 8" wide	No. 1	1,500	175	565	1,650	1,700,000	620,000
	No. 2	1,200	175	565	1,550	1,600,000	580,000
2"-4" thick and 10" wide	No. 1	1,300	175	565	1,600	1,700,000	620,000
	No. 2	1,050	175	565	1,500	1,600,000	580,000
2"-4" thick and 12" wide	No. 1	1,250	175	565	1,600	1,700,000	620,000
	No. 2	975	175	565	1,450	1,600,000	580,000
Multiplier[3] of values for loads <7 days		1.25	1.25	1.0	1.25	1.0	1.0
Multiplier[4] of values for moisture >19%		0.85*	0.97	0.67	0.8**	0.9	0.9

Notes:

1. Values shown are for visually graded Southern Pine, ref. [2].
2. Values for Southern Pine are already adjusted for size, $C_F = 1.0$.
3. See Tables 4-4 through 4-8 for other adjustment factors that may be applicable.
4. *denotes when $(F_b)(C_F) \leq 1,150$ lb per sq in., $C_M = 1.0$.
5. **denotes when $(F_b)(C_F) \leq 750$ lb per sq in., $C_M = 1.0$.
6. See the National Design Specification for Wood Construction for other grades and species of wood and for additional adjustments that may be appropriate for conditions where wood is used.

TABLE 4-2 Reference Design Values of Southern Pine with Less Than 19% Moisture

| Species and Nominal Size | Grade | Reference Design Value, lb per sq in | | | | | |
| | | Extreme Fiber in Bending F_b | Shear Parallel to Grain F_v | Compression Perpendicular to Grain $F_{c\perp}$ | Compression Parallel to Grain $F_{c//}$ | Modulus of Elasticity | |
						E	E_{min}
Douglas-Fir-Larch 2"–4" thick and 2" & wider	No. 1	1,000	180	625	1,500	1,700,000	620,000
	No. 2	900	180	625	1,350	1,600,000	580,000
Hem-Fir 2"–4" thick and 2" & wider	No. 1	975	150	405	1,350	1,500,000	550,000
	No. 2	850	150	405	1,300	1,300,000	470,000
Spruce-Pine-Fir 2"–4" thick and 2" & wider	No. 1	875	135	425	1,150	1,400,000	510,000
	No. 2	875	135	425	1,150	1,400,000	510,000
Multiplier[3] of values for loads <7 days		1.25	1.25	1.0	1.25	1.0	1.0
Multiplier[4] of values for moisture >19%		0.85*	0.97	0.67	0.8**	0.9	0.9

Notes:
1. Values shown are not size adjusted for the above species of lumber.
2. See Table 4-3a for applicable size adjustment factors.
3. See Tables 4-4 through 4-8 for other adjustment factors that may be applicable.
4. *denotes when $(F_b)(C_F) \leq 1{,}150$ lb per sq in., $C_M = 1.0$.
5. **denotes when $(F_b)(C_F) \leq 750$ lb per sq in., $C_M = 1.0$.
6. See the National Design Specification for Wood Construction for other grades and species of wood and for additional adjustments that may be appropriate for conditions where wood is used.

TABLE 4-3 Reference Design Values for Several Species of Lumber Less Than 19% Moisture

Width of Lumber	Size Adjustment Factor for Bending Stress, F_b		Size Adjustment for Compression Parallel to Grain $F_c //$
	Thickness		
	2-in. & 3-in.	4-in.	
2-in.	1.5	1.5	1.15
3-in.	1.5	1.5	1.15
4-in.	1.5	1.5	1.15
5-in.	1.4	1.4	1.1
6-in.	1.3	1.3	1.1
8-in.	1.2	1.3	1.05
10-in.	1.1	1.2	1.0
12-in.	1.0	1.1	1.0
14-in. & wider	0.9	1.0	0.9

TABLE 4-3a   Size Adjustment Factors C_F for Species of Lumber shown in Table 4-3

Load-Duration	C_D	Typical Design Load
Permanent	0.9	Dead Load
10 years	1.0	Occupancy Live Load
2 months	1.15	Snow Load
7 days	1.25	Construction Load
10 minutes	1.6	Wind/Earthquake Load
Impact	2.0	Impact Load

TABLE 4-4   Adjustment Factors C_D for Load-Duration

Bending Stress	Shear Stress	Compression Stress Perpendicular to Grain	Compression Stress Parallel to Grain	Modulus of Elasticity
0.85*	0.97	0.67	0.8**	0.9

Notes:
1. * denotes when $(F_b)(C_F) \leq 1,150$ lb per sq in., $C_M = 1.0$.
2. ** denotes when $(F_b)(C_F) \leq 750$ lb per sq in., $C_M = 1.0$.

TABLE 4-5   Adjustment Factor for Moisture Content Greater Than 19%

d/b Range	Lateral Constraints for Bending Stability
$d/b \leq 2$	No lateral support is required; includes 2×4, 3×4, 4×6, 4×8, and any members used flat, examples are 4×2, 6×2, 8×2
$2 < d/b \leq 4$	Ends shall be held in position, as by full depth solid blocking, bridging, hangers, nailing, bolting, or other acceptable means; examples are 2×6, 2×8, 3×8
$4 < d/b \leq 5$	Compression edge of the member shall be held in line for its entire length to prevent lateral displacement and ends of member at point of bearing shall be held in position to prevent rotation and/or lateral displacement; such as a 2×10.
$5 < d/b \leq 6$	Bridging, full depth solid blocking, or diagonal cross-bracing must be installed at intervals not to exceed 8 ft, compression edge must be held in line for its entire length, and ends shall be held in position to prevent rotation and/or lateral displacement; example is a 2×12
$6 < d/b \leq 7$	Both edges of the member shall be held in line for its entire length and ends at points of bearing shall be held in position to prevent rotation and/or lateral displacement; examples are deep and narrow rectangular members

Note: d and b are nominal dimensions.

TABLE 4-6 Lateral Constraints for Stability of Bending Members

Width (depth)	Adjustment Factor for Bending Stress, F_b	
	Thickness (breadth)	
	2-in. and 3-in.	4-in.
2-in.	1.0	–
3-in.	1.0	–
4-in.	1.1	1.0
5-in.	1.1	1.05
6-in.	1.15	1.05
8-in.	1.15	1.05
10-in. and wider	1.2	1.1

TABLE 4-7 Adjustment Factors C_{fu} for Flat Use

Bearing Length	0.5-in.	1.0-in.	1.5-in.	2.0-in.	3.0-in.	4.0-in.	5.0-in.	≥ 6.0-in.
Value of C_b	1.75	1.38	1.25	1.19	1.13	1.10	1.075	1.0

TABLE 4-8   Adjustment Factors C_b for Bearing Area

Adjustment Factor C_L for Beam Stability

The beam stability factor C_L in the NDS is a multiplier to reduce the reference design value of bending stress based on the depth-to-thickness ratio (d/b) of the beam. When the conditions in Table 4-6 are satisfied, the beam stability factor $C_L = 1.0$; therefore, no adjustment is required in the reference design value in order to determine the allowable bending stress. The (d/b) ratios in Table 4-6 are based on nominal dimensions. For conditions where the support requirements in Table 4-6 are not met, the designer should follow the requirements of the NDS to reduce the reference design value by the adjustment factor C_L in order to determine the allowable stress in bending.

It is often possible to design members to satisfy the conditions in Table 4-6 to allow the designer to utilize the full allowable bending stress. One example of providing lateral support for bending members is by placing decking on the top of beams. Although the beam is used to support the decking, the decking can be fastened along the compression edge of the beam throughout the length of the beam to provide lateral stability. There are other methods that can be used by the designer to satisfy the requirements of providing adequate lateral bracing of bending members.

Adjustment Factor C_P for Column Stability

The column stability factor C_P in the NDS is a multiplier to reduce the reference design value of compression stress parallel to grain based on the slenderness ratio (l_e/d) of the beam. Shores are column members with axial loads that induce compression stresses that act parallel to the grain. Reference design values are presented in Tables 4-2 and 4-3 for compression stresses parallel to grain. However, these compression stresses do not consider the length of a member, which may affect its stability and strength. A column must be properly braced to prevent lateral buckling because its strength is highly dependent on its effective length.

The slenderness ratio for a column is the ratio of its effective length divided by the least cross-sectional dimension of the column, l_e/d. It is used to determine the allowable load that can be placed on a column. The allowable load decreases rapidly as the slenderness ratio increases. For this reason, long shores should be cross braced in two directions with one or more rows of braces.

The following equation for calculating the column stability factor C_P takes into consideration the modes of failure, combinations of crushing and buckling of wood members subjected to axial compression parallel to grain.

$$C_P = \left[[1 + (F_{ce}/F_c^*)]/2c - \sqrt{[1 + (F_{ce}/F_c^*)]/2c]^2 - (F_{ce}/F_c^*)/c} \right]$$

where F_c^* = compression stress parallel to grain, lb per sq in., obtained by multiplying the reference design value for compression stress parallel to grain by all applicable adjustment factors except C_P; $F_c^* = F_{C//} (C_D \times C_M \times C_F \times C_i \times C_t)$

$F_{ce} = 0.822 \, E'_{min}/(l_e/d)^2$, lb per sq in., representing the impact of Euler buckling

E'_{min} = modulus of elasticity for column stability, lb per sq in., obtained by multiplying the reference design value of E_{min} by all applicable adjustment factors; $E'_{min} = E_{min}$ $(C_M \bullet C_T \bullet C_i \bullet C_t)$

l_e/d = slenderness ratio, ratio of effective length, in., to least cross-sectional dimension, in. Note: For wood columns, l_e/d should never exceed 50.

$c = 0.8$ for sawn lumber

Adjustment Factors C_{fu} for Flat Use

Bending values in Tables 4-2 and 4-3 are based on loads applied to the narrow face of dimension lumber. When the lumber is laid flat and loaded perpendicular to the wide face, the reference design value for bending may be increased by the flat use factor, C_{fu}. Table 4-7 gives values of the flat use factor for dimension lumber when it is laid flat.

Adjustment Factors C_b for Bearing Area

In Tables 4-2 and 4-3 the reference design values for compression perpendicular to grain ($F_{C\perp}$) apply to bearing on a wood member. For bearings less than 6 in. in length and not nearer than 3 in. from the end of a member, the bearing area factor (C_b) can be used to account for an effective increase in bearing length. The bearing length is the dimension of the contact area measured parallel to the grain. The bearing area factor can be calculated by the equation C_b = (bearing length + ⅜ in.)/(bearing length). Values for C_b are always greater than 1.0, as shown in Table 4-8.

Application of Adjustment Factors

The designer must assess the condition the lumber will be used and then determine applicable adjustment factors that should be applied to the reference design values. Knowledge of the specifications of the NDS, ref. [4], is necessary to properly design wood members. The following examples illustrate applying adjustment factors to reference design values to determine allowable stresses. Additional examples are given in Chapter 5.

Example 4-1

A 3×8 beam of No. 2 grade Hem-Fir will be used to temporarily support a load at a job site during construction. The beam will be dry, less than 19% moisture content, and the applied loads will be less than 7 days. The adjusted stresses for bending and shear stresses for this beam can be calculated as follows:

From Table 4-3, the reference design value for bending of a 3×8 No. 2 grade Hem-Fir for bending is 850 lb per sq in. The allowable bending stress is obtained by multiplying the reference design value by the adjustment factors for size and load-duration. Table 4-3a shows a size factor for a 3×8 beam in bending as $C_F = 1.2$ and Table 4-4 shows a load-duration factor as $C_D = 1.25$. Therefore, the adjusted bending stress can be calculated as:

F'_b = (reference bending design value) $\times C_F \times C_D$

$\quad = (850 \text{ lb per sq in.})(1.2)(1.25)$

$\quad = 1,275 \text{ lb per sq in.}$

Before determining the bending adequacy of this member to resist the applied loads, the support constraints related to the d/b ratio must be satisfied as discussed in Chapter 5. Also, the beam stability factor must be in accordance with the NDS.

For shear Table 4-3 shows the reference design value is 150 lb per sq in. and Table 4-4 shows a load-duration adjustment factor $C_D = 1.25$. Therefore, the adjusted shear stress can be calculated as:

F'_v = (reference shear design value) $\times C_D$

$\quad = (150 \text{ lb per sq in.})(1.25)$

$\quad = 187.5 \text{ lb per sq in.}$

Example 4-2

Lumber of 2×8 of No. 1 grade Southern Pine will be laid flat as formwork for underside of a concrete beam. The load-duration will be less than 7 days and the lumber will be used in a wet condition. Determine the adjusted stress for bending.

From Table 4-2, the reference bending stress is 1,500 lb per sq in. No size adjustment is necessary because the design values in Table 4-2 are already adjusted for size. Adjustments for load-duration, moisture, and flat use can be obtained as follows:

From Table 4-4, the adjustment factor for load-duration $C_D = 1.25$
From Table 4-5, the adjustment factor for wet condition $C_M = 0.85$
From Table 4-7, the adjustment for flat-use $C_{fu} = 1.15$

The adjusted bending stress can be calculated as follows:

$F'_b = $ (reference bending design value) $\times (C_D) \times (C_M) \times (C_{fu})$

$= (1,500 \text{ lb per sq in.})(1.25)(.0.85)(1.15)$

$= 1,832 \text{ lb per sq in.}$

Plywood

Plywood is used extensively for formwork for concrete, especially for sheathing, decking, and form linings. Among the advantages are smooth surfaces, availability in a variety of thicknesses and lengths, and ease of handling during construction.

As presented later in this chapter, the plywood industry manufactures a special plywood called Plyform specifically for use in forming concrete structures. Plywood should not be confused with Plyform.

Plywood may be manufactured with interior glue or exterior glue. It is necessary to use exterior-glue plywood for formwork. In all exterior-type plywood manufactured under standards established by APA—The Engineered Wood Association, waterproof glue is used to join the plies together to form multi-ply panels. Also, the edges of the panels may be sealed against moisture by the mill that makes the panels.

Plywood is available in 4-ft widths and 8-ft lengths with thicknesses from ¼ in. through 1⅛ in. Larger sizes are available, such as 5 ft wide and 12 ft long. Plywood is manufactured to precise tolerances, within 1⁄16 in. for the width and length of the panel. The panel thickness is within 1⁄16 in. for panels specified as ¾ in. or less in thickness, and plus or minus 3% of the specified thickness for panels thicker than ¾ in.

Plywood is made in panels consisting of odd numbers of plies, each placed at right angles to the adjacent ply, which accounts for the physical properties that make it efficient in resisting bending, shear, and deflection. Therefore, the position in which a panel is attached to the supporting members will determine its strength. For example, for a panel 4 ft wide and 8 ft long, the fibers of the surface plies are parallel to the 8-ft length. Such a panel, installed with the outer fibers perpendicular to the supporting members, such as studs or joists, is stronger than it would be if the panel were attached with the outer fibers parallel to the supporting members. This is indicated by the accompanying tables.

Plywood is graded by the veneers of the plies. Veneer Grade N or A is the highest grade level. Grade N is intended for natural finish, whereas Grade A is intended for a plane-table surface. Plywood graded as N or A has no knots or restricted patches. Veneer Grade B has a solid surface but may have small round knots, patches, and round plugs. Grade B is commonly used for formwork. The lowest grade of exterior-glue plywood is a veneer Grade C, which has small knots, knotholes, and patches.

Table 4-9 gives the section properties of plywood per foot of width. The effective section properties are computed taking into account the species of wood used for the inner and outer plies and the variables involved for each grade. Because this table represents a wide variety of grades, the section properties presented are generally the minima that can be expected.

Allowable Stresses for Plywood

The allowable stresses for plywood are shown in Table 4-10. When the allowable stresses are used from this table, the plywood must be manufactured in accordance with Voluntary Product Standard PS 1-95 and must be identified by the APA—The Engineered Wood Association.

The allowable stresses are divided into three levels, which are related to grade. Plywood with exterior glue, and with face and back plies containing only N, A, or C veneers, must use level 1 (S-1) stresses. Plywood with exterior glue, and with B, C plugged, or D veneers in either face or back, must use level 2 (S-2) stresses. All grades with interior or intermediate glue (IMG or exposure 2) shall use level 3 (S-3) stresses.

The woods that may be used to manufacture plywood under Voluntary Product Standard PS 1-95 are classified into five groups based on the elastic modulus in bending and important strength properties. Group 1 is the highest grade, and Group 5 is the lowest grade. Groups 1 and 2 are the common classifications used for formwork. A list of the species that are in each group classification is provided in ref. [2].

Plyform

The plywood industry produces a special product, designated as Plyform, for use in forming concrete. Plyform is exterior-type plywood limited to certain species of wood and grades of veneer to ensure high performance as a form material. The term is proprietary and may be applied only to specific products that bear the trademark of APA—The Engineered Wood Association. Products bearing this identification are available in two classes: Plyform Class I and Plyform Class II. Class I is stronger and stiffer than Class II. For special applications, an APA Structural I Plyform is available that is stronger and stiffer than Plyform Classes I and II. It is recommended for high pressures where the face grain is parallel to supports.

Thickness see Note #3 below	Approx Weight (psf)	t Effective Thickness for Shear, in.	Properties for Stress Applied Parallel to Face Grain				Properties for Stress Applied Perpendicular to Face Grain			
			A Cross Sectional Area, in.²/ft	I Moment of Inertia, in.⁴/ft	S_e Effective Section Modulus, in.³/ft	Ib/Q Rolling Shear Constant, in.²/ft	A Cross Sectional Area, in.²/ft	I Moment of Inertia, in.⁴/ft	S_e Effective Section Modulus. in.³/ft	Ib/Q Rolling Shear Constant, in.²/ft
Different Species Group see Note #3 below										
¼-S	0.8	0.267	0.996	0.008	0.059	2.010	0.348	0.001	0.009	2.019
⅜-S	1.1	0.288	1.307	0.027	0.125	3.088	0.626	0.002	0.023	3.510
½-S	1.5	0.425	1.947	0.077	0.236	4.466	1.240	0.009	0.087	2.752
⅝-S	1.8	0.550	2.475	0.129	0.339	5.824	1.528	0.027	0.164	3.119
¾-S	2.2	0.568	2.884	0.197	0.412	6.762	2.081	0.063	0.285	4.079
⅞-S	2.6	0.586	2.942	0.278	0.515	8.050	2.651	0.104	0.394	5.078
1-S	3.0	0.817	3.721	0.423	0.664	8.882	3.163	0.185	0.591	7.031
1⅛-S	3.3	0.836	3.854	0.548	0.820	9.883	3.180	0.271	0.744	8.428

1/4-S	0.8	0.342	1.280	0.012	0.083	2.009	0.626	0.001	0.013	2.723
3/8-S	1.1	0.373	1.680	0.038	0.177	3.086	1.126	0.002	0.033	4.927
1/2-S	1.5	0.545	1.947	0.078	0.271	4.457	2.232	0.014	0.123	2.725
5/8-S	1.8	0.717	3.112	0.131	0.361	5.934	2.751	0.045	0.238	3.073
3/4-S	2.2	0.748	3.848	0.202	0.464	6.189	3.745	0.108	0.418	4.047
7/8-S	2.6	0.778	3.952	0.288	0.569	7.539	4.772	0.179	0.579	5.046
1-S	3.0	1.091	5.215	0.479	0.827	7.978	5.693	0.321	0.870	6.981
1 1/8-S	3.3	1.121	5.593	0.623	0.955	8.841	5.724	0.474	1.098	8.377

Notes:

1. Courtesy, APA—The Engineered Wood Association, "Plywood Design Specification," 1997.
2. All section properties are for sanded panels in accordance with APA manufactured specifications.
3. The section properties presented here are specifically for face plies of different species group from inner plies (includes all Product Standard Grades except Structural I and Marine).
4. These section properties presented here are for Structural I and Marine.

TABLE 4-9 Effective Section Properties of Sanded Plywood with 12-in. Widths

Type of Stress		Species Group of Face Ply	Grade Stress Level				
			S-1		S-2		S-3
			Wet	Dry	Wet	Dry	Dry Only
Extreme Fiber Stress in Bending (F_b) Tension in Plane of Plies (F_t) Face Grain Parallel or Perpendicular to Span (at 45° to Face Grain, Use 1/6 F_t)	F_b and F_t	1	1,430	2,000	1,190	1,650	1,650
		2, 3	980	1,400	820	1,200	1,200
		4	940	1,330	780	1,110	1,110
Compression in Plane of Plies (F_c) Parallel or Perpendicular to Face Grain (at 45° to Face Grain, Use 1/3 F_c)	F_c	1	970	1,640	900	1,540	1,540
		2	730	1,200	680	1,100	1,100
		3	610	1,060	580	950	950
		4	610	1,000	580	950	950
Shear through the Thickness (F_v) Parallel or Perpendicular to Face Grain (at 45° to Face Grain, Use 2 F_v)	F_v	1	155	190	155	190	160
		2, 3	120	140	120	140	120
		4	110	130	110	130	115
Rolling Shear (in the Plane of Plies) Parallel or Perpendicular to Face Grain (at 45° to Face Grain, Use 1 1/3 F_s)	F_s	Marine and Structural I	63	75	63	75	–
		All Others (see note 3 below)	44	53	44	53	48

TABLE 4-10   Allowable Stresses, in Pounds per Square Inch, for Plywood

Type of Stress		Species Group of Face Ply	Grade Stress Level				
			S-1		S-2		S-3
			Wet	Dry	Wet	Dry	Dry Only
Modulus of Rigidity (or Shear Modulus) Shear in Plane Perpendicular to Plies (through the Thickness) (at 45° to Face Grain, Use 4G)	G	1	70,000	90,000	70,000	90,000	82,000
		2	60,000	75,000	60,000	75,000	68,000
		3	50,000	60,000	50,000	60,000	55,000
		4	45,000	50,000	45,000	50,000	45,000
Bearing (on Face) Perpendicular to Plane of Plies	$F_{c\perp}$	1	210	340	210	340	340
		2, 3	135	210	135	210	210
		4	105	160	105	160	160
Modulus of Elasticity in Bending in Plane of Plies Face Grain Parallel or Perpendicular to Span	E	1	1,500,000	1,800,000	1,500,000	1,800,000	1,800,000
		2	1,300,000	1,500,000	1,300,000	1,500,000	1,500,000
		3	1,100,000	1,200,000	1,100,000	1,200,000	1,200,000
		4	900,000	1,000,000	900,000	1,000,000	1,000,000

Notes:
1. Courtesy, APA—The Engineered Wood Association, see ref. [3] pages 12 and 13 for Guide.
2. To qualify for stress level S-1, gluelines must be exterior and only veneer grades N, A, and C (natural, not repaired) are allowed in either face or back. For stress level S-2, glueline must be exterior and veneer grade B, C-Plugged and D are allowed on the face or back. Stress level S-3 includes all panels with interior or intermediate (IMG) gluelines.
3. Reduce stresses 25% for 3-layer (4- or 5-ply) panels over ⅝ in. thick. Such layups are possible under PS 1-95 for APA rated sheathing, APA rated Sturd-I-Floor, Underlayment, C-C Plugged and C-D Plugged grades over ⅝ in. through ¾ in. thick.
4. Shear-through-the-thickness stresses for Marine and special Exterior grades may be increased 33%. See Section 3.8.1 of ref. [3] for conditions under which stresses for other grades may be increased.

TABLE 4-10   Allowable Stresses, in Pounds per Square Inch, for Plywood (*Continued*)

The panels are sanded on both sides and oiled at the mill unless otherwise specified. The face oiling reduces the penetration of moisture and also acts as a release agent when the forms are removed from the concrete. Although the edges are sealed at the mill, any exposed edges resulting from fabrication at a job should be sealed with appropriate oil before the plywood is used for formwork.

The section properties of Plyform Classes I and II, and Structural I Plyform are given in Table 4-11. The allowable stresses are shown Table 4-12.

High-Density Overlaid Plyform

Both classes of Plyform are available with a surface treatment of hard, smooth, semiopaque, thermosetting resin-impregnated materials that form a durable bond with the plywood. The plywood industry uses the name HDO (High-Density Overlay) to describe the overlaid surface. Overlaid Plyform panels have a tolerance of plus or minus $\frac{1}{32}$ in. for all thicknesses through $\frac{13}{16}$ in. Thicker panels have a tolerance of 5% over or under the specified thickness. The treated surface is highly resistant to abrasion, which has permitted in excess of 100 uses under favorable conditions and with good care. The abrasion-resistant surface does not require oiling. However, it may be desirable to lightly wipe the surface of the panels with oil or other release agents before each pour to ensure easy stripping of the panels.

Equations for Determining the Allowable Pressure on Plyform

Equations (4-1) through (4-7) can be used to calculate the maximum pressures of concrete on Plyform used for decking or sheathing. When computing the allowable pressure of concrete on Plyform, use the center-to-center distances between supports to determine the pressure based on the allowable bending stress in the fibers.

When computing the allowable pressure of concrete on Plyform as limited by the permissible deflection of the plywood, use the clear span between the supports plus $\frac{1}{4}$ in. for supports whose nominal thicknesses are 2 in. and the clear span between the supports plus $\frac{5}{8}$ in. for the supports whose nominal thicknesses are 4 in.

When computing the allowable pressure of concrete on Plyform as limited by the allowable unit shearing stress and shearing deflection of the plywood, use the clear span between the supports.

The recommended concrete pressures are influenced by the number of continuous spans. For face grain across supports, assume three continuous spans up to a 32-in. support spacing and two spans for greater spacing. For face grain parallel to supports, assume three

Thickness in.	Approx weight lb/ft²	Properties for Stress Applied Parallel to Face Grain (face grain across supports)			Properties for Stress Applied Perpendicular to Face Grain (face grain along supports)		
		I Moment of Inertia in.⁴/ft	S_e Effective Section Modulus in.³/ft	Ib/Q Rolling Shear Constant in.²/ft	I Moment of Inertia in.⁴/ft	S_e Effective Section Modulus in.³/ft	Ib/Q Rolling Shear Constant in.²/ft
Class I							
15/32	1.4	0.066	0.244	4.743	0.018	0.107	2.419
1/2	1.5	0.077	0.268	5.153	0.024	0.130	2.739
19/32	1.7	0.115	0.335	5.438	0.029	0.146	2.834
5/8	1.8	0.130	0.358	5.717	0.038	0.175	3.094
23/32	2.1	0.180	0.430	7.009	0.072	0.247	3.798
3/4	2.2	0.199	0.455	7.187	0.092	0.306	4.063
7/8	2.6	0.296	0.584	8.555	0.151	0.422	6.028
1	3.0	0.427	0.737	9.374	0.270	0.634	7.014
1 1/8	3.3	0.554	0.849	10.430	0.398	0.799	8.419
Class II							
15/32	1.4	0.063	0.243	4.499	0.015	0.138	2.434
1/2	1.5	0.075	0.267	4.891	0.020	0.167	2.727
19/32	1.7	0.115	0.334	5.326	0.025	0.188	2.812
5/8	1.8	0.130	0.357	5.593	0.032	0.225	3.074
23/32	2.1	0.180	0.430	6.504	0.060	0.317	3.781
3/4	2.2	0.198	0.454	6.631	0.075	0.392	4.049
7/8	2.6	0.300	0.591	7.990	0.123	0.542	5.997
1	3.0	0.421	0.754	8.614	0.220	0.812	6.987
1 1/8	3.3	0.566	0.869	9.571	0.323	1.023	8.388
Structural I							
15/32	1.4	0.067	0.246	4.503	0.021	0.147	2.405
1/2	1.5	0.078	0.271	4.908	0.029	0.178	2.725
19/32	1.7	0.116	0.338	5.018	0.034	0.199	2.811
5/8	1.8	0.131	0.361	5.258	0.045	0.238	3.073
23/32	2.1	0.183	0.439	6.109	0.085	0.338	3.780
3/4	2.2	0.202	0.464	6.189	0.108	0.418	4.047
7/8	2.6	0.317	0.626	7.539	0.179	0.579	5.991
1	3.0	0.479	0.827	7.978	0.321	0.870	6.981
1 1/8	3.3	0.623	0.955	8.841	0.474	1.098	8.377

Notes:
1. Courtesy, American Plywood Association, "Concrete Forming," 2004.
2. The section properties presented here are specifically for Plyform, with its special layup restrictions. For other grades, section properties are listed in the Plywood Design Specification, page 16 of ref. [3].

TABLE 4-11   Effective Section Properties of Plyform with 1-ft Widths

Item		Plyform Class I	Plyform Class II	Structural I Plyform
Bending stress, lb/in.²	F_b	1,930	1,330	1,930
Rolling shear stress, lb/in.²	F_s	72	72	102
Modulus of Elasticity— (psi, adjusted, use for bending deflection calculation)	E	1,650,000	1,430,000	1,650,000
Modulus of Elasticity— (lb/in.², unadjusted, use for shear deflection calculation)	E_e	1,500,000	1,300,000	1,500,000

Note:
1. Courtesy, APA—The Engineered Wood Association.

TABLE 4-12   Allowable Stresses for Plyform, lb per sq in.

spans up to 16 in. and two spans for 20 and 24 in. These are general rules as recommended by the APA. For specific applications, other span continuity relations may apply.

There are many combinations of frame spacings and plywood thicknesses that may meet the structural requirements for a particular job. However, it is recommended that only one thickness of plywood be used, and then the frame spacing varied for the different pressures. Plyform can be manufactured in various thicknesses, but it is good practice to base design on the most commonly available thicknesses.

Allowable Pressure Based on Fiber Stress in Bending

For pressure controlled by bending stress, use Eqs. (4-1) and (4-2).

$$w_b = 96F_b S_e / (l_b)^2 \qquad \text{for one or two spans} \qquad (4\text{-}1)$$

$$w_b = 120F_b S_e / (l_b)^2 \qquad \text{for three or more spans} \qquad (4\text{-}2)$$

where w_b = uniform pressure causing bending, lb per sq ft
F_b = allowable bending stress in plywood, lb per sq in.
S_e = effective section modulus of a plywood strip 12 in. wide, in.³/ft
l_b = length of span, center to center of supports, in.

Allowable Pressure Based on Bending Deflection

For pressure controlled by bending deflection, use Eqs. (4-3) and (4-4).

$$w_d = 2220EI\Delta_b/(l_s)^4 \qquad \text{for two spans} \qquad (4\text{-}3)$$

$$w_d = 1743EI\Delta_b/(l_s)^4 \qquad \text{for three or more spans} \qquad (4\text{-}4)$$

where Δ_b = permissible deflection of plywood, in.

$\quad$ w_d = uniform pressure causing bending deflection, lb per sq ft

$\quad$ l_s = clear span of plywood plus ¼ in. for 2-in. supports, and clear span plus ⅝ in. for 4-in. supports, in.

$\quad$ E = adjusted modulus of elasticity of plywood, lb per sq in.

$\quad$ I = moment of inertia of a plywood strip 12 in. wide, in.^{4}/ft

Allowable Pressure Based on Shear Stress

For pressure controlled by shear stresses, use Eqs. (4-5) and (4-6).

$$w_s = 19.2F_s(Ib/Q)/l_s \qquad \text{for two spans} \qquad (4\text{-}5)$$

$$w_s = 20F_s(Ib/Q)/l_s \qquad \text{for three spans} \qquad (4\text{-}6)$$

where $\quad$ w_s = uniform load on plywood causing shear, lb per sq ft

$\quad$ F_s = allowable rolling shear stress, lb per sq in.

$\quad$ Ib/Q = rolling shear constant, in.^{2}/ft of width

$\quad$ l_s = clear span between supports, in.

Allowable Pressure Based on Shear Deflection

For pressure controlled by shear deflection, use Eq. (4-7).

$$\Delta_s = w_s Ct^2(l_s)^2/1270E_eI \qquad (4\text{-}7)$$

where Δ_s = permissible deflection of the plywood, in.

$\quad$ w_s = uniform load causing shear deflection, lb per sq ft

$\quad$ C = a constant, equal to 120 for the face grain of plywood Perpendicular to the supports, and equal to 60 for the face grain of the plywood parallel to supports

$\quad$ t = thickness of plywood, in.

$\quad$ E_e = modulus of elasticity of the plywood, unadjusted for shear deflection, lb per sq in.

$\quad$ I = moment of inertia of a plywood strip 12 in. wide, in.^{4}/ft

Plywood Thick- ness, in.	Support Spacing, in.						
	4	**8**	**12**	**16**	**20**	**24**	**32**
15/32	2,715 (2,715)	885 (885)	355 (395)	150 (200)	– 115	– –	– –
1/2	2,945 (2,945)	970 (970)	405 (430)	175 (230)	100 (135)	– –	– –
19/32	3,110 (3,110)	1,195 (1,195)	540 (540)	245 (305)	145 (190)	– (100)	– –
5/8	3,270 (3,270)	1,260 (1,260)	575 (575)	265 (325)	160 (210)	– (100)	– –
23/32	4,010 (4,010)	1,540 (1,540)	695 (695)	345 (325)	210 (270)	110 (145)	– –
3/4	4,110 (4,110)	1,580 (1,580)	730 (730)	370 (410)	225 (285)	120 (160)	– –
1 1/8	5,965 (5,965)	2,295 (2,295)	1,370 (1,370)	740 (770)	485 (535)	275 (340)	130 (170)

Notes:
1. Courtesy APA—The Engineered Wood Association, "Concrete Forming," 2004.
2. Deflection limited to $l/360$th of the span, $l/270$th for values in parentheses.
3. Plywood continuous across two or more spans.

TABLE 4-13   Recommended Maximum Pressure on Plyform Class I—Values in Pounds per Square Foot with Face Grain across Supports

Tables for Determining the Allowable Concrete Pressure on Plyform

Tables 4-13 through 4-16 give the recommended maximum pressures of Class I and Structural I Plyform. Calculations for these pressures were based on deflection limitations of $l/360$ and $l/270$ of the span, or shear or bending strength, whichever provided the most conservative (lowest load) value.

Maximum Spans for Lumber Framing Used to Support Plywood

Tables 4-17 and 4-18 give the maximum spans for lumber framing members, such as studs and joists that are used to support plywood subjected to pressure from concrete. The spans listed in Table 4-17 are based on using No. 2 Douglas Fir or No. 2 Southern Pine. The spans listed in Table 4-18 are based on using No. 2 Hem-Fir.

Plywood Thickness, in.	Support Spacing, in.					
	4	8	12	16	20	24
15/32	1,385	390	110	–	–	–
	(1,385)	(390)	(150)	–	–	–
1/2	1,565	470	145	–	–	–
	(1,565)	(470)	(195)	–	–	–
19/32	1,620	530	165	–	–	–
	(1,620)	(530)	(225)	–	–	–
5/8	1,770	635	210	–	–	–
	(1,770)	(635)	(280)	120	–	–
23/32	2,170	835	375	160	115	–
	(2,170)	(835)	(400)	(215)	(125)	–
3/4	2,325	895	460	200	145	–
	(2,325)	(895)	(490)	(270)	(155)	(100)
1 1/8	4,815	1,850	1,145	710	400	255
	(4,815)	(1,850)	(1,145)	(725)	(400)	(255)

Notes:
1. Courtesy APA—The Engineered Wood Association, "Concrete Forming," 2004.
2. Deflection limited to $l/360$th of the span, $l/270$th for values in parentheses.
3. Plywood continuous across two or more spans.

TABLE 4-14 Recommended Maximum Pressure on Plyform Class I—Values in Pounds per Square Foot with Face Grain Parallel to Supports

Plywood Thickness, in.	Support Spacing, in.						
	4	8	12	16	20	24	32
15/32	3,560	890	360	155	115	–	–
	(3,560)	(890)	(395)	(205)	115	–	–
1/2	3,925	980	410	175	100	–	–
	(3,925)	(980)	(435)	(235)	(135)	–	–
19/32	4,110	1,225	545	245	145	–	–
	(4,110)	(1,225)	(545)	(305)	(190)	(100)	–
5/8	4,305	1,310	580	270	160	–	–
	(4,305)	(1,310)	(580)	(330)	(215)	100	–
23/32	5,005	1,590	705	350	210	110	–
	(5,005)	(1,590)	(705)	(400)	(275)	(150)	–
3/4	5,070	1,680	745	375	230	120	–
	(5,070)	(1,680)	(745)	(420)	(290)	(160)	–
1 1/8	7,240	2,785	1,540	835	545	310	145
	(7,240)	(2,785)	(1,540)	(865)	(600)	(385)	(190)

Notes:
1. Courtesy APA—The Engineered Wood Association, "Concrete Forming," 2004.
2. Deflection limited to $l/360$th of the span, $l/270$th for values in parentheses.
3. Plywood continuous across two or more spans.

TABLE 4-15 Recommended Maximum Pressures on Structural I Plyform—Values in Pounds per Square Foot with Face Grain Across Supports

Plywood Thickness, in.	Support Spacing, in.					
	4	8	12	16	20	24
15/32	1,970 (1,970)	470 (530)	130 (175)	– –	– –	– –
1/2	2,230 (2,230)	605 (645)	175 (230)	– –	– –	– –
19/32	2,300 (2,300)	640 (720)	195 (260)	– (110)	– –	– –
5/8	2,515 (2,515)	800 (865)	250 (330)	105 (140)	– (100)	– –
23/32	3,095 (3,095)	1,190 (1,190)	440 (545)	190 (255)	135 (170)	– –
3/4	3,315 (3,315)	1,275 (1,275)	545 (675)	240 (315)	170 (210)	– (115)
1 1/8	6,860 (6,860)	2,640 (2,640)	1,635 (1,635)	850 (995)	555 (555)	340 (355)

Notes:
1. Courtesy APA—The Engineered Wood Association, "Concrete Forming," 2004.
2. Deflection limited to $l/360$th of the span, $l/270$th for values in parentheses.
3. Plywood continuous across two or more spans.

TABLE 4-16 Recommended Maximum Pressures on Structural I Plyform—Values in Pounds per Square Foot with Face Grain Parallel to Supports

The allowable stresses are based on a load-duration less than 7 days and moisture content less than 19%. The deflections are limited to $l/360$ with maxima not to exceed ¼ in. Spans are measured center to center on the supports.

Use of Plywood for Curved Forms

Plywood is readily adaptable as form material where curved surfaces of concrete are desired. Table 4-19 lists the minimum bending radii for mill-run Plyform panels of the thicknesses shown when bent dry. Shorter radii can be developed by selecting panels that are free of knots or short grain and by wetting or steaming the plywood prior to bending.

Hardboard

Tempered hardboard, which is sometimes used to line the inside surfaces of forms, is manufactured from wood particles that are impregnated with a special tempering liquid and then polymerized by baking.

Equivalent Uniform Load lb per ft	Douglas Fir #2 or Southern Pine #2 Continuous over 2 or 3 Supports (1 or 2 Spans) Nominal Size Lumber							Douglas Fir #2 or Southern Pine #2 Continuous over 4 or more Supports (3 or more Spans) Nominal Size of Lumber						
	2×4	2×6	2×8	2×10	4×4	4×6	4×8	2×4	2×6	2×8	2×10	4×4	4×6	4×8
200	48	73	92	113	64	97	120	56	81	103	126	78	114	140
400	35	52	65	80	50	79	101	39	58	73	89	60	88	116
600	29	42	53	65	44	64	85	32	47	60	73	49	72	95
800	25	36	46	56	38	56	72	26	41	52	63	43	62	82
1000	22	33	41	50	34	50	66	22	35	46	56	38	56	73
1200	19	30	38	46	31	45	60	20	31	41	51	35	51	67
1400	18	28	35	43	29	42	55	18	28	37	47	32	47	62
1600	16	25	33	40	27	39	52	17	26	34	44	29	44	58
1800	15	24	31	38	25	37	49	16	24	32	41	27	42	55
2000	14	23	29	36	24	35	46	15	23	30	39	25	39	52
2200	14	22	28	34	23	34	44	14	22	29	37	23	37	48
2400	13	21	27	33	21	32	42	13	21	28	35	22	34	45
2600	13	20	26	31	20	31	41	13	20	27	34	21	33	43
2800	12	19	25	30	19	30	39	12	20	26	33	20	31	41
3000	12	19	24	29	18	29	38	12	19	25	32	19	30	39
3200	12	18	23	28	18	28	37	12	19	24	31	18	29	38

TABLE 4-17 Maximum Spans, in Inches, for Lumber Framing Using Douglas-Fir No. 2 or Southern Pine No. 2

Equivalent Uniform Load lb per ft	Douglas Fir #2 or Southern Pine #2 Continuous over 2 or 3 Supports (1 or 2 Spans) Nominal Size Lumber							Douglas Fir #2 or Southern Pine #2 Continuous over 4 or more Supports (3 or more Spans) Nominal Size of Lumber						
	2×4	2×6	2×8	2×10	4×4	4×6	4×8	2×4	2×6	2×8	2×10	4×4	4×6	4×8
3400	11	18	22	27	17	27	35	12	18	23	30	18	28	36
3600	11	17	22	27	17	26	34	11	18	23	30	17	27	35
3800	11	17	21	26	16	25	33	11	17	22	29	16	26	34
4000	11	16	21	25	16	24	32	11	17	22	28	16	25	33
4200	11	16	20	25	15	24	31	11	17	22	28	16	24	32
4400	10	16	20	24	15	23	31	10	16	22	27	15	24	31
4600	10	15	19	24	14	23	30	10	16	21	26	15	23	31
4800	10	15	19	23	14	22	29	10	16	21	26	14	23	30
5000	10	15	18	23	14	22	29	10	16	21	25	14	22	29

Notes:
1. Courtesy APA—The Engineered Wood Association, "Concrete Forming," 2004.
2. Spans are based on the 2001 NDS allowable stress values, $C_D = 1.25$, $C_r = 1.0$, $C_M = 1.0$.
3. Spans are based on dry, single-member allowable stresses multiplied by a 1.25 duration-of-load factor for 7-day loads.
4. Deflection is limited to $l/360$th of the span with ¼ in. maximum.
5. Spans are measured center-to-center on the supports.

TABLE 4-17 Maximum Spans, in Inches, for Lumber Framing Using Douglas-Fir No. 2 or Southern Pine No. 2 (*Continued*)

Equivalent Uniform Load lb per ft	Hem-Fir #2 Continuous over 2 or 3 Supports (1 or 2 Spans) Nominal Size Lumber							Hem-Fir #2 Continuous over 4 or more Supports (3 or more Spans) Nominal Size of Lumber						
	2 × 4	2 × 6	2 × 8	2 × 10	4 × 4	4 × 6	4 × 8	2 × 4	2 × 6	2 × 8	2 × 10	4 × 4	4 × 6	4 × 8
200	45	70	90	110	59	92	114	54	79	100	122	73	108	133
400	34	50	63	77	47	74	96	38	56	71	87	58	86	112
600	28	41	52	63	41	62	82	29	45	58	71	48	70	92
800	23	35	45	55	37	54	71	23	37	48	61	41	60	80
1000	20	31	40	49	33	48	64	20	32	42	53	37	54	71
1200	18	28	36	45	30	44	58	18	28	37	47	33	49	65
1400	16	25	33	41	28	41	54	16	26	34	43	29	45	60
1600	15	23	31	39	25	38	50	15	24	31	40	26	41	54
1800	14	22	29	37	23	36	48	14	22	30	38	24	38	50
2000	13	21	28	35	22	34	45	14	23	28	36	22	35	46
2200	13	20	26	33	20	32	42	13	21	27	34	21	33	43
2400	12	19	25	32	19	30	40	12	20	26	33	20	31	41
2600	12	19	25	30	18	29	38	12	20	25	32	19	30	39
2800	12	18	24	29	18	28	36	12	19	24	31	18	28	37
3000	11	18	23	28	17	26	35	11	18	24	30	17	27	36

TABLE 4-18 Maximum Spans, in Inches, for Lumber Framing Using Hem-Fir No. 2

Equivalent Uniform Load lb per ft	Hem-Fir #2 Continuous over 2 or 3 Supports (1 or 2 Spans) Nominal Size Lumber							Hem-Fir #2 Continuous over 4 or more Supports (3 or more Spans) Nominal Size of Lumber						
	2 × 4	2 × 6	2 × 8	2 × 10	4 × 4	4 × 6	4 × 8	2 × 4	2 × 6	2 × 8	2 × 10	4 × 4	4 × 6	4 × 8
3200	11	17	22	27	16	25	34	11	18	23	29	17	26	34
3400	11	17	22	27	16	25	32	11	17	22	29	16	25	33
3600	11	17	21	26	15	24	31	11	17	22	28	16	24	32
3800	10	16	21	25	15	23	31	10	17	22	28	15	24	31
4000	10	16	20	24	14	23	30	10	16	21	27	15	23	30
4200	10	15	10	24	14	22	29	10	16	21	27	14	22	30
4400	10	15	19	24	14	22	28	10	16	21	26	14	22	29
4600	10	15	19	23	13	21	28	10	15	20	26	14	21	28
4800	10	14	19	22	13	21	27	10	15	20	25	13	21	28
5000	10	14	18	22	13	20	27	10	15	20	24	13	21	27

Notes:

1. Courtesy American Plywood Association, "Concrete Forming," 2004.
2. Spans are based on the 2001 NDS allowable stress values, $C_D = 1.25$, $C_r = 1.0$, $C_M = 1.0$.
3. Spans are based on dry, single-member allowable stresses multiplied by a 1.25 duration-of-load factor for 7-day loads.
4. Deflection is limited to $l/360$th of the span with ¼ in. maximum.
5. Spans are measured center-to-center on the supports.

Table 4-18 Maximum Spans, in Inches, for Lumber Framing Using Hem-Fir No. 2 (*Continued*)

Plywood Thickness, in.	Minimum Bending Radii, ft	
	Across the Grain	Parallel to the Grain
¼	2	5
⁵⁄₁₆	2	6
³⁄₈	3	8
½	6	12
⁵⁄₈	8	16
¾	12	20

Courtesy, APA—The Engineered Wood Association.

TABLE 4-19   Minimum Bending Radii for Plyform Dry

The boards, which are available in large sheets, have a hard, smooth surface that produces a concrete whose surface is relatively free of blemishes and joint marks. The thin sheets can be bent to small radii, which is an advantage when casting concrete members with curved surfaces.

Table 4-20 gives the physical properties for tempered hardboard. Standard sheets are 4 ft wide by 6, 8, 12, and 16 ft long. Shorter lengths may be obtained by special order.

Table 4-21 gives the minimum bending radii for various thicknesses of hardboard when bending is performed under the specified conditions. Because the values given in the table are based on bending the hardboard around smooth cylinders, it is improbable that smooth bends to these radii can be obtained with formwork.

Thickness, in.	Weight, lb per sq ft	Tensile Strength, lb per sq in.	Modulus of Rupture, lb per sq in.	Punch Shearing Strength, lb per sq in.
⅛	0.71	5,100	10,000	5,800
³⁄₁₆	1.05	4,700	10,000	5,800
¼	1.36	4,300	9,300	4,500
⁵⁄₁₆	1.78	3,900	8,300	4,200

TABLE 4-20   Physical Properties of Tempered Hardboard

Thickness, in.	Cold, dry, smooth side		Cold, moist, smooth side	
	Out, in.	In, in.	Out, in.	In, in.
⅛	9	7	6	4
³⁄₁₆	16	14	9	6
¼	25	22	14	10
⁵⁄₁₆	35	30	20	16

TABLE 4-21   Minimum Bending Radii for Tempered Hardboard around Smooth Cylinders, in.

Fiber Form Tubes

Round fiber-tube forms are fabricated by several manufacturers. These fiber forms are available in lengths up to 50 ft with inside diameters ranging from 12 to 48 in., in increments of 6 in. Two types of waterproof coatings are used. One is a plasticized treatment for use where the forms are to be removed and a clean finish is specified, and the other is a wax treatment for use where the forms will not be removed or where the condition of the exposed surface of the concrete is not important. The latter treatment is less expensive than the former. These forms can be used only once.

The tubes are manufactured by wrapping successive layers of fiber sheets spirally and gluing them together to produce the desired wall thickness. When the forms are removed from the columns, they will leave a spiral effect on the surface of the concrete. Seamless tubes, which will not leave a spiral effect when they are removed, are also available at slightly higher costs. These tubes may be purchased precut to specified lengths at the mill, or they may be cut at the project.

Steel Forms

Steel forms are of two broad types: those that are prefabricated into standard panel sizes and shapes and those that are fabricated for special uses. For some projects either one or both types may be used. Among their uses are the following:

1. Concrete walls

2. Concrete piers, columns, and related items

3. Combined decking and reinforcing for concrete slabs

4. Built-in-place concrete conduit

5. Concrete tunnel linings and concrete dams

6. Precast concrete members

7. Architectural concrete

For certain uses, forms made of steel have several advantages over forms made of other materials. They can provide adequate rigidity and strength. They can be erected, disassembled, moved, and re-erected rapidly, provided suitable handling equipment is available for the large sections. They are economical if there are enough reuses. The smooth concrete surface may be an advantage for some projects.

Forms made of steel have some disadvantages. Unless they are reused many times, they are expensive. Also, unless special precautions are taken, steel forms offer little or no insulation protection to concrete placed during cold weather.

Corrugated metal sheets have been used extensively to form floor and roof slabs of buildings. Patented pans and domes are often used to form concrete decks of bridges and for structural floor slabs of buildings.

Aluminum Forms

Forms made from aluminum are in many respects similar to those made of steel. However, because of their lower density, aluminum forms are lighter than steel forms, and this is their primary advantage when compared with steel. Because the strength of aluminum in handling, tension, and compression is less than the strength of steel, it is necessary to use larger sections when forms are made of aluminum. Because wet concrete can chemically attack aluminum, it is desirable to use aluminum alloys in resisting corrosion from the concrete.

Support trusses fabricated with aluminum alloys have been effectively used for flying forms. These forms are lightweight and allow large lengths of deck forms to be moved easily. Cast aluminum alloy molds have also been used successfully to form ornamental concrete products. Aluminum wall forms have also been used to produce textures on the surfaces of concrete walls.

Plastic Forms

Fiberglass plastic forms can be used for unique shapes and patterns in concrete. In addition to their ability to form unusual shapes, plastic forms are lightweight, easy to handle and strip, and they eliminate rust and corrosion problems.

The forms are constructed in much the same manner that is commonly used in the hand forming of boats. Because of careful temperature and humidity controls that must be exercised at all times during the manufacture of plastic forms, fabrication of the forms is performed under factory conditions.

There are several manufacturers of plastic forms. Most of these forms are fiberglass reinforced for column forms and dome pan forms, which are custom-made forms for special architectural effects.

Form Liners

Designers of architectural concrete frequently specify a particular finish on the surface of the concrete. Special surface finishes can be achieved by attaching form liners to the inside faces of forms. The form liners can be used to achieve an extremely smooth surface, or to achieve a particular textured finish on the concrete surface. For example, for walls the texture may simulate a pattern of bricks or natural grains of wood. A variety of types and shapes of form liners are available to produce the desired finish on the concrete surface.

There are various types of materials that may be suitable as form liners, depending on the desired finish. This includes plywood, hardboard, coatings, and plastics. Liners may be attached to the sides of forms with screws, staples, or nails. In some situations, the liners may be attached with an adhesive that bonds the liner to the steel or wood forms.

Plastic liners may be flexible or rigid. A rubberlike plastic that is made of urethanes is flexible enough that it can be peeled away from the concrete surface, revealing the desired texture. Polyvinyl chloride (PVC) sheets are rigid with sufficient stiffness for self-support. The sheets are available in 10-ft lengths. A releasing agent should be applied to the form liner to ensure uniformity of the concrete surface and to protect the form liner for possible reuse.

Nails

There are three types of nails that are normally used in formwork: common wire nails, box nails, and double-headed nails. Common wire nails are used for attaching formwork members or panels for multiple uses, when nails are not required to be removed in stripping the forms. The common wire nail is the most frequently used nail for fastening form members together. A box nail has a thinner shank and head than a common wire nail, which makes it more useful for built-in-place forms. Box nails pull loose more easily than common or double-headed nails.

Double-headed nails are frequently used in formwork when it is desirable to remove the nail easily, such as for stripping forms. The first head permits the nail to be driven fully into the wood; the shaft of the nail extends a fraction of an inch beyond the first head to a second head. This head, protruding slightly outside the surface of the lumber, permits the claw of the hammer or bar to remove the nail easily when the form is being stripped from the concrete.

The nails most frequently used with wood and plywood forms are the common wire type. The sizes and properties of these types of nails are shown in Table 4-22. Where they are used to fasten form members together, the allowable loads are based on the resistance to withdrawal, the resistance to lateral movement, or a combination of

Size of Nail, penny-weight	Length, in.	Diameter, in.	Bending Yield Strength, lb per sq in.
6d	2	0.113	100,000
8d	2½	0.131	100,000
10d	3	0.148	90,000
12d	3¼	0.148	90,000
16d	3½	0.162	90,000
20d	4	0.192	80,000
30d	4½	0.207	80,000
40d	5	0.225	80,000
50d	5½	0.244	70,000

TABLE 4-22   Sizes and Properties of Common Wire Nails

the two resistances. The resistances of nails are discussed in the following sections.

Withdrawal Resistance of Nails

The resistance of a nail to direct withdrawal from wood is related to the density or specific gravity of the wood, the diameter of the nail, the depth of its penetration, and its surface condition. The NDS, published by the American Forest & Paper Association (AF&PA), is a comprehensive guide for the design of wood and its fastenings. The association's Technical Advisory Committee has continued to study and evaluate new data and developments in wood design. The information given in the following tables is from ref. [1].

Table 4-23 provides the nominal withdrawal design values for nails and species of wood commonly used for formwork. Withdrawal values are given only for nails driven in side grains [1]. Nails are not to be loaded in withdrawal from the end grain of wood. The nominal values are provided in pounds per inch of penetration for a single nail driven in the side grain of the main member with the nail axis perpendicular to the wood fibers. Tabulated values should be multiplied by the applicable adjustment factors to obtain allowable design values.

Lateral Resistance of Nails

For wood-to-wood connections, the lateral resistance of nails depends on the nail diameter and depth of penetration, the bending yield strength of the nail, the thickness and species of the wood, and the dowel bearing strength of the wood.

Size of Nail, penny-weight	Diameter, in.	Species of Wood and Specific Gravity		
		Southern Pine S.G. = 0.55	Douglas Fir-Larch S.G. = 0.50	Hem-Fir S.G. = 0.43
4d	0.099	31	24	17
6d	0.113	35	28	19
8d	0.131	41	32	22
10d	0.148	46	36	25
12d	0.148	46	36	25
16d	0.162	50	40	27
20d	0.192	59	47	32
30d	0.207	64	50	35
40d	0.225	70	55	38
50d	0.244	76	60	41

Note:
1. See National Design Specification for Wood Construction for additional adjustments that may be appropriate for a particular job condition.

TABLE **4-23** Withdrawal Design Values for Common Wire Nails Driven into Side Grain of Main Member, lb per in. of Penetration

The NDS presents four equations for calculating the lateral design values of nails in wood-to-wood connections. The allowable design value is the lowest result of the four equations. These equations include the variables described in the preceding sections. Table 4-24 provides the nominal lateral design values for common wire nails and several species of wood that are commonly used for formwork in dry conditions, less than 19% moisture content. Design values for other species of wood and other types of fasteners can be found in ref. [1].

The values given in Table 4-24 are based on nails driven in the side grain with the nail axis perpendicular to the wood fibers. Design values are based on nails driven in two members of identical species of seasoned wood with no visual evidence of splitting. The nominal lateral design values of nails are based on nails that are driven in the side grain of the main member, with the nail axis perpendicular to the wood fibers. When nails are driven in the end grain, with the nail axis parallel to the wood fibers, nominal lateral design values should be multiplied by 0.67. The values are based on a nail penetration into the main member of 12 times the nail diameter. The minimum nail penetration into the main member should be six times the nail diameter. There should be sufficient edge and end distances, as well as spacings of nails to prevent splitting of the wood.

Side Member Thickness in.	Common Wire Nail Size/ Dimensions			Species and Properties of Wood		
				Southern Pine S.G. = 0.55	Douglas Fir-Larch S.G. = 0.50	Hem-Fir S.G. = 0.43
	Size	Diameter in.	Length in.	Dowel bearing strength = 5,500	Dowel bearing strength = 4,650	Dowel bearing strength = 3,500
¾	8d	0.131	2½	104	90	73
	10d	0.148	3	121	105	85
	12d	0.148	3¼	121	105	85
	16d	0.162	3½	138	121	99
	20d	0.192	4	157	138	114
	30d	0.207	4½	166	147	122
	40d	0.225	5	178	158	132
	50d	0.244	5½	182	162	136
1½	16d	0.162	3½	154	141	122
	20d	0.192	4	185	170	147
	30d	0.207	4½	203	186	161
	40d	0.225	5	224	205	178
	50d	0.244	5½	230	211	183

Notes:
1. See NDS, ref. [1], for values for additional thicknesses and species, and for additional adjustments that may be required for a particular condition.
2. Dowel bearing strength is given in lb per sq in.

TABLE 4-24 Lateral Design Values for Common Nails Driven into Side Grain of Two Members with Identical Species, lb

Toe-Nail Connections

Toe-nails are sometimes used to fasten two members of wood. Toe-nails are driven at an angle of approximately 30° with the member and started approximately one-third the length of the nail from the member end.

When toe-nails are used to connect members, the tabulated withdrawal values shown in Table 4-23 should be multiplied by 0.67 and the nominal lateral values in Table 4-24 multiplied by 0.83.

Connections for Species of Wood for Heavy Formwork

In the preceding sections, values were given for nails to fasten Southern Pine, Douglas Fir-Larch, and Hem-Fir species of wood. These are the materials that are commonly used for formwork. However, for some projects it is necessary to build formwork from larger members, with species of wood that have higher densities and strengths, and with fasteners that are stronger than common nails.

Screws, bolts, or timber connectors are often used to fasten larger members or members with high-density wood. The following sections present information for these types of fasteners for Southern Pine, Douglas Fir-Larch, and Hem-Fir wood.

Lag Screws

Lag screws are sometimes used with formwork, especially where heavy wood members are fastened together. These screws are available in sizes varying from ¼ to 1¼ in. shank diameter with lengths from 1 to 12 in. The larger diameters are not available in the shorter lengths.

Lag screws require prebored holes of the proper sizes. The lead hole for the shank should be of the same diameter as the shank. The diameter of the lead hole for the threaded part of the screw should vary with the density of the wood and the diameter of the screw. For lightweight species of wood with a specific gravity less than 0.50, the diameter for the threaded portion should be 40 to 50% of the shank diameter. For species with a specific gravity between 0.50 and 0.60, the diameter of the threaded portion should be 60 to 75% of the shank diameter. For dense hardwoods with a specific gravity greater than 0.60, the diameter of the threaded portion should be 65 to 85% of the shank diameter.

During installation, some type of lubricant should be used on the lag screw or in the lead hole to facilitate insertion and prevent damage to the lag screw. Lag screws should be turned, not driven, into the wood.

Withdrawal Resistance of Lag Screws

The resistance of a lag screw to direct withdrawal from wood is related to the density or specific gravity of the wood and the unthreaded shank diameter of the lag screw. Table 4-25 provides the withdrawal design values for a single lag screw installed in species of wood that are commonly used for formwork, including Southern Pine, Douglas Fir-Larch, and Hem-Fir. The values for the lag screws are given in pounds per inch of thread penetration into the side grain of the main member. The length of thread penetration in the main member shall

Size of Lag Screw Unthreaded Shank Dia., in.	Species and Specific Gravity of Wood		
	Southern Pine S.G. = 0.55	Douglas Fir-Larch S.G. = 0.50	Hem-Fir S.G. = 0.43
¼	260	225	179
⁵⁄₁₆	307	266	212
⅜	352	305	243
⁷⁄₁₆	395	342	273
½	437	378	302
⅝	516	447	357
¾	592	513	409
⅞	664	576	459

Note:
1. See NDS [1] for additional adjustments that may be required.

TABLE 4-25 Lag Screws Withdrawal Design Values into Side Grain of Main Member, lb per in. of Thread Penetration

not include the length of the tapered tip. For lag screws installed in the end grain, the values shown in Table 4-25 should be multiplied by 0.75.

Lateral Resistance of Lag Screws

For wood-to-wood connections, the lateral resistance of lag screws depends on the depth of penetration, the diameter and bending yield strength of the screw, and the dowel bearing strength of the wood. The NDS provides equations for calculating the lateral design values of lag screws in wood-to-wood connections and for wood-to-metal connections. The lowest result of the equations is the lateral design value. Design values are based on dry wood, less than 19% moisture content, and a normal load-duration.

Table 4-26 provides design values for lateral resistance of lag screws in wood-to-wood connections for several thicknesses and species of wood. These values are for lag screws inserted into the side grain of the main member with the lag screw axis perpendicular to the wood fibers. The values are based on a penetration, not including the length of the tapered tip, into the main member of approximately eight times the shank diameter. If the penetration is less than this amount, an appropriate adjustment must be made in accordance with the NDS requirements. The values in Table 4-26 are based on an edge distance, end distance, and spacing of lag screws of four times the diameter of the lag screw.

Side Member Thickness, in.	Lag Screw, unthreaded Shank diameter, in.	Species and Specific Gravity of Wood								
		Southern Pine S.G. = 0.55 Lateral Value, lb			Douglas Fir-Larch S.G. = 0.50 Lateral Value, lb			Hem-Fir S.G. = 0.43 Lateral Value lb,		
		//	s⊥	m⊥	//	s⊥	m⊥	//	s⊥	m⊥
¾	¼	150	110	120	140	100	110	130	90	100
	⁵⁄₁₆	180	120	130	170	110	120	150	100	110
	⅜	180	120	130	170	110	120	150	100	110
1	¼	160	120	120	150	120	120	140	100	110
	⁵⁄₁₆	210	140	150	190	130	140	170	110	130
	⅜	210	130	150	200	120	140	170	100	120
1½	¼	160	120	120	150	120	120	140	110	110
	⁵⁄₁₆	210	150	150	200	140	140	180	130	130
	⅜	210	150	150	200	140	140	190	130	130
	⁷⁄₁₆	320	220	230	310	200	210	290	170	190
	½	410	250	290	390	220	270	350	190	240
	⅝	600	340	420	560	310	380	500	280	340
	¾	830	470	560	770	440	510	700	360	450
	⅞	1,080	560	710	1,020	490	660	930	390	580

D	2½								
1/4	160	120	120	150	120	120	140	110	110
5/16	210	150	150	200	140	140	180	130	130
3/8	210	150	150	200	140	140	190	130	130
7/16	320	230	230	310	210	210	290	190	190
1/2	410	290	290	390	270	270	360	240	240
5/8	670	430	440	640	390	420	590	330	380
3/4	1,010	550	650	960	500	610	890	430	550
7/8	1,370	690	880	1,280	630	550	1,130	550	730

Notes:

1. Courtesy, American Forest & Paper Association.
2. See NDS, [1], for values for additional diameters of lag screws and for additional thicknesses and species of wood.
3. See NDS, [1], for additional adjustments that may be required for a particular condition.
4. Symbols: // – represents values for loads parallel to grain of wood in both members.
   s⊥ – represents values for side members with loads perpendicular to grain and main members parallel to grain.
   m⊥ – represents values for main members with loads perpendicular to grain and side members parallel to grain.

TABLE 4-26 Lateral Design Values for Lag Screws for Single Shear in Two Members with Both Members of Identical Species, lb

Timber Connectors

The strength of joints between wood members used for formwork can be increased substantially by using timber connectors. Three types of connectors are available: split ring, two shear plates, and one shear plate. The split ring is used with its bolt or a lag screw in single shear. The two-shear plate connector consists of two shear plates used back to back in the contact faces of a wood-to-wood connection with their bolt or lag screw in single shear.

The one-shear-plate connector is used with its bolt or lag screw in single shear with a steel strap for wood-to-metal connections. The split ring and the two-shear plate connectors are the most common types of connectors used with formwork.

When lag screws are used in lieu of bolts, the hole for the unthreaded shank should be the same diameter as the shank. The diameter of the hole for the threaded portion of the lag screw should be approximately 70% of the shank diameter. When bolts are used, a nut is installed, with washers placed between the outside wood member and the bolt head and between the outside wood and the nut.

When timbers are joined with metal or other types of connectors, the strength of the joint depends on the type and size of the connector, the species of wood, the thickness and width of the member, the distance of the connector from the end of the timber, the spacing of the connectors, the direction of application of the load with respect to the direction of the grain of the wood, the duration of the load, and other factors.

Split-Ring Connectors

Split-ring connectors are usually restricted to situations where the formwork is fabricated in a yard, remote from the jobsite, where a drill press is readily available. Table 4-27 lists the design unit values for one split ring connector and bolt installed in the side grain for joining two members of seasoned wood in single shear. The values are based on adequate member thicknesses, edge distances, side distances, and spacing [1].

Tabulated nominal design values for split-ring and shear-plate connectors are based on the assumption that the faces of the members are brought into contact with the connector units, and allow for seasonal variations after the wood has reached the moisture content normal for the condition of service. When split-ring or shear-plate connectors are installed in wood that is not seasoned to the moisture content of the normal service condition, the connections should be tightened by turning down the nuts periodically until moisture equilibrium is attained.

Shear Plate Dia. (in.)	Bolt Dia. (in.)	Number of Faces of Member with Connectors on Same Bolt	Net Thickness of Member (in.)	Loaded Parallel to Grain Design Value, lb		Loaded Perpendicular to Grain Design Value, lb	
				Southern Pine & Douglas Fir-Larch	Hem-Fir	Southern Pine & Douglas Fir-Larch	Hem-Fir
2½	½	1	1" minimum	2,270	1,900	1,620	1,350
			1½" or thicker	2,730	2,290	1,940	1,620
		2	1½" minimum	2,100	1,760	1,500	1,250
			2" or thicker	2,730	2,290	1,940	1,620
4	¾	1	1" minimum	3,510	2,920	2,440	2,040
			1½" thick	5,160	4,280	3,590	2,990
			1⅝" or thicker	5,260	4,380	3,660	3,050
		2	1½" minimum	3,520	2,940	2,450	2,040
			2" thick	4,250	3,540	2,960	2,460
			2½" thick	5,000	4,160	3,480	2,890
			3" or thicker	5,260	4,380	3,660	3,050

Notes:
1. Courtesy American Forest and Paper Association.
2. See NDS, ref. [1], for additional adjustments that may be appropriate for a particular application.

TABLE 4-27   Split Ring Connector Unit Design Values, in Pounds, for One Split-Ring and Bolt in Single Shear

Shear-Plate Connectors

Table 4-28 shows the design unit values for one-shear plate connectors installed in the side grain for joining two members of seasoned wood in single shear. The values are based on adequate member thicknesses, edge distances, side distances, and spacing [1].

Split-Ring Dia. (in.)	Bolt Dia. (in.)	Number of Faces of Member with Connectors of on Same Bolt	Net Thickness of Member (in.)	Loaded Parallel to Grain Design Value, lb		Loaded Perpendicular to Grain Design Value, lb	
				Southern Pine and Douglas Fir-Larch	Hem-Fir	Southern Pine and Douglas Fir-Larch	Hem-Fir
2⅝	¾	1	1½" minimum	2,670	2,200	1,860	1,550
			1½" minimum	2,080	1,730	1,450	1,210
		2	2" thick	2,730	2,270	1,910	1,580
			2½" or thick	2,860	2,380	1,990	1,650
4	¾ or ⅞	1	1½" minimum	3,750	3,130	2,620	2,170
			1¾" or thicker	4,360	3,640	3,040	2,530
		2	1¾" minimum	2,910	2,420	2,020	1,680
			2" thick	3,240	2,700	2,260	1,880
			2½" thick	3,690	3,080	2,550	2,140
			3" thick	4,140	3,450	2,880	2,400
			3½" or thicker	4,320	3,600	3,000	2,510

Notes:
1. Courtesy American Forest and Paper Association.
2. See NDS, ref. [1], for additional adjustments that may be appropriate for a particular application.

TABLE 4-28   Shear Plate Connector Unit Design Values, lb, for One Shear Plate Unit and Bolt in Single Shear

Split-Ring and Shear-Plate Connectors in End Grain

When split-ring and shear-plate connectors are installed in a surface that is not parallel to the general direction of the grain of the member, adjustments in the values tabulated in Tables 4-27 and 4-28 must be made in accordance with NDS [1].

Penetration Requirements of Lag Screws

When lag screws are used, in lieu of bolts, for split-ring or shear-plate connectors, the nominal design values tabulated in Tables 4-27 and 4-28 must be multiplied by the appropriate penetration depth factor [1].

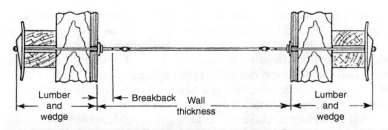

Lumber and wedge — Breakback — Wall thickness — Lumber and wedge

FIGURE 4-2   Form snap tie for concrete walls. (*Source: Dayton Superior Corporation*)

Form Ties

Form ties are placed between wall forms to resist the lateral pressure against wall forms resulting from concrete. Form ties serve two purposes: they hold the forms apart prior to placing the concrete, and they resist the lateral pressure of the concrete after it is placed.

There are many varieties of ties available, which may consist of narrow steel bands, plain rods, rods with hooks or buttons on the ends, or threaded rods, with suitable clamps or nuts to hold them in position. Although most ties are designed to break off inside the concrete, there are some that are tapered to allow them to be pulled out of the wall after the forms are removed. After the form ties are removed, the holes can be filled with a suitable grout material.

In ordering form ties, it is necessary to specify the thickness of the wall, sheathing, studs, and wales. Ties are available for wall thicknesses from 4 to 24 in. Figure 4-2 shows a standard snap tie that is commonly used in formwork for walls. The tension load of a standard snap tie is generally 3,000 lb. Coil ties are also available, which have tension load capacities that exceed 15,000 lb. Form ties are further discussed in Chapter 9.

Manufacturers specify the safe loads that can be resisted by their ties, which generally range from 3,000 to 5,000 lb. Special high strength ties are available for loads that exceed 50,000 lb. ACI Committee 347 recommends a safety factor of 2.0 for form ties. When high-strength ties are used, the bearing capacity of the tie clamp against the supporting member, such as the wale, must have adequate strength to keep the member from crushing.

Concrete Anchors

Anchors are mechanical devices that are installed in the concrete to support formwork or to provide a means of lifting a concrete member, such as tilt-up wall panels. Anchors placed in the form before the placement of con-crete are sometimes referred to as "inserts." Generally anchors of this type are used for lifting the concrete member.

Anchors may also be installed after the concrete is placed and cured. A hole is drilled into the concrete to allow installation of a threaded mechanical device, such as a helical coil or an expandable nut-type device. After the threaded mechanical device is installed, a bolt or all-threaded rod can be installed to fasten formwork, to secure an existing concrete structure to a newly placed concrete member, or to attach rigging for lifting the concrete member.

Self-threading anchors are available for light-duty applications. The ultimate strength of these fasteners will vary, depending on the diameter and depth of embedment of the anchor, and on the strength of concrete in which it is installed. When properly installed in concrete, the ultimate pullout load will range from 300 to 1,800 lb, and the ultimate shear strength may vary from 800 to 1,600 lb.

There are a variety of types of anchors available from manufacturers. The strength of an anchor is provided by each manufacturer. ACI Committee 347 recommends a safety factor of 2.0 for concrete anchors used in the application of formwork, supporting form weight and concrete pressures, and for precast concrete panels when used as formwork. A safety factor of 3.0 is recommended for formwork supporting the weight of forms, concrete, construction live loads, and impact.

References

1. "National Design Specification for Wood Construction," ANSI/AF&PA NDS-2005, American Forest & Paper Association, Washington, DC, 2005.
2 "Design Values for Wood Construction," Supplement to the National Design Specification, National Forest Products Association, Washington, DC, 2005.
3. "Plywood Design Specification," APA—The Engineered Wood Association, Tacoma, WA, 1997.
4. "Concrete Forming," APA—The Engineered Wood Association, Tacoma, WA, 2004.
5. "Plywood Design Specification," APA—The Engineered Wood Association, r Tacoma, WA, 2004.
6. "Manufacturer's Catalogue," Dayton Superior Corporation, Parsons, KS.
7. "Manufacturer's Brochure," The Sonoco Products Company, Hartsville, SC.

CHAPTER 5

Design of Wood Members for Formwork

General Information

The terms *timber*, *lumber*, and *wood* are often used interchangeably in the construction industry. In general, timber is used to describe larger size members, whereas lumber is used for smaller size members. In either case, both timber and lumber are made from wood.

Wood is an ideal material for construction. It is lightweight, which makes it easy to handle at the jobsite. It is also easily fabricated to almost any configuration by cutting and fastening together various combinations of standard size lumber.

Design of wood members involves selection of the size and grade of lumber to withstand the applied loads and deflections. The applied loads produce stresses in a member; including bending, shear, compression, and bearing. The selected member must also have enough rigidity to prevent excessive deformations and deflection.

Arrangement of Information in This Chapter

The beginning sections in this chapter present the basic equations for analyzing beams, including analysis of bending moments and shear forces. The basic equations for calculating stresses and deflection of beams are also presented and illustrated. Following the sections concerned with the analysis, Table 5-2 summarizes the equations commonly used in the analysis of beams for formwork. Table 5-2 is valuable and frequently referenced in this book to calculate bending stress, shear stress, and deflection during the process of analyzing and designing formwork and temporary structures.

In subsequent sections allowable span lengths are calculated based on the bending, shear, and deflection of beams and of plywood. These are the equations commonly used for the design of formwork and temporary structures during construction. Tables 5-3 and 5-4 provide a summary of the equations for the reader to reference for design purposes, followed by a presentation and discussion of compression stresses in axially loaded wood columns. Throughout this book, Tables 5-3 and 5-4 are frequently referenced in example problems to illustrate the design of wood members for formwork and temporary structures.

Lumber versus Timber Members

Sawn lumber is classified into two classifications; *dimension lumber* and *timbers*. Dimension lumber is the smaller sizes of structural lumber, whereas *timbers* are the larger sizes of lumber.

Dimension lumber includes the range of sizes from 2×2 through 4×16. Thus, dimension lumber is lumber that have nominal thicknesses of 2 to 4 in. Thickness refers to the smaller dimension of the piece of wood and width refers to the larger dimension. Timbers are the larger sizes and have a 5-in. minimum nominal dimension. Thus, the smallest size timber is 6×6, and all members larger than 6×6 are classified as timber.

The species of trees used for structural wood are classified as hardwoods and softwoods. These terms represent the classifications of trees and not the physical properties of the wood. Hardwoods are broad-leaved trees, whereas softwoods have narrow, needlelike leaves, like evergreens. The majority of structural lumber comes from the softwood category.

Lumber is graded by species groups, reference NDS Supplement 2005. Douglas Fir-Larch and Southern Pine are two very common species groups that are widely used in structural applications. They are relatively dense and their structural properties exceed many hardwoods. For Douglas Fir-Larch the species of trees that may be included in combination are Douglas Fir and Western Larch. For Southern Pine, the species of trees that may be included in combination are Loblolly Pine, Longleaf Pine, Shortleaf Pine, and Slash Pine.

The designer must be aware that more than one set of grading rules can be used to grade some commercial species groups. For example, Douglas Fir-Larch can be graded under West Coast Lumber Inspection Bureau (WCLIB) rules or under Western Wood Products Association (WWPA) rules. There are some differences in reference design values between the two sets of rules. The tables of reference design values in the NDS Supplement have the grading rules clearly identified, such as WCLIB and/or WWPA. These differences in reference design values occur only in large-size members known as timbers.

The reference design values are the same under both sets of grading rules for dimension lumber. The designer should use the lower reference design value in designing structural members because he/she does not have control over which set of grading rules will be used.

Loads on Structural Members

A load is a force exerted on a structural member, such as a beam, column, or slab. The loads may be concentrated or uniformly distributed. A concentrated load is a point load at a particular location on the member. For example, a column may have single concentrated load acting along the axis of the column, or a beam may have one load applied transverse to the axis of the beam. Often, there are several concentrated loads acting on a beam. For example, a girder in a building receives concentrated loads at the points where the floor beams are placed on the girder.

A uniformly distributed load is a load of uniform magnitude per unit of length that extends over a portion or the entire length of the member. A floor joist that supports floor decking is an example of a member supporting a uniformly distributed load. Concrete placed in a wall form produces uniformly distributed pressures against the formwork sheathing, which in turn produces a uniformly distributed load along the studs that support the sheathing. For some structural members, there is a combination of concentrated loads and uniformly distributed loads.

Equations Used in Design

It is the responsibility of the designer to assess the conditions and to determine the proper analysis and design equations. Design procedures must comply with all codes, specifications, and regulatory requirements.

The equations presented in the following sections are the traditional basic equations for bending, shear, and compression. The designer must ensure that the appropriate adjustments are made in the equations based on the conditions that are unique to each job.

Loads on structural members produce bending, shear, and compression stresses. These stresses must be kept within an allowable limit to ensure safety. In addition to strength requirements, limits frequently are specified for the maximum permissible deflection of structural members. Stresses and deflections are analyzed by the use of basic equations shown in subsequent sections of this chapter.

The following symbols and units will be used in the equations presented in this chapter:

M = bending moment, in.-lb
c = distance from neutral axis of beam to extreme fiber in bending, in.

f = applied stress, $\text{lb}/\text{in.}^2$
F = allowable stress, $\text{lb}/\text{in.}^2$
I = moment of inertia of beam section, in.^4
b = width of beam section, in.
d = depth of beam section, in.
S = section modulus, in.^3
V = shear force, lb
P = concentrated point load, lb
A = cross-sectional area, in.^2
E = modulus of elasticity for deflection calculations, $\text{lb}/\text{in.}^2$
E_{min} = modulus of elasticity for beam and column stability calculations, $\text{lb}/\text{in.}^2$
$F_{c//}$ = compression stress parallel to grain, $\text{lb}/\text{in.}^2$
$F_{c\perp}$ = compression stress perpendicular to grain, $\text{lb}/\text{in.}^2$
Ib/Q = rolling shear constant, in.^2, used to design plywood
W = total uniformly distributed load, lb
w = uniformly distributed load, lb per lin ft
L = beam span or column length, ft
l = beam span or column length, in.
Δ = deflection, in.

In designing and fabricating wood members, there is frequent intermixing of units of measure. For example, the span length for a long beam may be defined in feet, whereas the span length of plywood sheathing is often defined in inches. Because different units of measure are used frequently, many of the equations in this chapter show both units. As presented above, span lengths in feet are denoted by the symbol L, whereas span lengths expressed in inches are denoted by the symbol l: thus $l = 12L$. Similarly, the total load on a uniformly loaded beam is $W = wL$, which is equivalent to $wl/12$. These simple illustrations are presented here so the reader will be cognizant of the symbols and units of measure in subsequent sections of this book.

Analysis of Bending Moments in Beams with Concentrated Loads

When a load is applied in a direction that is transverse to the long axis of a beam, the beam will rotate and deflect. This flexure of the beam causes bending stresses, compression on one edge, and tension on the opposite edge of the beam. The flexure in a beam is caused by bending moments, the product of force times distance. The transverse load also creates shear forces in the beam, resulting in shear stresses in the transverse and longitudinal directions. Shear stresses and deflection of beams will be discussed in subsequent sections of this chapter.

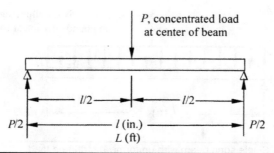

FIGURE 5-1 Simple span beam with concentrated load.

For a simple beam supported at each end, with a concentrated load P at its center, the reaction at each end of the beam will be $P/2$, as illustrated in Figure 5-1. The maximum bending moment due to the concentrated load will occur at the center of the beam and can be calculated as the product of force times distance: $M = P/2 \times l/2 = Pl/4$. The following equations can be used to calculate the maximum bending moment for a simple span beam for the stated loads.

For one concentrated load P acting at the center of the beam,

$$M = Pl/4 \text{ in.-lb} \tag{5-1}$$

For two equal loads P acting at the third points of the beam,

$$M = Pl/3 \text{ in.-lb} \tag{5-2}$$

For three equal loads P acting at the quarter points of the beam,

$$M = Pl/2 \text{ in.-lb} \tag{5-3}$$

For one concentrated load P acting at a distance x from one end of the beam, the maximum bending moment will be:

$$M = [P(l - x)x]/l \text{ in.-lb} \tag{5-4}$$

Equation (5-4) can be used to determine the maximum bending moment in a beam resulting from two or more concentrated loads acting at known locations by adding the moment produced by each load at the critical point along the beam. The critical point is that point at which the combined moments of all loads is a maximum.

Analysis of Bending Moments in Beams with Uniformly Distributed Loads

For a simple beam, supported at each end, with a uniform load distributed over its full length, the total vertical load on the beam wL is divided between the two supports at the end of the beam, as illustrated in Figure 5-2.

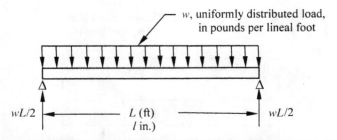

Figure 5-2 Single-span beam with uniformly distributed load.

The maximum bending moment, which will occur at the center of the beam, can be calculated as follows:

$$M = [wL/2 \times L/2] - [wL/2 \times L/4]$$
$$= wL^2/4 - wL^2/8$$
$$= wL^2/8 \text{ ft-lb}$$

Multiplying the preceding equation by 12 to convert the units of bending moment from ft-lb to in.-lb, and recognizing that $W = wL = wl/12$, the maximum bending moment for a uniformly distributed load on a simply supported beam can be calculated by the following equation:

$$M = 12wL^2/8$$
$$= Wl/8$$
$$= wl^2/96 \text{ in.-lb} \tag{5-5}$$

If the beam is continuous over three or more equally spaced supports, the maximum bending moment will be:

$$M = 12wL^2/10$$
$$= Wl/10$$
$$= wl^2/120 \text{ in.-lb} \tag{5-6}$$

Bending Stress in Beams

For a member in flexure, the applied bending stress must not exceed the allowable bending design stress $f_b < F_b$. For a beam subjected to a bending moment M, the applied bending stress can be calculated from the following equation:

$$f_b = Mc/I = M/S \tag{5-7}$$

For a solid rectangular beam, the moment of inertia is

$$I = bd^3/12 \tag{5-8}$$

and because $c = d/2$ for a rectangular beam, the section modulus is

$$S = I/c = bd^2/6 \qquad (5\text{-}9)$$

from which

$$f_b = M/S = 6M/bd^2 \qquad (5\text{-}10)$$

Equations (5-7) and (5-10) are used to determine the applied bending stress in a beam of known size for a known bending moment.

For design purposes, the bending moment M and the allowable bending stress F_b are known, with the required size of the beam to be determined. Therefore, substituting the allowable bending stress F_b for the applied bending stress f_b, Eq. (5-7) can be written for design purposes in the form:

$$S = M/F_b \qquad (5\text{-}11)$$

For a rectangular section, substituting $bd^2/6$ for S, we get

$$bd^2 = 6M/F_b \qquad (5\text{-}12)$$

When analyzing a beam, the magnitude of the load uniformly distributed along the beam or the magnitude and location of the concentrated load or loads are known. Also, the length of the span and the type of supports (whether simple or continuous) are known. Therefore, the bending moment M can be calculated. Also, for a given grade and species of lumber, the allowable stress F_b is known. Using this information, the required size (width and depth) of a beam can be determined as illustrated in the following sections.

Stability of Bending Members

Equations (5-7) through (5-12) for bending stresses are based on beams that are adequately braced and supported. The allowable bending stresses for beams are based on beams that have adequate lateral bracing and end supports to prevent lateral buckling of members subject to bending. If the beam is not adequately braced and supported, the reference design value is reduced in order to determine the allowable bending stress.

The beam stability factor C_L in the NDS is a multiplier to reduce the reference design value of bending stress based on the depth-to-thickness ratio (d/b) of the beam. When the conditions in Table 5-1 are satisfied, the beam stability factor $C_L = 1.0$; therefore, no adjustment is required in the reference design value in order to determine the allowable bending stress. The d/b ratios in Table 5-1 are based on

d/b Range	Lateral Constraints for Bending Stability
$d/b \leq 2$	No lateral support is required; includes 2×4, 3×4, 4×6, 4×8, and any members used flat, examples are 4×2, 6×2, 8×2.
$2 < d/b \leq 4$	Ends shall be held in position, as by full depth solid blocking, bridging, hangers, nailing, bolting, or other acceptable means; examples are 2×6, 2×8, 3×8.
$4 < d/b \leq 5$	Compression edge of the member shall be held in line for its entire length to prevent lateral displacement and ends of member at point of bearing shall be held in position to prevent rotation and/or lateral displacement; such as a 2×10.
$5 < d/b \leq 6$	Bridging, full depth solid blocking, or diagonal cross-bracing must be installed at intervals not to exceed 8 feet, compression edge must be held in line for its entire length, and ends shall be held in position to prevent rotation and/or lateral displacement; example is a 2×12.
$6 < d/b \leq 7$	Both edges of the member shall be held in line for its entire length and ends at points of bearing shall be held in position to prevent rotation and/or lateral displacement; examples are deep and narrow rectangular members.

Note: d and b are nominal dimensions

TABLE 5-1   Lateral Constraints for Stability of Bending Members

nominal dimensions. For conditions where the support requirements in Table 5-1 are not met, the designer should follow the requirements of the NDS to reduce the reference design value by the adjustment factor C_L in order to determine the allowable stress in bending.

It is often possible to design members to satisfy the conditions in Table 5-1 to allow the designer to utilize the full allowable bending stress. One example of providing lateral support for bending members is by placing decking on the top of beams. Although the beam is used to support the decking, the decking can be fastened along the compression edge of the beam throughout the length of the beam to provide lateral stability. There are other methods that can be used by the designer to satisfy the requirements of providing adequate lateral bracing of bending members.

Examples of Using Bending Stress Equations for Designing Beams and Checking Stresses in Beams

The following examples illustrate methods of designing and checking bending stresses in rectangular wood beams for specified loading conditions. The beams will have adequate lateral bracing and end supports ($C_L = 1.0$) to allow the reference design values to be used for the allowable bending stress.

Example 5-1

Determine the minimum-size wood beam, 8 ft long, required to support three equal concentrated loads of 190 lb spaced at 2, 4, and 6 ft from the ends of the beam. The beam will be used in a dry condition with normal load-duration and temperature. Adequate bracing and end supports will be provided.

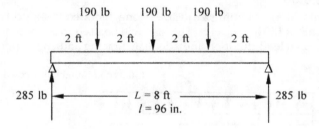

Because the loads are acting at the quarter points of the beam, Eq. (5-3) will be used to calculate the applied bending moment.

$$M = Pl/2$$
$$= (190 \text{ lb} \times 96 \text{ in.})/2$$
$$= 9{,}120 \text{ in.-lb}$$

Consider a 2×6 S4S (surfaced four sides) lumber of No. 2 grade Southern Pine. Table 4-2 indicates an allowable bending stress of 1,250 lb per sq in. From Eq. (5-11), the required section modulus S can be calculated:

$$S = M/F_b$$
$$= (9{,}120 \text{ in.-lb})/(1{,}250 \text{ lb per sq in.})$$
$$= 7.29 \text{ in.}^3$$

Thus, the required section modulus of the beam is 7.29 in.^{3}. From Table 4-1, the section modulus of a 2×6 S4S beam is 7.56 in.^{3}, which is greater than the required 7.29 in.^{3}. Therefore, a 2×6 S4S No. 2 grade Southern Pine is satisfactory provided it satisfies the requirements for the allowable unit stress in shear and the limitations on deflection.

Because the nominal dimension d/b ratio of 6/2 is 3.0, which is in the range of $2 < d/b \le 4$ in Table 5-1, the ends of the beam must be held in position to prevent displacement or rotation. The beam must have this end condition in order to satisfy beam stability and to justify using the allowable bending stress of 1,250 lb per sq in.

An alternate method to design a beam is to assume an allowable bending stress F_b and calculate the required section modulus S using Eq, (5-11), $S = M/F_b$. Then, searching Table 4-1, an available section modulus is selected that is greater than the calculated required section modulus. However, the allowable stress F_b will vary depending on the depth of the beam. Therefore, after the section modulus is selected from Table 4-1, the designer must then verify that the originally assumed allowable stress F_b is correct for the particular depth of the member that was selected from Table 4-1.

Example 5-2

Determine the minimum-size joist, 9 ft long, required to support a uniform load of 150 lb per lin ft. The joist will be used in a dry condition with normal load-duration and adequate bracing and end supports.

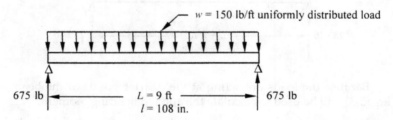

$w = 150$ lb/ft uniformly distributed load

675 lb ← $L = 9$ ft → 675 lb

$l = 108$ in.

For this condition of a uniformly distributed load over a single-span beam, the maximum bending moment can be calculated from Eq. (5-5) as follows:

$$\text{Using Eq. (5-5), } M = wl^2/96$$
$$= (150 \text{ lb per ft})(108 \text{ in.})^2/96$$
$$= 150(108)^2/96$$
$$= 18{,}225 \text{ in.-lb}$$

Consider a joist of No. 2 Douglas Fir-Larch with 8 in. nominal depth. Determine the allowable bending stress as follows:

Table 4-3, reference a design value for bending stress = 900 lb per sq in.

Table 4-3a, size adjustment for 8-in. lumber $C_F = 1.2$
Allowable bending stress $F_b = 900$ lb per sq in. × 1.2
$$= 1{,}080 \text{ lb per sq in.}$$

From Eq. (5-11) the required section modulus $S = M/F_b$

$$= (18{,}225 \text{ in.-lb})/(1{,}080 \text{ lb per sq in.})$$

$$= 16.88 \text{ in.}^3$$

Consider 8-in. nominal depths from Table 4-1:

For a 2×8, $S = 10.88 < 16.88$, therefore not acceptable

For a 3×8, $S = 18.13 > 16.88$, therefore acceptable

Therefore, a 3×8 member is satisfactory for bending. Because the nominal dimension d/b ratio of $8/3$ is 2.7, which is in the range of $2 < d/b \le 4$ in Table 5-1, the ends of the beam must be held in position to prevent displacement or rotation. The beam must have this end condition in order to satisfy beam stability and to justify using the allowable bending stress of 1,080 lb per sq in. Before the final selection is made, this member should be checked for the allowable unit stress in shear and the permissible deflection by methods that are presented in the following sections.

Example 5-3

A 2×4 beam of No. 2 grade Hem-Fir is supported over multiple supports that are spaced at 30 in. on center. The beam must carry a 500 lb per lin ft uniformly distributed load over the entire length. The beam will be used in a dry condition with normal load-duration. Check the bending stress in the beam and compare it to the allowable stress.

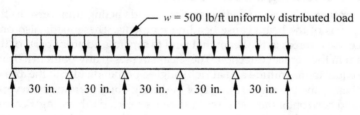

The maximum bending moment for a beam with multiple supports can be calculated as:

$$\text{From Eq. (5-6), } M = wl^2/120$$

$$= 500 \text{ lb/ft}(30 \text{ in.})^2/120$$

$$= 3{,}750 \text{ in.-lb}$$

The section modulus for a 2×4 beam from Table 4-1 is 3.06 in.^{3}, or the section modulus for bending can be calculated as follows:

$$\text{From Eq. (5-9), } S = bd^2/6$$

$$= 1.5 \text{ in.}(3.5 \text{ in.})^2/6$$

$$= 3.06 \text{ in.}^3$$

The applied bending stress can be calculated as follows:

$$\text{From Eq. (5-7)}, f_b = M/S$$
$$= 3750 \text{ in.-lb}/3.06 \text{ in.}^3$$
$$= 1225 \text{ lb per sq in.}$$

Determine the allowable bending stress for a No. 2 grade Hem-Fir as follows:

Table 4-3, reference design value for bending stress = 850 lb per sq in.
Table 4-3a, size adjustment factor for a 2×4 beam, $C_F = 1.5$
Allowable bending stress $F_b = (850 \text{ lb per sq in.})(C_F)$
$$= (850 \text{ lb per sq in.})(1.5)$$
$$= 1,275 \text{ lb per sq in.}$$

Because the allowable bending stress F_b of 1,275 lb per sq in. is greater than the applied bending stress f_b of 1225 lb per sq in., the No. 2 grade Hem-Fir beam is satisfactory for bending. The d/b ratio of a 2×4 is $4/2 = 2.0$; therefore, no lateral support is required. Although the 2×4 is adequate for bending, it must be checked for shear and deflection as shown in the following sections.

Horizontal Shearing Stress in Beams

As presented in the preceding sections, loads acting transverse to the long axis of the beam cause bending moments. These loads also produce shear forces in the beam that tend to separate adjacent parts of the beam in the vertical direction. The shear stresses at any point in a beam are equal in magnitude and at right angles, perpendicular to the axis of the beam, and parallel to the axis of the beam. Thus, there are both vertical and horizontal shear stresses in a beam subjected to bending. Because the strength of wood is stronger across grain than parallel to grain, the shear failure of a wood member is higher in the horizontal direction (parallel to grain) along the axis of the beam. Thus, in the design of wood beams for shear, the horizontal shear stress is considered.

The maximum applied horizontal shear stress in a rectangular wood beam is calculated by the following equation:

$$f_v = 3V/2bd \tag{5-13}$$

For a single-span beam with one support at each end and a uniformly load distributed over its full length, the maximum total shear will occur at one end, and it will be:

$$V = wL/2$$
$$= wl/24 \text{ lb} \tag{5-14}$$

For a beam continuous over more than two equally spaced supports, with a uniform load distributed over its full length, the maximum total shear is frequently determined as an approximate value from the following equation:

$$V = 5wL/8$$
$$= 5wl/96 \text{ lb} \qquad (5\text{-}15)$$

For a simple beam with one or more concentrated loads acting between the supports, the maximum total shear will occur at one end of the beam, and it will be the reaction at the end of the beam. If the two reactions are not equal, the larger one should be used.

Example 5-4

Consider the 2 × 6 beam selected in Example 5-1 that supports three equally concentrated loads of 190 lb each. Check this beam for the unit stress in horizontal shear.

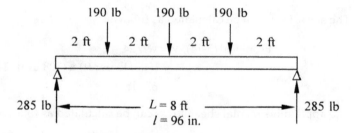

The maximum shear force can be calculated as:

$$V = [3 \times 190 \text{ lb}]/2$$
$$= 285 \text{ lb}$$

From Eq. (5-13), the applied horizontal shear stress is

$$f_v = 3V/2bd$$
$$= [3 \times 285 \text{ lb}]/[2 \times 1.5\text{-in.} \times 5.5\text{-in.}]$$
$$= 51.8 \text{ lb per sq in.}$$

Table 4-2 indicates an allowable horizontal shear stress for a 2 × 6 No. 2 grade Southern Pine as 175 lb per sq in., which is greater than the applied shear stress of 51.8 lb per sq in. Therefore, the 2 × 6 beam will be satisfactory in horizontal shear.

Example 5-5

Consider the beam previously selected to resist the bending moment in Example 5-2. It was determined that a 3 × 8 S4S beam of No. 2

Douglas Fir-Larch would be required for bending. Will this beam satisfy the requirements for the allowable unit stress in horizontal shear?

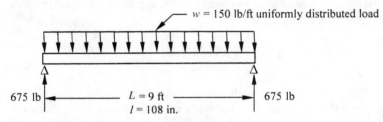

The uniform load and dimensions of the beam are:

$$w = 150 \text{ lb per lin ft}$$
$$L = 9 \text{ ft, or } l = 108 \text{ in.}$$
$$b = 2.5 \text{ in.}$$
$$d = 7.25 \text{ in.}$$

The shear force can be calculated as follows:

From Eq. (5-14), $V = wl/24$
$$= [(150 \text{ lb per ft}) \times (108 \text{ in.})]/24$$
$$= 675 \text{ lb}$$

The applied horizontal shear stress can be calculated as follows:

From Eq. (5-13), $f_v = 3V/2bd$
$$= [3 \times 675 \text{ lb})]/[2 \times 2.5 \text{ in.} \times 7.25 \text{ in.}]$$
$$= 55.9 \text{ lb per sq in.}$$

Table 4-3 indicates an allowable shear stress for a 3×8 beam of No. 2 grade Douglas Fir-Larch as 180 lb per sq in., which is greater than the required 55.9 lb per sq in. Therefore, the 3×8 beam will be satisfactory for shear.

Modified Method of Determining the Unit Stress in Horizontal Shear in a Beam

When calculating the shear force V in bending members with uniformly distributed loads, the NDS allows neglecting all loads within a distance d from the supports provided the beam is supported by full bearing on one surface and loads are applied to the opposite surface. For this condition, the shear force at a reaction or support point of single-span and multiple-span beams that sustain a uniformly distributed load will be as follows.

For a single-span beam with a uniform load of w over its entire length of L:

Total vertical load $= wL$
All loads within a distance d from the supports $= w(2d/12)$
Therefore, the net total vertical load $= wL - w(2d/12)$

For this condition, the shear force V is

$$V = (wL/2) - w(d/12)$$
$$= (wL/2)[1 - 2d/l] \qquad (5\text{-}16)$$

Substituting $L = l/12$, Eq. (5-16) for shear in a single-span beam can be written in the following form:

$$V = wl[1 - 2d/l]/(24) \qquad (5\text{-}16a)$$

From Eq. (5-13), the total shear for a beam is

$$V = 2f_v bd/3 \qquad (5\text{-}16b)$$

Combining Eqs. (5-16a) and (5-16b) and solving for f_v gives:

$$2f_v bd/3 = wl[1 - 2d/l]/24$$
$$f_v = 3wl[1 - 2d/l]/[2 \times 24bd]$$
$$= wl[1 - 2d/l]/16bd$$
$$= w[l - 2d]/16bd \qquad (5\text{-}17)$$

For multiple-span beams the shear force V can be calculated as follows:

$$V = (5wL/8) - [(5w/8) \times (2d/12)]$$
$$= (5wL/8)[1 - 2d/1] \qquad (5\text{-}18)$$

Substituting $L = l/12$, Eq. (5-18) for shear in a multiple-span beam can be written in the following form:

$$V = 5wl[1 - 2d/l]/96 \qquad (5\text{-}18a)$$

From Eq. (5-13), the total shear in a beam is

$$V = 2f_v bd/3 \qquad (5\text{-}18b)$$

Combining Eqs. (5-18a) and (5-18b) and solving for f_v gives:

$$2f_v bd/3 = 5wl[1 - 2d/l]/96$$
$$f_v = 15wl[1 - 2d/l]/192bd$$
$$= 15w[l - 2d]/192bd \qquad (5\text{-}19)$$

Example 5-6

A 6-ft-long rectangular 4 × 6 beam will be used for temporary support of formwork. The duration of the load will be less than 7 days and the beam will be in a dry condition, moisture content less than 19%. The beam will be grade No. 2 Spruce-Pine-Fir whose actual dimensions are 3½ by 5½ in. If the beam supports a uniformly distributed load of 450 lb per lin ft over a single span, determine the values of the shear force V and the applied horizontal shear stress f_v using the modified method of determining shear. Also, check the bending capacity for this beam.

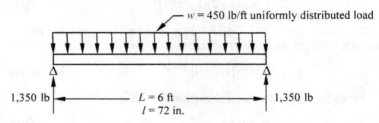

$w = 450$ lb/ft uniformly distributed load

1,350 lb ← $L = 6$ ft → 1,350 lb
$l = 72$ in.

Check Shear Capacity

From Eq. (5-16), the modified shear force for a single-span beam with a uniformly distributed load can be calculated by Eq. (5-16) as follows:

$$V = wL/2[1 - 2d/l]$$
$$= [(450 \text{ lb/ft} \times 6 \text{ ft})/2][1 - 2(5.5 \text{ in.})/72 \text{ in.}]$$
$$= (1,350 \text{ lb})[1 - 0.153]$$
$$= 1,143.5 \text{ lb}$$

From Eq. (5-17), the applied shear stress can be calculated:

$$f_v = w[l - 2d]/16bd$$
$$= 450 \text{ lb/ft}[72 \text{ in.} - 2(5.5 \text{ in.})]/[16(3.5 \text{ in.})(5.5 \text{ in.})]$$
$$= 27,450/308$$
$$= 89.1 \text{ lb per sq in.}$$

Determine the allowable shear stress for a 4 × 6 No. 2 grade Spruce-Pine-Fir with a 7-day load-duration as follows:

Table 4-3, reference design value for shear stress = 135 lb per sq in.
Table 4-4, adjustment factor for 7-day load-duration, $C_D = 1.25$

Allowable shear stress $F_v = (135 \text{ lb per sq in.})(C_D)$
$$= (135 \text{ lb per sq in.})(1.25)$$
$$= 168.8 \text{ lb per sq in.}$$

The allowable shear stress of 168.8 lb per sq in. is greater than the applied shear stress of 89.1 lb per sq in.; therefore, the beam is satisfactory for shear stress.

Check Bending Capacity

For a uniformly distributed load on a single span 4 × 6 beam, the bending moment and stress can be calculated as follows:

From Eq. (5-5), $M = wl^2/96 = [(450 \text{ lb per ft}) \times (72 \text{ in.})^2]/96$
$$= 24,300 \text{ in.-lb}$$

From Eq. (5-10), $f_b = M/S = 6M/bd^2 = [6(24,300)]/[(3.5) \times (5.5)^2]$
$$= 1,376.8 \text{ lb per sq in.}$$

Determine the allowable bending stress for a 4 × 6 No. 2 grade Spruce-Pine-Fir with a 7-day load-duration as follows:

Table 4-3, reference design value in bending = 875 lb per sq in.
Table 4-3a, size adjustment factor for 4 × 6, $C_F = 1.3$
Table 4-4, adjustment factor for 7-day load-duration, $C_D = 1.25$

Allowable bending stress, $F_b = (875 \text{ lb per sq in.})(C_F)(C_D)$
$$= (875 \text{ lb per sq in.})(1.3)(1.25)$$
$$= 1,421.9 \text{ lb per sq in.}$$

The allowable bending stress of 1,421.9 lb per sq in. is greater than the applied bending stress of 1,376.8 lb per sq in. Therefore, the 4 × 6 beam is adequate for bending. No lateral support is required because the d/b ratio is less than 2.

Example 5-7

Consider the beam previously selected to resist the bending moment in Example 5-3. It was determined that a 2 × 4 beam of No. 2 grade Hem-Fir would be required for bending. Check the shear capacity of this beam using the modified method of determining the unit shear stress.

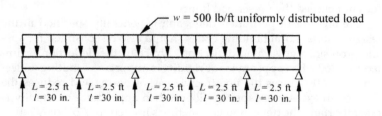

From Eq. (5-18), the modified shear force for a multiple-span beam with a uniformly distributed load can be calculated as follows:

$$V = 5wL/8[1 - 2d/l]$$
$$= [5(500 \text{ lb/ft} \times 2.5 \text{ ft})/8][1 - 2(3.5 \text{ in.})/30 \text{ in.})]$$
$$= (781.2 \text{ lb})[1 - 0.233]$$
$$= 598.9 \text{ lb}$$

From Eq. (5-19), the applied shear stress can be calculated:

$$f_v = 15w[l - 2d]/192bd$$
$$= 15(500 \text{ lb/ft})[(30 \text{ in.} - 2(3.5 \text{ in.})]/[192(1.5 \text{ in.})(3.5 \text{ in.})]$$
$$= 172,500/1,008$$
$$= 171.1 \text{ lb per sq in.}$$

Table 4-3 shows an allowable shear stress F_v as 150 lb per sq in., which is less than the calculated applied shear stress of 171.1 lb per sq in. Therefore, the beam is not satisfactory for shear stress.

Example 5-3 showed the 2×4 beam adequate for bending using the No. 2 grade Hem-Fir, but this example shows it is not adequate for shear. A different grade and species of lumber could be considered. For example, a No. 2 grade Southern Pine has an allowable shear stress of 175 lb per sq in., or a No. 2 grade Douglas Fir-Larch, which has an allowable shear stress of 180 lb per sq in. Either of these grades and species would be adequate for shear. The beam should also be checked for deflection as shown in the following sections.

Deflection of Beams

When a beam is subjected to a load, there is a change in its shape. This change is called deformation. Regardless of the magnitude of the load, some deformation always occurs. When a force causes bending moments in a beam, the deformation is called deflection. A beam may have adequate strength to resist applied bending and shear stresses, but it may not have adequate rigidity to resist deflections that are created by the loads applied on the beam. Some amount of deflection exists in all beams, and the designer must ensure that the deflection does not exceed the prescribed limits.

The amount of permitted deflection is generally specified in the design criteria. Local building codes should also be consulted for specific provisions governing deflection. Typically, the deflection is limited to $l/360$ when appearance or rigidity is important. Limiting deflection to $l/360$ reduces the unattractive sag of beams and reduces the effect of an excessively springy floor. When appearance or rigidity is less important, the deflection is sometimes limited to $l/270$ or $l/240$.

Most specifications for concrete structures limit the deflection of formwork members in order to prevent excessive wave effects on the surface of the concrete. For formwork members where appearance is important, the permissible deflection may be limited to $l/360$, or as $\frac{1}{16}$ in. where l is the distance between the centers of supports. An example of important appearance is sheathing for flooring where it is desirable to have a smooth concrete surface without waves between the supporting joints of the sheathing. For limitations in deflections, the distance between supports for which the two will be the same is obtained by equating the two values.

$$l/360 = \frac{1}{16}$$
$$l = 22.5 \text{ in.}$$

For formwork members where strength is the primary concern and appearance is secondary, the allowable deflection may be specified as $l/270$, or $\frac{1}{8}$ in., where l is the distance between the centers of supports. For example, the inside surface of a parapet wall around a roof will only be exposed to maintenance workers, rather than the general public. Thus, appearance may not be important. For limitations in deflections, the distance between supports for which the two will be the same is obtained by equating the two values.

$$l/270 = \frac{1}{8} \text{ in.}$$
$$l = 34 \text{ in.}$$

For spans less than 34 in. the $l/270$ limit is more rigid, and for spans greater than 34 in. the $\frac{1}{8}$ in. is more rigid. For long structural members, some designers limit the deflection to $\frac{1}{4}$ in.

Thus, $l/360$ is a more rigid deflection requirement than $l/270$. The selection of these two limiting deflection criteria depends on the relative importance of appearance and strength. If appearance is not important, the less stringent requirement will allow economy in the construction costs of formwork.

Deflection of Beams with Concentrated Loads

A beam may have one or more concentrated loads applied at various locations on the beam. The loads cause the beam to deflect downward. The amount of resistance to deflection is called rigidity. For a beam the amount of rigidity is determined by the physical properties of the beam, EI. The amount of deflection depends on the magnitude and location of the applied loads, length and size of beam, grade and species of the beam, and the number and location of supports underneath the beam.

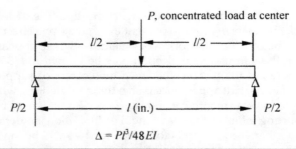

FIGURE 5-3 Single-span beam with concentrated load at the center.

Deflection of Single-Span Beams with Concentrated Loads

For a single-span beam supported at each end, with a concentrated load P at its center, the reaction at each end of the beam will be $P/2$ as illustrated in Figure 5-3. The concentrated load will cause a bending moment of $Pl/4$ as discussed previously in this chapter. The maximum deflection will occur at the center of the beam, and can be calculated by Eq. (5-20).

$$\Delta = Pl^3/48EI \tag{5-20}$$

When the concentrated load is not located at the center of the single span beam, as shown in Figure 5-4, the deflection can be calculated by Eq. (5-21).

$$\Delta = Pb[l^2/16 - b^2/12]/EI \tag{5-21}$$

When there are three concentrated load as shown on the single-span beam in Figure 5-5, the deflection can be calculated by Eq. (5-22).

$$\Delta = P\,[l^3 + 6al^2 - 8a^3]/48EI \tag{5-22}$$

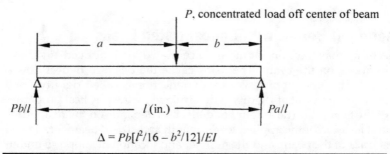

FIGURE 5-4 Single-span beam with concentrated load off center.

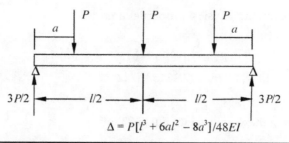

$$\Delta = P[l^3 + 6al^2 - 8a^3]/48EI$$

Figure 5-5   Single-span beam with three concentrated loads.

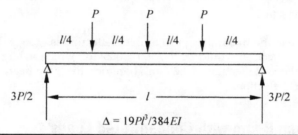

$$\Delta = 19Pl^3/384EI$$

Figure 5-6   Single-span beam with concentrated loads at quarter points.

For a beam with three equal loads P acting at the quarter points of the beam as shown in Figure 5-6, the maximum deflection can be calculated from Eq. (5-23).

$$\Delta = 19Pl^3/384EI \tag{5-23}$$

Example 5-8

In Example 5-1, it was determined that the bending moment required the use of a 2×6 beam, 8 ft long, to support three concentrated loads of 190 lb each, located 2, 4, and 6 ft, respectively, from the end of the beam. The actual dimensions of this beam are 1½ in. by 5½ in. Determine the deflection of the beam considering the three concentrated loads using Eq. (5-23).

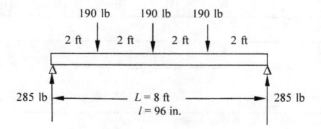

For this beam, the physical properties are:

$l = 96$ in.

$= 1,600,000$ lb per sq in.

$I = bd^3/12 = [1.5 \text{ in.} \times (7.25 \text{ in.})^3]/12$

$= 20.8$ in.^{4}

Substituting these values into Eq. (5-23) gives:

$\Delta = 19Pl^3/384EI$

$= [19(190 \text{ lb})(96 \text{ in.})^3]/[384(1,600,000 \text{ lb per sq in.})(20.8 \text{ in.}^4)]$

$= 0.25$ in.

Suppose the maximum permissible deflection is $l/360$. The permissible deflection would be $96/360 = 0.27$ in., which is larger than the calculated deflection of 0.25 in. Therefore, the 2×6 beam is adequate for deflection.

Multiple-Span Beam with Concentrated Loads

Figure 5-7 illustrates a beam, such as a wale, that is continuous over several supports, the form ties, with equally spaced concentrated loads acting on it. The concentrated loads act on the wale through the studs. The same loading condition will exist where floor decking is supported on joists, which in turn are supported by stringers, and where the stringers are supported by vertical shores. For this type of loading, the positions of the concentrated loads P can vary considerably, which will result in some rather complicated calculations to determine the maximum deflection. If the concentrated loads are transformed into a uniformly distributed load, having the same total value as the sum of the

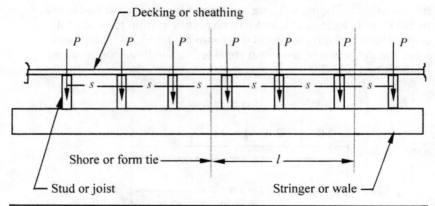

FIGURE 5-7   Concentrated loads on a beam.

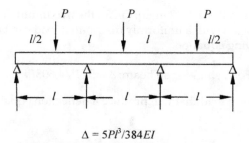

$$\Delta = 5Pl^3/384EI$$

FIGURE 5-8   Multiple-span beam with concentrated loads.

concentrated loads, the maximum deflection can usually be approximated from Eq. (5-24) with sufficient accuracy for form design.

For a beam with three equal loads P acting a multiple-span beam as shown in Figure 5-8, the maximum deflection can be calculated from Eq. (5-24).

$$\Delta = 5Pl^3/384EI \qquad (5\text{-}24)$$

Deflection of Beams with Uniform Loads

A beam may have a uniformly distributed load applied over a portion or the entire length of the beam. The amount of deflection depends on the magnitude of the distributed load, length and size of beam, grade and species of the beam, and the number and location of supports underneath the beam.

Single-Span Beams with Uniformly Distributed Loads

For a uniformly loaded beam supported at each end as shown in Figure 5-9, the maximum deflection at the center of the beam can be calculated by Eq. (5-25):

$$\Delta = 5Wl^3/384EI \qquad (5\text{-}25)$$

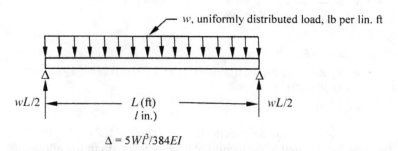

$$\Delta = 5Wl^3/384EI$$

FIGURE 5-9   Single-span beam with uniformly distributed load.

Substituting $W = wl/12$ in Eq. (5-25), the maximum deflection of a single span beam with a uniformly distributed load can be calculated by the following equation.

$$\text{For a single-span beam; } \Delta = 5wl^4/4{,}608EI \qquad (5\text{-}26)$$

The uniform load that will produce a deflection equal to Δ will be as follows:

$$\text{For a single-span beam; } w = 4{,}608EI\Delta/5l^4$$
$$= 921.6EI\Delta/l^4 \qquad (5\text{-}27)$$

$$\text{For } \Delta = l/270, \ w = 4{,}608EI/1350l^3 \qquad (5\text{-}27a)$$

$$\text{For } \Delta = l/360, \ w = 4{,}608EI/1800l^3 \qquad (5\text{-}27b)$$

$$\text{For } \Delta = \tfrac{1}{8} \text{ in., } w = 4{,}608EI/40l^4 \qquad (5\text{-}27c)$$

$$\text{For } \Delta = \tfrac{1}{16} \text{ in., } w = 4{,}608EI/80l^4 \qquad (5\text{-}27d)$$

Example 5-9

In Example 5-2, it was determined that the bending moment required a No. 2 grade Douglas Fir-Larch 3×8 beam to support a uniformly distributed load of 150 lb per lin ft over the 9 ft length of the beam. Determine the deflection of this beam. Because the beam is a single-span beam with a uniform load, the deflection is calculated using Eq. (5-26). From Table 4-1, the moment of inertia $I = 79.39$ in.[4]

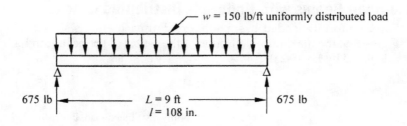

$\Delta = 5wl^4/4{,}608EI$

$\quad = [5(150 \text{ lb per lin ft})(108 \text{ in.})^4]/[4{,}608(1{,}600{,}000 \text{ lb per sq in.})$
$\quad \quad (79.39 \text{ in.}^4)]$

$\quad = 0.17 \text{ in.}$

Assume the allowable deflection is $l/360 = 108/360 = 0.30$ in. Because the calculated deflection of 0.17 in. is less than the allowable deflection of 0.30 in., then this beam is acceptable for deflection.

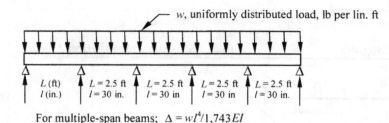

For multiple-span beams; $\Delta = wl^4/1{,}743EI$

FIGURE 5-10   Multiple-span beam with uniformly distributed loads.

Deflection of Multiple-Span Beams with Uniformly Distributed Loads

Figure 5-10 shows a uniformly loaded beam extending over several supports, such as studs supporting sheathing or joists supporting decking. The approximate value of maximum deflection in the end spans will be as shown in Eq. (5-28).

$$\text{For multiple-span beams; } \Delta = wl^4/1{,}743EI \qquad (5\text{-}28)$$

The uniform load that will produce a deflection equal to Δ will be as follows:

$$\text{For multiple-span beams; } w = 1{,}743EI\Delta/l^4 \qquad (5\text{-}29)$$
$$\text{For } \Delta = l/270,\ w = 1{,}743EI/270l^3 \qquad (5\text{-}29a)$$
$$\text{For } \Delta = l/360,\ w = 1{,}743EI/360l^3 \qquad (5\text{-}29b)$$
$$\text{For } \Delta = \tfrac{1}{8}\text{ in., } w = 1{,}743EI/40l^4 \qquad (5\text{-}29c)$$
$$\text{For } \Delta = \tfrac{1}{16}\text{ in., } w = 1{,}743EI/80l^4 \qquad (5\text{-}29d)$$

Table for Bending Moment, Shear, and Deflection for Beams

As illustrated in the preceding sections, there are many combinations of loads and locations of loads on beams used in formwork. Numerous books are available that provide the maximum bending moment, shear, and deflection for various applications of load, combinations of loads, span lengths, and end conditions.

There are several methods of analysis that can be used to determine the maximum shear force, bending moment, and deflection in beams. For complicated forming systems, it may be necessary to provide a precise analysis of the structural members, including shear and moment diagrams, to determine the appropriate values of shear and moment.

Table 5-2 gives the maximum bending moment, shear, and deflection for beams subjected to the indicated load and support conditions.

Load Diagram	Shear Force V, lb	Bending Moment M, in.-lb	Deflection Δ, in.
1. P, lb ; l	P	Pl	$Pl^3/3EI$
2. $l/2$ P $l/2$; l	$P/2$	$Pl/4$	$Pl^3/48EI$
3. a P b ; l	Pa/l	Pab/l	$Pb[l^2/16 - b^2/12]/EI$
4. P a P a ; l	P	Pa	$Pa[3l^2/4 - a^2]/6EI$
5. P a P P a ; $l/2$ $l/2$	$3P/2$	$P[l/4 + a]$	$P[l^3 + 6al^2 - 8a^3]/48EI$
6. $l/2$ P l P l P $l/2$; l l l	$5P/8$	s$3Pl/16$	$5Pl^3/384EI$
7. w, lb/ft ; l	$wl/12$	$wl^2/24$	$wl^4/96EI$
8. w, lb/ft ; l	$wl/24$	$wl^2/96$	$5wl^4/4{,}608EI$
9. w, lb/ft ; l l l	$5wl/96$	$wl^2/120$	$wl^4/1{,}743EI$

TABLE 5-2   Maximum Bending Moment, Shear, and Deflection for Beams

Many of the equations shown in this table were derived and illustrated in previous sections. The symbols and units were also defined earlier in this chapter. The span length l and deflection Δ are measured in inches, and the unit of measure for the uniformly distributed load w is lb per lin ft. The concentrated load P and shear force V are measured in pounds.

Calculating Deflection by Superposition

Table 5-2 gives equations for various load conditions for calculating bending stress, shear stress, and deflection. These equations are commonly used to design wood members for formwork and temporary structures. However, sometimes there are other load conditions that may be applied to the beams shown in Table 5-2. The method of superposition is a technique of determining the total deflection of a beam by superimposing the deflections caused by several simple loadings. In essence, the total deflection is determined by adding the deflections caused by several loads acting separately on a beam.

The method of superposition is restricted to beams with the same length and end conditions. Also, this method is restricted to the theory of small deformations, which means the effect produced by each load must be independent of that produced by other loads. Each load must not cause an excessive change in the original shape or length of the beam. Also, the point of maximum deflection caused by the separate loadings must occur at the same location on the beam.

The technique of superposition is applicable for combining the types of loads shown in Table 5-2. The following examples illustrate the method of superposition for calculating deflection.

Example 5-10

Use the method of superposition to calculate the total deflection of the beam shown below. The single-span beam has four concentrated loads, two 8-lb loads, and two 120-lb loads. A review of Table 5-2 shows there is no beam with this load condition. However, Beam 4 in Table 5-2 is a single-span beam with two loads located at a distance a from each end of the beam. Using Beam 4 in Table 5-2, the deflection due to the two 80-lb loads can be calculated and then the deflection due to the two 120-lb loads can be calculated. Then, by method of superposition, the two deflections can be added together to obtain the total deflection due to all four of the loads.

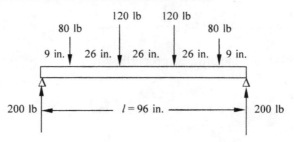

For this beam the physical properties are

$$l = 96 \text{ in.}$$
$$E = 1,600,000 \text{ lb per sq in.}$$
$$I = 20.8 \text{ in.}^4$$

Using Beam 4 in Table 5-2 to calculate the deflection from the two 80-lb loads:

$$\Delta = Pa\,[3(l^2)/4 - (a)^2]/6EI$$
$$= 80(9)[3(96)^2/4 - (9)^2]/6(1,600,000)(20.8)$$
$$= 0.0245 \text{ in.}$$

Using Beam 4 in Table 5-2 to calculate the deflection from the two 120-lb loads:

$$\Delta = Pa\,[3(l^2)/4 - (a)^2]/6EI$$
$$= 120(35)[3(96)^2/4 - (35)^2]/6(1,600,000)(20.8)$$
$$= 0.1196 \text{ in.}$$

By superposition, the total deflection is obtained by adding the two deflections as follows:

Total deflection = deflection from 80-lb loads + deflection from 120-lb loads

$$\Delta = 0.0245 \text{ in.} + 0.1196 \text{ in.}$$
$$= 0.144 \text{ in.}$$

Example 5-11

The beam shown below has both a uniformly distributed load of 150 lb per lin ft and two 200-lb concentrated loads. Use the method of superposition to calculate the total deflection of the beam.

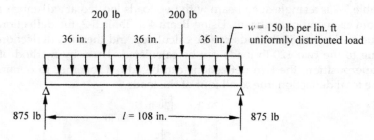

For this beam the physical properties are

$$l = 108 \text{ in.}$$
$$E = 1,600,000 \text{ lb per sq in.}$$
$$I = 79.39 \text{ in.}^4$$

Using Beam 4 in Table 5-2 to calculate the deflection from the two 200-lb loads:

$$\Delta = Pa\,[3(l^2)/4 - (a)^2]/6EI$$
$$= 200(36)[3(108)^2/4 - (36)^2]/6(1,600,000)(79.39)$$
$$= 0.07 \text{ in.}$$

Using Beam 8 in Table 5-2 to calculate the deflection from the 150 lb per lin ft uniformly distributed load:

$$\Delta = 5wl^4/4,608EI$$
$$= 5(150)(108)^4/4,608(1,600,000)(79.39)$$
$$= 0.17 \text{ in.}$$

By superposition the total deflection is obtained by adding the two deflections as follows:

Total deflection = deflection from 200-lb loads + deflection from the 150-lb per lin ft uniformly distributed load:

$$\Delta = 0.07 \text{ in.} + 0.17 \text{ in.}$$
$$= 0.24 \text{ in.}$$

Allowable Span Length Based on Moment, Shear, or Deflection

As illustrated in the preceding sections, many calculations are required to determine the adequacy of the strength of structural members. It requires considerable time to analyze each beam element to determine the applied and allowable stresses for a given load condition, and to compare the calculated deflection with the permissible deflection. For each beam element, a check must be made for bending, shear, and deflection. Thus, it is desirable to develop a method that is convenient for determining the adequacy of the strength and rigidity of form members.

For design purposes it is often useful to rewrite the previously derived stress and deflection equations in order to calculate the permissible span length of a member in terms of member size, allowable stress, and loads on the member. The equations derived earlier in this chapter can be rewritten to show the span length *l* in terms of the bending and shear stresses and deflection.

The following sections show the method of determining the length of span that is permitted for the stated conditions. The equations presented in these sections are especially useful in designing formwork.

Allowable Span Length for Single-Span Members with Uniformly Distributed Loads

For a single-span beam with a uniformly distributed load over its entire length (Beam 8 in Table 5-2) the allowable span length for bending, shear, and deflection can be determined as follows.

Combining the applied bending moment $M = wl^2/96$ with the allowable bending stress $F_b = M/S$, and rearranging terms, provides the following equation for allowable span length due to bending for a single-span beam:

$$l_b = [96F_bS/w]^{1/2} \text{ in.} \tag{5-30}$$

Combining the applied shear force $V = wl/24$ with the allowable shear stress $F_v = 3V/2bd$, the allowable span length due to shear for a single-span beam is provided by the following equation:

$$l_v = 16F_vbd/w \tag{5-31}$$

Combining the modified shear force $V = wl/24(1 - 2d/l)$ with the allowable shear stress $F_v = 3V/2bd$ provides the following equation for the allowable span length for modified shear:

$$l_v = 16F_vbd/w + 2d \tag{5-32}$$

The general equation for calculating the allowable span length based on the deflection of a single-span wood beam can be obtained by rearranging the terms in Eq. (5-26) as follows:

$$l_\Delta = [4,608EI\Delta/5w]^{1/4} \tag{5-33}$$

If the permissible deflection is $l/360$, substituting $l/360$ for Δ in Eq. (5-26) and rearranging terms, the allowable span length for single-span beams will be:

$$l_\Delta = [4,608EI/1800w]^{1/3} \tag{5-33a}$$

If the permissible deflection is $\frac{1}{16}$ in., substituting $\frac{1}{16}$ in. for Δ in Eq. (5-26) for a single-span beam, the allowable span length based on a permissible deflection of $\frac{1}{16}$ in. is

$$l_\Delta = [4,608EI/80w]^{1/4} \tag{5-33b}$$

Allowable Span Length for Multiple-Span Members with Uniformly Distributed Loads

For a multiple-span beam with a uniformly distributed load over its entire length (Beam 9 in Table 5-2), the allowable span length can be determined as described in the following paragraphs.

Combining the applied bending moment $M = wl^2/120$ with the allowable bending stress $F_b = M/S$ and rearranging terms, the following equation can be used to calculate the allowable span length due to bending for multiple-span wood beams:

$$l_b = [120F_bS/w]^{1/2} \text{ in.} \tag{5-34}$$

Combining the shear force $V = 5wl/96$ with the allowable shear stress $F_v = 3V/2bd$ and rearranging terms, the allowable span length due to shear in a multiple-span beam will be:

$$l_v = 192F_vbd/15w \tag{5-35}$$

Combining the modified shear force $V = 5wl/96[1 - 2d/l]$ with the allowable shear stress $F_v = 3V/2bd$, the allowable span length due to modified shear in multiple-span beams is

$$l_v = 192F_vbd/15w + 2d \tag{5-36}$$

The general equation for calculating the allowable span length based on the deflection of wood beams with multiple-span lengths can be obtained by rearranging the terms in Eq. (5-23) as follows:

$$l_\Delta = [1{,}743EI\Delta/w]^{1/4} \tag{5-37}$$

If the permissible deflection is $l/360$, substituting $l/360$ for Δ in Eq. (5-28) for multiple-span beams, the allowable span length due to a deflection of $l/360$ is

$$l_\Delta = [1{,}743EI/360w]^{1/3} \tag{5-37a}$$

If the permissible deflection is $\frac{1}{16}$ in., substituting $\frac{1}{16}$ in. for Δ in Eq. (5-28) for multiple-span beams, the allowable span length based on a permissible deflection of $\frac{1}{16}$ in. is

$$l_\Delta = [1{,}743EI/16w]^{1/4} \tag{5-37b}$$

The techniques described in this section can be used for other load and support conditions of beams to determine the allowable length for a specified bending stress, shear stress, or permissible deflection.

Stresses and Deflection of Plywood

Plywood is used extensively as sheathing in formwork, to support lateral pressures against wall or column forms or vertical pressures on floor forms. The loads on plywood are usually considered as being uniformly distributed over the entire surface of the plywood. Thus, the equations for calculating stresses and deflections in plywood are based on uniform loads in pounds per square foot (lb per sq ft), whereas in previous sections the uniform loads were considered as pounds per linear foot (lb per lin ft).

The Plywood Design Specification [4] recommends three basic span conditions for computing the uniform load capacity of plywood panels. The spans may be single-span, two-span, or three-span.

When plywood is placed with the face grain across (perpendicular to) the supports, the three-span condition is used for support spacings up to and including 32 in. The two-span condition is used when the support spacing is greater than 32 in. In designing or checking the strength of plywood, it should be noted that when the face grain of plywood is placed across (perpendicular to) supports, the physical properties under the heading "stress applied parallel to face grain" should be used (see Table 4-9 for plywood and Table 4-11 for Plyform).

When plywood face grain is placed parallel to the supports, the three-span condition is assumed for support spacings up to and including 16 in. The two-span condition is assumed for face grain parallel to supports when the support spacing is 20 or 24 in. The single-span condition is used for spans greater than 24 in. It should be noted that when the face grain of plywood is placed parallel to the supports, the physical properties under the heading "stress applied perpendicular to face grain" should be used (see Table 4-10 for plywood and Table 4-12 for Plyform).

In designing plywood, it is common to assume a 1-ft-wide strip that spans between the supporting joists of plywood sheathing for floors or between the studs that support plywood sheathing for wall forms. The physical properties for plywood are given for 1 ft widths of plywood.

APA—The Engineered Wood Association recommends using the center-to-center span length for calculating the bending stress. For calculating the rolling shear stress and shear deflection, the clear distance between supports should be used. For bending deflection, the recommendations are to use the clear span plus ¼ in. for 2 in. nominal framing, and the clear span plus ⅝ in. for 4 in. framing. Figure 5-11 shows these various span lengths.

Allowable Pressure on Plywood Based on Bending Stress

The allowable pressure of plywood based on bending can be calculated using the conventional engineering equations presented in previous sections of this book. However, because plywood is fabricated by alternate layers of wood that are glued together, the previously presented stress equations must be used with the physical properties of plywood. For example, in calculating the section modulus for bending, an effective section modulus S_e is used. Values for the effective section modulus and other physical properties of plywood per foot of width can be found in Table 4-9 for plywood and in Table 4-11 for Plyform.

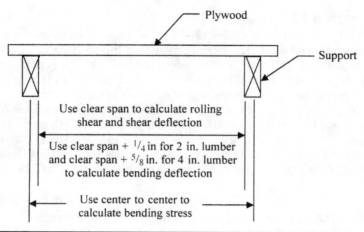

Figure 5-11   Effective span length for design of plywood.

The allowable pressure based on fiber stress bending of plywood was presented in Chapter 4. The following equations are restated here for clarity:

$$w_b = 96F_bS_e/(l_b)^2 \text{ for one or two spans} \tag{5-38}$$

$$w_b = 120F_bS_e/(l_b)^2 \text{ for three or more spans} \tag{5-39}$$

where w_b = allowable pressure of concrete on plywood, lb per sq ft
F_b = allowable bending stress in plywood, lb per sq in.
S_e = effective section modulus of a 1-ft-wide plywood strip, in.^{3}/ft
l_b = center-to-center distance between supports, in.

Equations (5-38) and (5-39) can be rewritten to calculate the allowable span length in terms of the uniformly distributed pressure, the allowable bending stress, and the section properties of plywood. The equations written in this manner are useful for determining the joist spacing necessary to support a given concrete pressure on the forms. Rearranging the terms in the equations, the span lengths can be calculated as follows:

$$l_b = [96F_bS_e/w_b]^{1/2} \text{ for one or two spans} \tag{5-40}$$

$$l_b = [120F_bS_e/w_b]^{1/2} \text{ for three or more spans} \tag{5-41}$$

Using the equations in this manner allows the designer to determine the permissible pressure that can be applied on plywood panels. The following examples illustrate the bending stress equations for plywood.

Example 5-12

Class I Plyform with a thickness of 1 in. is proposed as sheathing for a concrete wall form. The lateral pressure of the freshly placed concrete against the Plyform is 900 lb per sq ft. The Plyform will be placed across grain, so the bending stress is applied parallel to face grain. Determine the maximum spacing of the studs based on bending.

From Table 4-11, the value of the effective section modulus for stress applied parallel to face grain is $S_e = 0.737$ in.^{3}/ft, and from Table 4-12 the indicated allowable bending stress for Class I Plyform is 1,930 lb per sq in. Substituting these values into Eq. (5-41), the permissible spacing based on bending can be calculated as follows:

$$l_b = [120F_b S_e / w_b]^{1/2}$$
$$= [120(1,930 \text{ lb per sq in.})(0.737 \text{ in.}^3/\text{ft})/(900 \text{ lb per sq ft})]^{1/2}$$
$$= 13.8 \text{ in.}$$

For constructability, a 12-in. stud spacing may be selected that provides easier fabrication of the forms to match the 8 ft lengths of the Plyform. The shear stress and the deflection due to bending and shear should also be checked for the 1-in.-thick Plyform to ensure adequate shear strength and rigidity for the deflection criteria. Shear and deflection of plywood are presented in the following sections.

Example 5-13

Standard ¾-in.-thick plywood panels of the type shown in Table 4-9 are to be used as sheathing for a concrete form. The plywood will be installed across 2-in. nominal-thickness joists, spaced at 16 in. on centers. The plywood will be installed across grain, so the bending stress will be applied parallel to face grain. Assume a Group II species with an S-1 stress level rating in dry conditions and determine the permissible pressure, based on bending, that can be applied to the plywood panels.

From Table 4-9, the effective section modulus $S_e = 0.412$ in.^{3}/ft and from Table 4-10 the indicated allowable bending stress $F_b = 1,400$ lb per sq in. The allowable pressure can be calculated from Eq. (5-39) as follows:

$$w_b = 120F_b S_e / (l_b)^2$$
$$= 120(1,400 \text{ lb per sq in.})(0.412 \text{ in.}^3/\text{ft})/(16.0 \text{ in.})^2$$
$$= 270 \text{ lb per sq ft}$$

The allowable uniformly distributed load of 270 lb per sq ft is based on bending stress. The plywood should also be checked for shear and deflection to ensure that it has adequate strength and rigidity to sustain this pressure.

Example 5-14

Class I Plyform sheathing, ¾ in. thick, is to be used for forming a concrete floor. The combined total dead and live load pressure on the form is 175 lb per sq ft. The Plyform will be placed with the face grain across multiple supporting joists, three or more spans. The spacing of joists will be 24 in. center to center. Is this ¾-in.-thick Plyform adequate to sustain the 175 lb per sq ft uniformly distributed pressure?

From Table 4-11, the value of the effective section modulus for stress applied parallel to face grain is $S_e = 0.455$ in.^{3}/ft, and from Table 4-12 the allowable bending stress for Class I Plyform is 1,930 lb per sq in. The allowable uniformly distributed pressure can be calculated from Eq. (5-39) as follows:

$$w_b = 120 F_b S_e / (l_b)^2$$
$$= 120(1,930 \text{ lb per sq in.})(0.455 \text{ in.}^3/\text{ft})/(24 \text{ in.})^2$$
$$= 183 \text{ lb per sq ft}$$

Because the allowable pressure of 183 lb per sq ft is greater than the applied 175 lb per sq ft pressure, the ¾-in.-thick Class I Plyform is satisfactory for bending stress. The margin of safety for bending can be calculated as follows:

$$\text{Margin of safety} = 183/175$$
$$= 1.05, \text{ or } 5\%$$

Although the ¾-in. Class I Plyform is adequate for bending, it must also be checked for shear and deflection.

Allowable Pressure on Plywood Based on Rolling Shear Stress

The general engineering equation for calculating the shear stress in a member due to flexure is

$$f_v = VQ/Ib$$

For plywood the equation for rolling shear resistance is written in the form:

$$F_s = V/[Ib/Q] \qquad (5\text{-}42)$$

When plywood is subjected to flexure, shear stresses are induced in the plies and layers of glue of the plywood. Because some of the plies in plywood are at right angles to others, certain types of load subject them to stresses that tend to make them roll, which is defined as rolling shear. The term "Ib/Q (in.^{2}/ft)" is called the rolling shear constant. It can be obtained from tables of the physical properties (see Table 4-9 for plywood and Table 4-11 for Plyform). The value of F_s is the allowable rolling shear stress that can be obtained from Table 4-10 for plywood and from Table 4-12 for Plyform.

The value of shear V depends on the loads that are applied to the member and the number of supports for the member. The load on plywood is generally a uniformly distributed load. The spans may be single or multiple. As presented earlier, the traditional engineering equation for calculating the shear in a single-span beam with a uniformly distributed load is as follows:

$$\text{For a single span, } V = W/2$$
$$= wl/24 \text{ lb}$$

When plywood is used as sheathing, the Plywood Design Specification recommends the following equations for calculating the allowable uniformly distributed pressure:

$$\text{For a single span, } w_s = 24F_s(Ib/Q)/l_s \tag{5-43}$$
$$\text{For two spans, } w_s = 19.2F_s(Ib/Q)/l_s \tag{5-44}$$
$$\text{For three or more spans, } w_s = 20F_s(Ib/Q)/l_s \tag{5-45}$$

where w_s = allowable pressure of concrete on plywood, lb per sq ft
F_s = allowable stress in rolling shear of plywood, lb per sq in.
Ib/Q = rolling shear constant of plywood, in.^{2} per ft of width
l_s = clear distance between supports, in.

Equations (5-43) through (5-45) can be rewritten to calculate the allowable span length in terms of the uniformly distributed load, the allowable shear stress, and the rolling shear constant of plywood. Rearranging terms in the equations, the allowable span length due to rolling shear stress for plywood can be calculated as follows:

$$\text{For a single span, } l_s = 24F_s(Ib/Q)/w_s \tag{5-46}$$
$$\text{For two spans, } l_s = 19.2F_s(Ib/Q)/w_s \tag{5-47}$$
$$\text{For three or more spans, } l_s = 20F_s(Ib/Q)/w_s \tag{5-48}$$

Example 5-15

Consider the ¾-in.-thick plywood panel in Example 5-13 that is to be supported by 2-in. joists, spaced at 16 in. on centers. Calculate the allowable pressure based on rolling shear stress.

For rolling resistance shear, the clear span is calculated as the net distance between supports, reference Figure 5-11. Thus, for 2-in. nominal joists spaced at 16 in. on centers, the clear span will be:

$$l_s = 16.0 \text{ in.} - 1.5 \text{ in.}$$
$$= 14.5 \text{ in.}$$

From Table 4-9, for a 3/4-in.-thick plywood panel installed across supports, the rolling shear constant $Ib/Q = 6.762$ in.^{2}/ft. From Table 4-10, the allowable rolling shear stress $F_s = 53.0$ lb per sq in. The allowable pressure can be calculated from Eq. (5-45).

$$w_s = 20F_s(Ib/Q)/l_s$$
$$= 20(53.0)(6.762)/(14.5)$$
$$= 494 \text{ lb per sq ft}$$

For the 3/4-in. plywood, the allowable pressure of 494 lb per sq ft due to shear is greater than the allowable pressure of 270 lb per sq ft due to bending, therefore bending governs the panel for this particular condition.

Allowable Pressure on Plywood Based on Deflection Requirements

Plywood used for formwork must have adequate rigidity to resist deflection of the plywood between supporting joists or studs. As presented in previous sections, the permissible deflection is usually specified as $l/270$ for structural concrete formwork and $l/360$ for architectural concrete. To prevent waves in the surface of concrete that is formed with plywood, it is common to limit the deflection to $l/360$, ⅞ but not greater than ⅟₁₆ in.

There are two types of deflections that apply to plywood, bending deflection and shear deflection. The APA has conducted numerous tests to evaluate the deflection of plywood. Plywood sheathing is generally used where the loads are uniformly distributed and the span lengths are normally less than 30 to 50 times the thickness of the plywood. For small l/d ratios, tests have shown that shear deformation accounts for only a small percentage of the total deflection.

Allowable Pressure on Plywood due to Bending Deflection

The following equations are recommended by the Plywood Design Specification for calculating deflection due to bending:

For a single span, $\Delta_d = w_d l_d^4/921.6EI$ (5-49)

For two spans, $\Delta_d = w_d l_d^4/2220EI$ (5-50)

For three or more spans, $\Delta_d = w_d l_d^4/1743EI$ (5-51)

where Δ_d = deflection of plywood due to bending, in.
w_d = uniform pressure of concrete on plywood, lb per sq ft

l_d = clear span of plywood plus ¼ in. for 2-in. supports, and clear span plus ⅝ in. for 4-in. supports, in.

E = modulus of elasticity of plywood, lb per sq in.

I = effective moment of inertia of a 1-ft-wide strip of plywood, in.^{4}

Rearranging the terms in Eqs. (5-49) through (5-51), the allowable pressure of concrete on plywood can be calculated in terms of the span length, physical properties of plywood, and permissible deflection.

$$\text{For a single span, } w_d = 921.6EI\Delta_d/l_d^4 \tag{5-52}$$
$$\text{For two spans, } w_d = 2{,}220EI\Delta_d/l_d^4 \tag{5-53}$$
$$\text{For three or more spans, } w_d = 1{,}743EI\Delta/l_d^4 \tag{5-54}$$

Equations (5-52) through (5-54) can be used to calculate the allowable concrete pressure on plywood with different span conditions for given deflection criteria. For example, if the permissible deflection is $l/360$, then by substituting the value of $l/360$ for Δ_d in the equation, the designer of formwork can calculate the allowable pressure on the plywood for the stipulated deflection criterion.

Equations (5-52) through (5-54) can be rewritten to calculate the allowable span length for a given concrete pressure and deflection. The equations for calculating the allowable the span length for plywood based on bending deflection are

$$\text{For a single span, } l_d = [921.6EI\Delta_d/w_d]^{1/4} \tag{5-55}$$
$$\text{For two spans, } l_d = [2{,}220EI\Delta_d/w_d]^{1/4} \tag{5-56}$$
$$\text{For three or more spans, } l_d = [1{,}743EI\Delta_d/w_d]^{1/4} \tag{5-57}$$

As discussed previously, the deflection criterion is usually limited to $l/270$ or $l/360$, or sometimes not to exceed a certain amount, such as ⅛ in. or ¹⁄₁₆ in. The value of permissible deflection can be substituted into the preceding equations to determine the allowable span length based on the permissible bending deflection criterion.

The following equations can be used to calculate the allowable span length for plywood sheathing based on bending deflection for the stated deflection criterion:

Single spans,

$$\text{For } \Delta_d = l/360, l_d = [921.6EI/360w_d]^{1/3} \tag{5-55a}$$
$$\text{For } \Delta_d = \text{¹⁄₁₆ in., } l_d = [921.6EI/16w_d]^{1/4} \tag{5-55b}$$

Two spans,

$$\text{For } \Delta_d = l/360, \; l_d = [2{,}220EI/360w_d]^{1/3} \qquad (5\text{-}56a)$$
$$\text{For } \Delta_d = \tfrac{1}{16} \text{ in., } l_d = [2{,}220EI/16w_d]^{1/4} \qquad (5\text{-}56b)$$

Three or more spans,

$$\text{For } \Delta_d = l/360, \; l_d = [1{,}743EI/360w_d]^{1/3} \qquad (5\text{-}57a)$$
$$\text{For } \Delta_d = \tfrac{1}{16} \text{ in., } l_d = [1{,}743EI/16w_d]^{1/4} \qquad (5\text{-}57b)$$

Example 5-16

Compute the bending deflection of a ¾-in.-thick Plyform Class I panel that must resist a concrete pressure of 250 lb per sq ft. The plywood will be installed across multiple 2-in.-thick joists, spaced at 18 in. on centers. The physical properties are shown in Table 4-11, and the allowable stresses are shown in Table 4-12. From Eq. (5-51), calculate the bending deflection.

Reference Figure 5-10, effective span length is clear span + ¼ in.:

$$l_s = 18.0 \text{ in.} - 1.5 \text{ in.} + \tfrac{1}{4} \text{ in.} = 16.75 \text{ in.}$$
$$\Delta_d = w_d l_d^{\,4}/1{,}743EI$$
$$= (250 \text{ lb per sq ft})(16.75 \text{ in.})^4/(1{,}743)(1{,}650{,}000 \text{ lb per sq in.})$$
$$(0.199 \text{ in.}^4)$$
$$= 0.035 \text{ in.}$$

Allowable Pressure on Plywood Based on Shear Deflection

The shear deflection may be closely approximated for all span conditions by the following equation. This equation applies to all spans—single, double, and multiple spans:

$$\Delta_s = w_s Ct^2 l_s^{\,2}/1{,}270E_e I \qquad (5\text{-}58)$$

where Δ_s = shear deflection of plywood due to shear, in.

w_s = uniform pressure on plywood, lb per sq ft

C = constant = 120 for face grain of plywood perpendicular to supports;

= 60 for face grain of the plywood parallel to the supports

t = effective thickness of plywood for shear, in.

l_s = clear distance between supports, in.

E_e = modulus of elasticity of plywood, unadjusted for shear deflection, lb per sq in.

I = effective moment of inertia of a 1-ft-wide strip of plywood, in.[4]

Rearranging the terms in Eq. (5-58), the allowable pressure of concrete on plywood can be calculated in terms of the permissible deflection and the physical properties of concrete.

$$w_s = 1{,}270 E_e I \Delta_s / C t^2 l_s^2 \qquad (5\text{-}59)$$

Equation (5-58) can also be rewritten to calculate the allowable span length in terms of the permissible deflection and the physical properties of plywood.

$$l_s = [1{,}270 E_e I \Delta_s / w_s C t^2]^{1/2} \qquad (5\text{-}60)$$

Substituting the deflection criteria of $l/360$ and $\frac{1}{16}$ in. for Δ_s in Eq. (5-58), the allowable span length for shear deflection of plywood can be calculated as follows:

For $\Delta_s = l/360$, $l_s = 1{,}270 E_e I / 360 w_s C t^2$ (5-60a)

For $\Delta_s = \frac{1}{16}$ in., $l_s = [1{,}270 E_e I / 16 w_s C t^2]^{1/2}$ (5-60b)

Plywood is generally used in applications where the loads are considered uniformly distributed over the plywood and the spans are normally 30 to 50 times the thickness of the plywood. Tests have shown that shear deformation accounts for only a small percentage of the total deflection when the span to thickness is in the range of $30 \le l/t \le 50$ (see ref. [4]). However, for shorter l/t ratios (15 to 20 or lower), the shear deflection should be calculated separately and added to the bending deflection.

Example 5-17

Consider the ¾-in.-thick Plyform Class I panel in Example 5-16 that is to be supported by 2-in. joists, spaced at 18 in. on centers. Calculate the shear deflection for the uniformly distributed load of 250 lb per sq ft.

From Figure 5-10, effective span length for shear deflection is the clear span:

$$l_s = 18.0 \text{ in.} - 1.5 \text{ in.}$$
$$= 16.5 \text{ in.}$$

From Table 4-11 for physical properties,

Moment of inertia, $I = 0.199$ in.[4]

From Table 4-12 for allowable stresses,

Modulus of elasticity for shear deflection, $E = 1{,}500{,}000$ lb per sq in.

From Eq. (5-58) the shear deflection can be calculated as:

$$\Delta_s = w_s C t^2 l_s^2 / 1{,}270 E_e I$$
$$= 250(120)(3/4)^2(16.5)^2 / 1{,}270(1{,}500{,}000)(0.199)$$
$$= 0.012 \text{ in.}$$

The calculated shear deflection of 0.012 in. is small compared to the bending deflection of 0.035 in. that was calculated in Example 5-14. As previously discussed, shear deflection often accounts to a small percentage of the total deflection. For this Plyform panel, the total deflection is:

Total deflection = bending deflection + shear deflection

$$= 0.035 \text{ in.} + 0.012 \text{ in.}$$
$$= 0.047 \text{ in.}$$

Assume the maximum permissible deflection is $l/360 = 0.05$. The calculated deflection is 0.047, which less than 0.05; therefore, the ¾ in. Plyform is satisfactory for bending and shear deflection.

Tables of Equations for Calculating Allowable Span Lengths for Wood Beams and Plywood Sheathing

The previous sections have provided numerous equations and examples of calculating bending stress, shear stress, and deflection of wood beams and plywood sheathing. As noted previously, for form design it is often desirable to calculate the allowable span length of a member based on bending, shear, and deflection.

Table 5-3 summarizes the equations that are most often used for calculating the allowable span lengths for wood beams with uniformly distributed loads. Table 5-4 summarizes the equations for calculating the allowable span lengths of plywood sheathing for uniformly distributed concrete pressure. The symbols and units in the equations have been presented in preceding sections of this chapter.

Compression Stresses and Loads on Vertical Shores

Shores must safely support all dead and live loads from the formwork system above the shores. Special care must be taken in calculating stresses and allowable loads on columns, because many failures in formwork have been due to inadequate shoring and bracing of forms.

The maximum load that a vertical shore can safely support varies with the following factors:

1. Allowable unit stress in compression parallel to grain

2. Net area of shore cross section

3. Slenderness ratio of shore

Span Condition	Equations for Calculating Allowable Span Lengths, in Inches, of Wood Beams			
			Bending Deflection	
	Bending l_b, in.	Shear l_v, in.	l_Δ, in. for $\Delta = l/360$	l_Δ, in. for $\Delta = \frac{1}{16}$ in.
Single Spans	$[96F_bS/w]^{1/2}$ Eq. (5-30)	$16F_vbd/w + 2d$ Eq. (5-32)	$[4,608EI/1,800/w]^{1/3}$ Eq. (5-33a)	$[4,608EI/80w]^{1/4}$ Eq. (5-33b)
Multiple Spans	$[120F_bS/w]^{1/2}$ Eq. (5-34)	$192F_vbd/15w + 2d$ Eq. (5-36)	$[1,743EI/360w]^{1/3}$ Eq. (5-37a)	$[1,743EI/16w]^{1/4}$ Eq. (5-37b)

Notes:
1. Physical properties of wood beams can be obtained from Table 4-1.
2. See Tables 4-2 and 4-3 for allowable stresses with adjustments from Tables 4-3a through 4-8.
3. Units for stresses and physical properties are in pounds and inches.
4. Units for w are lb per lin ft along the entire length of beam.

TABLE 5-3 Equations for Calculating Allowable Span Lengths, in Inches, for Wood Beams based on Bending, Shear, and Deflection for Beams with Uniformly Distributed Loads

Span Condition	Equations for Calculating Allowable Span Lengths, in Inches, for Plywood and Plyform				Shear Deflection
			Bending Deflection		
	Bending l_b, in.	Shear l_s, in.	l_d, in. for $\Delta = l/360$	l_d, in. for $\Delta = 1/16$ in.	l_s, in. for $\Delta = l/360$
Single Spans	$[96F_bS_e/w_b]^{1/2}$ Eq. (5-40)	$24F_s(Ib/Q)/w_s$ Eq. (5-46)	$\left[\dfrac{921.6EI}{360w_d}\right]^{1/3}$ Eq. (5-55a)	$\left[\dfrac{921.6EI}{16w_d}\right]^{1/4}$ Eq. (5-55b)	$\dfrac{1,270E_eI}{360w_sCt^2}$ Eq. (5-60a)
Two Spans	$[96F_bS_e/w_b]^{1/2}$ Eq. (5-40)	$19.2F_s(Ib/Q)/w_s$ Eq. (5-47)	$\left[\dfrac{2,220EI}{360w_d}\right]^{1/3}$ Eq. (5-56a)	$\left[\dfrac{2,220EI}{16w_d}\right]^{1/4}$ Eq. (5-56b)	$\dfrac{1,270E_eI}{360w_sCt^2}$ Eq. (5-60a)
Three or more Spans	$[120F_bS_e/w_b]^{1/2}$ Eq. (5-41)	$20F_s(Ib/Q)/w_s$ Eq. (5-48)	$\left[\dfrac{1,743EI}{360w_d}\right]^{1/3}$ Eq. (5-57a)	$\left[\dfrac{1,743EI}{16w_d}\right]^{1/4}$ Eq. (5-57b)	$\dfrac{1,270E_eI}{360w_sCt^2}$ Eq. (5-60a)

Notes:
1. See Tables 4-9 and 4-11 for physical properties of plywood and Plyform, respectively.
2. Reference Tables 4-10 and 4-12 for allowable stresses of plywood and Plyform, respectively.
3. Units for stresses and physical properties are in pounds and inches.
4. Units for w are lb per lin ft for a 12-in.-wide strip of plywood.

TABLE 5-4 Equations for Calculating Allowable Span Lengths, in Inches, for Plywood and Plyform based on Bending, Shear, and Deflection due to Uniformly Distributed Pressure

Shores are column members with axial loads that induce compression stresses that act parallel to the grain. Values of compression stresses are presented in Tables 4-2 and 4-3 for stresses parallel to grain. However, these compression stresses do not consider the length of a member, which may affect its stability and strength. A shore must be properly braced to prevent lateral buckling because its strength is highly dependent on its unbraced length.

The slenderness ratio for a shore is the ratio of its unbraced length divided by the least cross-sectional dimension of the shore, l_u/d. It is used to determine the allowable load that can be placed on a shore. The allowable load on a shore decreases rapidly as the slenderness ratio increases. For this reason, long shores should be cross-braced in two directions with one or more rows of braces.

Figure 5-12 illustrates the calculations of l_u/d for a 4×6 rectangular column member. For this 4×6-in. column, it is necessary to calculate two values of l_u/d to determine the value that governs for buckling.

For strong-axis buckling, the unbraced length is 9 ft, or 108 in., and the dimension $d = 5\frac{1}{2}$ in. Therefore, $l_u/d = 108/5.5 = 19.6$.

For weak-axis buckling, the unbraced length is 6 ft, or 72 in., and the dimension $d = 3\frac{1}{2}$ in. Therefore, $l_u/d = 72/3.5 = 20.6$.

Because 20.6 is greater than 19.6, the value of 20.6 must be used in calculating the allowable compression stress in the member. Thus, there can be different bracing patterns for a given size compression member. For each particular bracing pattern, the governing l_u/d ratio must be determined.

In Figure 5-12 the bracing is shown as a small member to illustrate the concept of unbraced length. Bracing must be of a sufficient size and must be securely attached to a rigid support in order to prevent lateral movement of the compression member.

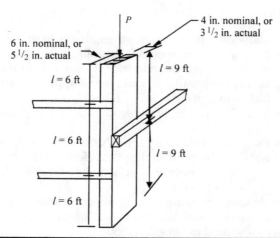

FIGURE 5-12   Illustration of unbraced lengths for compression members.

Square shores are often preferred to rectangular ones because the minimum dimension of the cross section is the same about both the x-x and y-y axes. This eliminates the requirement of calculating slenderness ratios about two axes to determine the governing value. Using square shores in lieu of rectangular ones also reduces the possibility of orienting the shores improperly during erection of the forms. Installing incorrect bracing patterns of shores can adversely affect the safety of shoring.

Solid wood columns may be classified into three categories according to their length: short columns, intermediate columns, and long columns. Short columns are defined by small l_u/d ratios, the ratio of unbraced length to the least cross-sectional dimension. Columns are generally considered short when the l_u/d ratio is less than 10. The failure of short columns is primarily by crushing of the wood. Therefore, the limiting stress in a short column is the allowable compression stress parallel to grain.

For intermediate l_u/d ratios, the failure is generally a combination of crushing and buckling. For long columns, large l_u/d ratios, the failure is by lateral buckling of the member. The instability of long wood columns is often referred to as long-column buckling. The equation for determining the critical buckling load of long columns was derived by Francis Euler and is discussed in many publications related to column and compression members. For rectangular wood members, the limiting value is $l_u/d = 50$. Columns with ratios greater than this value are not permitted in formwork.

Equation (5-61) is the recommended equation to determine the allowable compression stress in a wood column with loads applied parallel to grain. This single equation takes into consideration modes of failure, combinations of crushing, and buckling. Because a single equation is used, many terms are included:

$$F_C = F_c^* \left[[1 + (F_{ce}/F_c^*)]/2c \sqrt{\left[[1 + (F_{ce}/F_c^*)]/2c \right]^2 - (F_{ce}/F_c^*)/c} \right] \qquad (5\text{-}61)$$

where F_c = allowable compression stress, lb per sq in.

F_c^* = compression stress parallel to grain, lb per sq in., obtained by multiplying the reference design value for compression stress parallel to grain by all applicable adjustment factors except C_P; $F_c^* = F_{C//} (C_D \times C_M \times C_F \times C_i \times C_t)$

$F_{ce} = 0.822E'_{min}/(l_e/d)^2$, lb per sq in., representing the impact of Euler buckling

E'_{min} = modulus of elasticity for column stability, lb per sq in., obtained by multiplying the reference design value of E_{min} by all applicable adjustment factors; $E'_{min} = E_{min} (C_M \cdot C_T \cdot C_i \cdot C_t)$

l_e/d = slenderness ratio, ratio of effective length, in., to least cross-sectional dimension, in. *Note:* For wood columns l_e/d should never exceed 50

c = 0.8 for sawn lumber

The allowable load on a column is equal to the allowable compression stress multiplied by the cross-sectional area of the member. When determining the maximum allowable load on a shore, the net area of the cross section of the shore should be used:

$$P_c = F_c A \qquad (5\text{-}62)$$

where P_c = allowable compression load, lb
F_c = allowable compression stress, lb per sq in.
A = net cross-sectional area, bd, in.^{2}

Example 5-18

A 4×6 S4S rectangular shore of No. 2 Southern Pine has a reference compression stress parallel to grain of 1,600 lb per sq in. and a reference modulus of elasticity for column stability of 580,000 lb per sq in. It has an effective length of 6 ft about both the strong x-x axis and the weak y-y axis. Calculate the allowable compression load.

The shore will be used for a condition that requires no adjustments in modules of elasticity or the reference value of compression stress parallel to grain, that is, dry condition, normal temperature and load-duration, no incising, and so on. Therefore, the adjustment factors for reference values of compression stress and modulus of elasticity will be equal to 1.0: ($F^*_c = F_{c//}$ and $E'_{min} = E_{min}$).

Because the shore is braced in both axes, the slenderness ratio will be governed by the least cross-sectional dimension, $3\frac{1}{2}$ in.

The slenderness ratio, $l_e/d = 72/3.5$
$$= 20.57$$

$F_{ce} = 0.822\, E'_{min}/(l_e/d)^2$
$$= 0.822(580{,}000)/(20.57)^2$$
$$= 1{,}126.6 \text{ lb per sq in.}$$

The allowable compression stress can be calculated:

$$F_C = F^*_c \left[[1+(F_{ce}/F^*_c)]/2 - \sqrt{\left[[1+(F_{ce}/F^*_c)]/2c \right]^2 - (F_{ce}/F^*_c)/c} \right]$$

$$F_C = 1{,}600 \left[[1+(1{,}126.6/1{,}600)]/(0.8) \right.$$

$$\left. - \sqrt{\left[[1+1{,}126.6/1{,}600]/2(0.8) \right]^2 - (1{,}126.6)/0.8} \right]$$

$$= 1{,}600 \left[1.065 - \sqrt{(1.065)^2 - 0.704/0.8} \right]$$

$$= 1{,}600 \left[1.065 - \sqrt{1.134 - 00.880} \right]$$

$$= 1,600 \, (1.065 - 0.504)$$
$$= 1,600 \, (0.561)$$
$$= 897.6 \text{ lb per sq in.}$$

Using Eq. (5-62), the allowable compression load can be calculated by multiplying the allowable compression stress times the actual cross-sectional area of the member:

$$P_c = F_c A$$
$$= (897.6 \text{ lb per sq in.}) \times [(3\frac{1}{2} \text{ in.}) \times (5\frac{1}{2} \text{ in.})]$$
$$= (897.6 \text{ lb per sq in.}) \times (19.25 \text{ sq in.})$$
$$= 17,279 \text{ lb}$$

Table for Allowable Loads on Wood Shores

Table 5-5 shows allowable compression loads in rectangular solid wood shores for the specified condition. The loads in this table are calculated using Eqs. (5-61) and (5-62) for dry condition wood having a moisture content of less than 19%. The calculated values are based on sawn wood; therefore, the value of $c = 0.8$ is used. If shores are used that have design values different than those shown in Table 5-5, Eqs. (5-61) and (5-62) must be used to determine the allowable load on the shore. For a particular job condition, it may be necessary to adjust the values used in calculating the allowable loads shown in Table 5-5.

Shores of 4×4 S4S lumber are commonly used because screw jacks are readily available for this size lumber. Therefore, Table 5-5 shows values for 4×4 lumber. Screw jacks are not commonly available for 4×6 or 6×6 lumber. However, the 4×6 and 4×4 sizes of lumber are sometimes used for reshoring. The most common method of shoring is patented shores or scaffolding. Manufacturers of shoring and scaffolding have devoted considerable effort to developing shoring systems that are reliable and easily erected and removed. Towers can be erected of scaffolding to the desired height and the required strength, provided the manufacturer's recommendations are followed. Chapter 6 provides additional information on shoring and scaffolding.

Bearing Stresses Perpendicular to Grain

When a wood member is placed on another wood member, bearing stresses are created that tend to crush the fibers of the wood. For example, joists rest, or bear, on stringers. Similarly, stringers rest on posts or shores, and studs bear against wales. When a joist is placed

$P_c = F_c A$ where F_c is given by the following equation				
$F_c = F^*{}_c \left[[1 + (F_{ce}/F^*{}_c)]/2c - \sqrt{\left[[1 + (F_{ce}/F^*{}_c)]/2c \right]^2 - (F_{ce}/F^*{}_c)/c} \right]$				
Effective length, ft	$F^*{}_c = 1{,}650$ $E'{}_{min} = 580{,}000$ 4 × 4 S4S	$F^*{}_c = 1{,}552$ $E'{}_{min} = 580{,}000$ 4 × 4 S4S	$F^*{}_c = 1{,}495$ $E'{}_{min} = 470{,}000$ 4 × 4 S4S	$F^*{}_c = 1{,}322$ $E'{}_{min} = 510{,}000$ 4 × 4 S4S
4	16,484	15,755	14,433	13,537
5	13,846	13,425	11,811	11,608
6	11,076	10,883	9,333	9,462
7	8,792	8,681	7,252	7,581
8	7,020	6,948	5,761	6,090
9	5,519	5,656	4,655	4,952
10	4,688	4,672	3,819	4,098
11	3,919	3,897	3,206	3,428
12	3,318	3,338	2,697	2,904
13	2,851	2,841	2,315	2,487
14	2,462	2,461	1,999	2,162
15	l_e/d>50	l_e/d>50	l_e/d>50	l_e/d>50
Effective Length, ft	$F^*{}_c = 1{,}600$ $E'{}_{min} = 580{,}000$ 4 × 6 S4S	$F^*{}_c = 1{,}485$ $E'{}_{min} = 580{,}000$ 4 × 6 S4S	$F^*{}_c = 1{,}430$ $E'{}_{min} = 470{,}000$ 4 × 6 S4S	$F^*{}_c = 1{,}265$ $E'{}_{min} = 510{,}000$ 4 × 6 S4S
4	25,334	17,646	22,831	20,551
5	21,442	14,922	18,202	17,796
6	17,267	12,258	14,396	14,639
7	13,727	9,952	11,315	11,810
8	10,993	8,154	9,014	9,499
9	8,891	6,711	7,278	7,758
10	7,337	5,599	5,991	6,391
11	6,163	4,747	5,006	5,370
12	5,388	4,047	4,240	4,523
13	4,476	3,481	3,635	3,891
14	3,891	3,010	3,139	3,401
15	l_e/d>50	l_e/d>50	l_e/d>50	l_e/d>50

Notes:
1. Calculated values are based on Eqs. (5-61) and (5-62).
2. Values are calculated using $c = 0.8$ for sawn lumber, and minimum cross-sectional dimension $d = 3\frac{1}{2}$ in.
3. No adjustments have been made for moisture, temperature, load-duration, or other conditions; additional adjustments may be necessary for a particular job.
4. Values of $F^*{}_c$ and $E'{}_{min}$ are in lb per sq in.
5. Lumber is considered in dry condition, moisture content < 19%.
6. For wood members the slenderness ratio l_e/d should not exceed 50.

TABLE 5-5 Allowable Load of Rectangular Solid Wood 4 × 4 and 4 × 6 S4S Columns Members, Based on Stipulated Values of $F^*{}_c$ and $E'{}_{min}$

Size of shore and horizontal member, S4S	Net contact bearing area between members, sq in.	Allowable Compression Stress Perpendicular to Grain, lb per sq in.			
		$F_{c\perp} = 405$	$F_{c\perp} = 425$	$F_{c\perp} = 565$	$F_{c\perp} = 625$
2 × 4	5.25	2,126	2,231	2,966	3,281
3 × 4	8.75	3,543	3,718	4,943	5,468
4 × 4	12.25	4,961	5,206	6,921	7,656
4 × 6	19.25	7,796	8,181	10,876	12,031

TABLE 5-6 Allowable Loads that May Be Transmitted from Horizontal Wood Members to Vertical Wood Shores, lb

on a stringer, compression stresses perpendicular to grain are created. The contact area between two members is the bearing area. As shown in Tables 4-2 and 4-3, the allowable compression stress perpendicular to grain is less than the allowable compression stress parallel to grain. Therefore, it is necessary to check the bearing stress of wood members.

The values given in Table 5-6 are the maxima that the shores will support without danger of buckling. However, these values do not consider the contact bearing stresses at the top or bottom of the shore. Because a shore cannot support a load greater than the one that can be transmitted to it at the top or bottom, it may be necessary to reduce the loads on some shores to values less than those given in the table. For example, if the allowable unit compressive stress of a wood member that rests on a shore is 565 lb per sq in., then the maximum load on a shore that supports the wood member will be limited to the product of the net area of contact between the wood member and the shore, multiplied by 565 lb per sq in.

The area of contact between a 4 × 4 S4S horizontal wood member and the top of a 4 × 4 S4S shore will be 12.25 sq in. The maximum load that can be transmitted to the shore will be 12.25 sq in. × 565 lb per sq in. = 6,921 lb. If this load is exceeded, it is probable that the underside surface of the horizontal wood member in contact with the shore will be deformed permanently. This may endanger the capacity of transferring loads between the horizontal wood member and the shore.

Table 5-6 gives the allowable loads that may be transmitted to shores from horizontal wood members, based on the allowable unit compressive stresses perpendicular to the grain of the wood member. The same loads must be transmitted from the bottoms of the shores to the bases on which the shores rest.

In Table 5-6, the area of contact between a horizontal wood member and a shore is determined with the narrower face of the wood member bearing on the major dimension of the shore if the two faces of a shore are of unequal dimensions. For example, the area of contact between a 3 × 4 S4S horizontal wood member and a 4 × 6 S4S shore is assumed to be 2½ in. by 3½ in., or 8.75 sq in. For an allowable compression stress perpendicular to grain of 565 lb per sq in., the allowable load would be (8.75 sq in.)(565 lb per sq in.) = 4,943 lb.

Design of Forms for a Concrete Wall

The use of the equations and principles previously developed will be illustrated by designing the forms for a concrete wall. The design will include the sheathing, studs, wales, and form ties (refer to Figure 5-13).

In the design of wall forms, the thickness and grade of the plywood are often selected based on the availability of materials. For this design example, a ¾-in.-thick plywood from the Group I species with S-2 stress rating has been selected. The lumber will be No. 2 grade S4S Southern Pine with no splits. The following design criteria will apply:

1. Height of wall, 12 ft 0 in.
2. Rate of filling forms, 6 ft per hr
3. Temperature, 80°F
4. Concrete unit weight, 150 lb per cu ft
5. Type I cement will be used without retarders

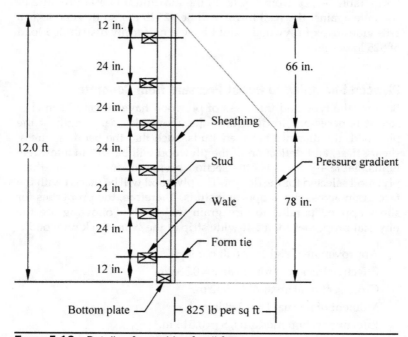

FIGURE 5-13 Details of one side of wall form.

6. Dry condition for wood, <19% moisture content

7. Deflection limited to $l/360$, but not greater than $\frac{1}{16}$ in.

8. All form lumber will be 2-in.-thick S4S lumber, No. 2 grade Southern Pine with no splits, (Table 4-2)

9. Short load-duration, less than 7 days

10. Sheathing will be ⅞-in. plywood from Group I species (Table 4-9), with a level S-2 stress rating, (Tables 4-9 and 4-10)

The design will be performed in the following sequence: determine the maximum lateral pressure of the concrete on the wall forms based on temperature and rate of filling the forms; determine the spacing of studs based on the strength and deflection of plywood sheathing; determine the spacing of wales based on the strength and deflection of studs; check the bearing strength between studs and wales; determine the span length of wales between ties based on the strength and deflection of wales; and determine the required strength of the form ties. Figure 5-14 illustrates the span lengths and spacing of the components of the wall form. Tables 5-3 and 5-4 provide the equations that are necessary to design the wall forms.

Lateral Pressure of Concrete on Forms

From Table 3-3, or from Eq. (3-2), the maximum lateral pressure of concrete against the wall forms will be 825 lb per sq ft. Therefore, a 1-ft-wide strip of plywood would have a uniformly distributed load of 825 lb per lin ft.

Plywood Sheathing to Resist Pressure from Concrete

Because the type and thickness of plywood have already been chosen, it is necessary to determine the allowable span length of the plywood, the distance between studs, such that the bending stress, shear stress, and deflection of the plywood will be within allowable limits. Table 4-9 provides the section properties for the ⅞-in.-thick plywood selected for the design. The plywood will be placed with the face grain across the supporting studs. Therefore, the properties for stress applied parallel to face grain will apply. Following are the physical properties for a 1-ft-wide strip of the ⅞-in.-thick plywood:

Approximate weight = 2.6 lb per sq ft

Effective thickness for shear, $t = 0.586$ in.

Cross-sectional area, $A = 2.942$ in.^{2}

Moment of inertia, $I = 0.278$ in.^{4}

Effective section modulus, $S_e = 0.515$ in.^{3}

Rolling shear constant, $Ib/Q = 8.050$ in.^{2}

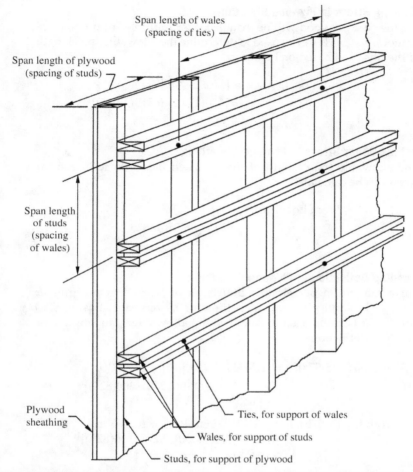

Span length of wales
(spacing of ties)

Span length of plywood
(spacing of studs)

Span length
of studs
(spacing
of wales)

Plywood
sheathing

Ties, for support of wales

Wales, for support of studs

Studs, for support of plywood

FIGURE 5-14   Span length and spacing of components of wall form.

For dry conditions, Table 4-10 indicates the allowable stresses for the plywood of Group I species and S-2 stress rating for normal duration of load. However, values for stresses are permitted to be increased by 25% for short-duration load conditions, less than 7 days, as discussed in Chapter 4 (Table 4-4). The 25% increase applies to bending and shear but does not apply to the modulus of elasticity. Therefore, the allowable stresses for the design of the plywood are obtained by multiplying the tabulated stresses by 1.25 as follows:

Allowable bending stress, $F_b = 1.25(1,650) = 2,062$ lb per sq in.

Allowable rolling shear stress, $F_s = 1.25 (53) = 66$ lb per sq in.

Modulus of elasticity, $E = 1,800,000$ lb per sq in.

Bending Stress in Plywood Sheathing

Because the plywood will be installed over multiple supports, three or more, Eq. (5-41) can be used to determine the allowable span length of the plywood based on bending.

$$\text{From Eq. (5-41), } l_b = [120F_bS_e/w_b]^{1/2}$$
$$= [120(2{,}062)(0.515)/(825)]^{1/2}$$
$$= 12.4 \text{ in.}$$

Rolling Shear Stress in Plywood Sheathing

The allowable span length of the plywood based on rolling shear stress can be calculated.

$$\text{From Eq. (5-48), } l_s = 20F_s(Ib/Q)/w_s$$
$$= 20(66)(8.050)/825$$
$$= 12.8 \text{ in.}$$

Bending Deflection in Plywood Sheathing

For a maximum permissible deflection of $l/360$, but not greater than $\frac{1}{16}$ in., and plywood with three or more spans, Eqs. (5-57a) and (5-57b) can be used to calculate the allowable span lengths for bending deflection.

$$\text{From Eq. (5-57a), } l_d = [1{,}743EI/360w_d]^{1/3} \text{ for } \Delta = l/360$$
$$= [1{,}743(1{,}800{,}000)(0.278)/360(825)]^{1/3}$$
$$= 14.3 \text{ in.}$$
$$\text{From Eq. (5-57b), } l_d = [1{,}743EI/16w_d]^{1/4} \text{ for } \Delta = \frac{1}{16} \text{ in.}$$
$$= [1{,}743(1{,}800{,}000)(0.278)/16(825)]^{1/4}$$
$$= 16.0 \text{ in.}$$

Summary for the ¾-in.-Thick Plywood Sheathing

For bending, the maximum span length of the plywood = 12.4 in.

For shear, the maximum span length of the plywood = 12.8 in.

For deflection, the maximum span length of the plywood = 16.0 in.

For this design, bending controls the maximum span length of the plywood sheathing. The studs that support the plywood must be placed at a spacing that is no greater than 12.4 in. For uniformity and constructability, choose a maximum stud spacing of 12.0 in., center to center.

Studs for Support of Plywood

The studs support the plywood and the wales must support the studs. Therefore, the maximum allowable spacing of wales will be governed by the bending, shear, and deflection in the studs.

Consider using 2 × 4 S4S studs whose actual size is 1½ by 3½ in. From Table 4-1, the physical properties can be obtained for 2 × 4 S4S lumber:

Area, $A = 5.25$ in.^{2}
Section modulus, $S = 3.06$ in.^{3}
Moment of inertia, $I = 5.36$ in.^{4}

The lumber selected for the forms is No. 2 grade Southern Pine with no splits. From Table 4-2, using the values that are increased by 25% for short-duration load conditions, the allowable stresses are as follows:

Allowable bending stress, $F_b = 1.25(1,500) = 1,875$ lb per sq in.
Allowable shear stress, $F_v = 1.25(175) = 218$ lb per sq in.
Modulus of elasticity, $E = 1,600,000$ lb per sq in.

Bending in 2 × 4 S4S Studs

Each stud will support a vertical strip of sheathing 12-in. wide, which will produce a uniform load of 825 lb per sq ft × 1.0 ft = 825 lb per lin ft on the lower portion of the stud. Because the studs will extend to the full height of the wall, the design should be based on continuous beam action for the studs. The allowable span length of studs, the distance between wales, based on bending can be determined as follows:

From Eq. (5-34), $l_b = [120F_bS/w]^{1/2}$
$= [120(1,875)(3.06)/825]^{1/2}$
$= 28.9$ in.

Shear in 2 × 4 S4S Studs

The allowable span length based on shear in the studs can be calculated as follows.

From Eq. (5-36), $l_v = 192F_vbd/15w + 2d$
$= 192(218)(1.5)(3.5)/15(825) + 2(3.5)$
$= 24.8$ in.

Deflection in 2 × 4 S4S Studs

The permissible deflection is $l/360$, but not greater than $\frac{1}{16}$ in. Using Eqs. (5-37a) and (5-37b), the allowable span lengths due to deflection can be calculated as follows:

From Eq. (5-37a), $l_\Delta = [1,743EI/360w]^{1/3}$ for $\Delta = l/360$
$= 1,743(1,600,000)(5.36)/360(825)]^{1/3}$
$= 36.9$ in.

From Eq. (5-37b), $l_\Delta = [1{,}743EI/16w]^{1/4}$ for $\Delta = \frac{1}{16}$ in.
$$= [1{,}743(1{,}600{,}000)(5.36)/16(825)]^{1/4}$$
$$= 32.6 \text{ in.}$$

Summary for the 2 × 4 S4S Studs

For bending, the maximum span length of studs = 28.9 in.

For shear, the maximum span length of studs = 24.8 in.

For deflection, the maximum span length of studs = 32.6 in.

For this design, shear governs the maximum span length of the studs. Because the wales must support the studs, the maximum spacing of the wales must be no greater than 24.8 in. Therefore, the wales will be spaced at 24 in. on center for this design.

Wales for Support of

For this design, double wales will be used. Wales for wall forms of this size are frequently made of two-member lumber whose nominal thickness is 2 in., separated by short pieces of 1-in.-thick blocks.

Determine the unit stress in bearing between the studs and the wales. If a wale consists of two 2-in. nominal thickness members, the contact area between a stud and a wale will be:

$$A = 2 \times [1.5 \text{ in.} \times 1.5 \text{ in.}]$$
$$= 4.5 \text{ sq. in.}$$

The total load in bearing between a stud and a wale will be the unit pressure of the concrete acting on an area based on the stud and wale spacing. For this design, the area will be 12 in. long by 24 in. high. Therefore, the pressure acting between the stud and wale will be:

$$P = 825 \text{ lb per sq ft}[1.0 \text{ ft} \times 2.0 \text{ ft}]$$
$$= 1{,}650 \text{ lb}$$

The calculated unit stress in bearing, perpendicular to grain, between a stud and a wale will be:

$$f_{c\perp} = P/A$$
$$= 1{,}650 \text{ lb}/4.5 \text{ sq in.}$$
$$= 367 \text{ lb per sq in.}$$

From Table 4-2, the indicated allowable unit compression stress perpendicular to the grain for No. 2 grade Southern Pine is 565 lb per sq in., which is greater than the applied stress of 367 lb per sq in. Therefore, a 24-in. spacing of wales will be satisfactory.

Size of Wale Based on Selected 24-in. Spacing

Although the loads transmitted from the studs to the wales are concentrated, it is generally sufficiently accurate to treat them as uniformly distributed loads, having the same total values as the concentrated loads, when designing formwork for concrete walls. However, in some critical situations it may be desirable to design the forms using concentrated loads as they actually exist. Assuming a uniformly distributed load from a pressure of 825 lb per sq ft on the wales that are spaced at 24 in. (2.0 ft), the value of w can be calculated as follows:

$$w = 825 \text{ lb per sq ft (2.0 ft)}$$
$$= 1,650 \text{ lb per lin ft}$$

With 24-in. spacing and 825 lb per sq ft pressure on the wales, consider double 2×4 S4S wales.

From Table 4-1, for a double 2×4, $A = 2(5.25 \text{ in.}^2) = 10.5 \text{ in.}^2$
$$S = 2(3.06 \text{ in.}^3) = 6.12 \text{ in.}^3$$
$$I = 2(5.36 \text{ in.}^4) = 10.72 \text{ in.}^4$$

From Table 4-2, for a 2×4, $F_b = 1.25(1,500) = 1,875 \text{ lb per sq in.}$
$$F_v = 1.25(175) = 218 \text{ lb per sq in.}$$
$$E = 1,600,000 \text{ lb per sq in.}$$

Bending in Wales

Based on bending, the span length of the wales must not exceed the value calculated.

From Eq. (5-34), $l_b = [120F_bS/w]^{1/2}$
$$= [120(1,875)(6.12)/1,650]^{1/2}$$
$$= 28.8 \text{ in.}$$

Shear in Wales

Based on shear, the span length of the wales must not exceed the value calculated.

From Eq. (5-36), $l_v = 192F_vbd/15w + 2d$
$$= 192(218)(10.5)/15(1,650) + 2(3.5)$$
$$= 24.7 \text{ in.}$$

Deflection in Wales

Based on the deflection criterion of less than $l/360$, the span length of the wales must not exceed the following calculated value:

$$\text{From Eq. (5-37a), } l_\Delta = [1{,}743EI/360w]^{1/3}$$
$$= [1{,}743(1{,}600{,}000)(10.72)/360(1{,}650)]^{1/3}$$
$$= 36.9 \text{ in.}$$

Based on a deflection criterion of less than $\frac{1}{16}$ in., the span length of the wales must not exceed the following calculated value:

$$\text{From Eq. (5-37b), } l_\Delta = [1{,}743EI/16w]^{1/4}$$
$$= [1{,}743(1{,}600{,}000)(10.72)/16(1{,}650)]^{1/4}$$
$$= 32.9 \text{ in.}$$

Summary for the Double 2 × 4 Wales

For bending, the maximum span length of wales = 28.8 in.

For shear, the maximum span length of wales = 24.7 in.

For deflection, the maximum span length of wales = 32.6 in.

For this design, shear governs the maximum span length of the wales. Because the wales are supported by the ties, the spacing of the ties must not exceed 24.7 in. However, the ties must have adequate strength to support the load from the wales. For uniformity and constructability, choose a maximum span length of 24 in. for wales and determine the required strength of the ties to support the wales.

Strength Required of Ties

Based on the strength of the wales in bending, shear, and deflection, the wales must be supported every 24 in. by a tie. A tie must resist the concrete pressure that acts over an area consisting of the spacing of the wales and the span length of the wales, 24 in. by 24 in., or 2 ft by 2 ft. The load on a tie can be calculated as follows:

$$\text{Area} = 2.0 \text{ ft} \times 2.0 \text{ ft}$$
$$= 4.0 \text{ sq ft}$$
$$\text{Tie tension load} = 825 \text{ lb per sq ft (4.0 sq ft)}$$
$$= 3{,}300 \text{ lb}$$

The ties for the wall must have a safe tension load of 3,300 lb. A standard-strength 3,000-lb working load tie is not adequate. Therefore a heavy-duty 4,000-lb working load tie is required.

An alternative method is to select a 3,000-lb tie and calculate the allowable spacing of the ties to support the wale. For a 24 in., or 2.0 ft, spacing of wales, the equivalent uniformly distributed load along a wale will be 825 lb per sq ft multiplied by 2.0 ft, or 1,650 lb per lin ft. The maximum spacing of the ties based on the strength of a 3,000 lb safe working load tie is as follows:

$$l = (3,000 \text{ lb})/(1,650 \text{ lb per ft}) = 1.8 \text{ ft.}$$
$$l = 21.6 \text{ in.}$$

For a 3,000 lb tie, the spacing must not exceed 21.6 in. For this design, 4,000-lb ties will be used and placed at 24 in. center to center, because the 24-in. spacing is convenient for layout and ease of construction.

Design Summary of Forms for Concrete Wall

The allowable stresses and deflection criteria used in the design are based on the following lumber values.

For the 2 × 4 S4S No. 2 grade Southern Pine with no splits:
Allowable bending stress = 1,875 lb per sq in.

Allowable shear stress = 218 lb per sq in.

Modulus of elasticity = 1,600,000 lb per sq in.

Permissible deflection less than $l/360$, but not greater than 1/16 in.

For the ¾-in.-thick plywood from Group I species with S-1 stress rating:

Allowable bending stress = 2,062 lb per sq in.

Allowable shear stress = 66 lb per sq in.

Modulus of elasticity = 1,800,000 lb per sq in.

Permissible deflection less than $l/360$, but not greater than 1/16 in.

The forms for the concrete wall should be built to the following conditions:

Item	Nominal size and spacing
Sheathing	⅞-in.-thick plywood
Studs	2 × 4 S4S at 12 in. on centers
Wales	double 2 × 4 S4S at 24 in. on centers
Form ties	4,000-lb capacity at 24 in.

Figure 5-15 summarizes the design, including the size and spacing of studs, wales, and form ties.

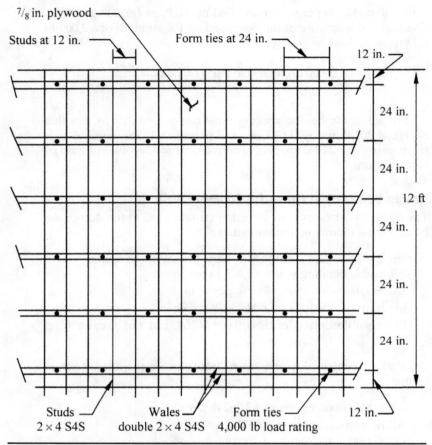

7/8 in. plywood

Studs at 12 in.

Form ties at 24 in.

12 in.

24 in.

24 in.

12 ft

24 in.

24 in.

24 in.

12 in.

Studs
2 × 4 S4S

Wales
double 2 × 4 S4S

Form ties
4,000 lb load rating

12 in.

Figure 5-15   Summary of design for concrete wall.

Minimum Lateral Force for Design of Wall Form Bracing Systems

The lateral pressure of freshly placed concrete on wall forms is carried by the ties. However, the wall forms must be braced adequately to resist any eccentric loads or lateral forces due to wind. The minimum lateral force applied at the top of wall forms is shown in Table 5-7. The force may act in either direction as illustrated in Figure 5-16.

Bracing for Wall Forms

To ensure safety and alignment, forms must be adequately braced to resist any lateral movement. Braces must be securely attached at both ends to prevent displacement and rotation. Figure 5-17 shows the

Wall Height above Grade h, ft	ACI 347 Minimum 100 lb per ft or 15 lb per sq ft Wind	Wind Force Prescribed by Local Code, lb per sq ft			
		10	20	25	30
4 or less	30	20	40	50	60
6	45	30	60	75	90
8	100	100	100	100	120
10	100	100	100	125	150
12	100	100	120	150	180
14	105	100	140	175	210
16	120	100	160	200	240
18	135	100	180	225	270
20	150	100	200	250	300
20 or more	$7.5h$	$5.0h$	$10.0h$	$12.5h$	$15.0h$
Walls below grade					
8 ft or less	Brace to maintain alignment				
More than 8 ft	100 lb per sq ft minimum, or brace for any known lateral forces that are greater				

Notes:
1. Reprinted from "Formwork for Concrete," ACI Special Publication No. 4, the American Concrete Institute.
2. Wind force prescribed by local code shall be used whenever it would require a lateral force for design greater than the minimum above.

TABLE 5-7   Minimum Lateral Force for Design of Wall Form Bracing, lb per lin ft

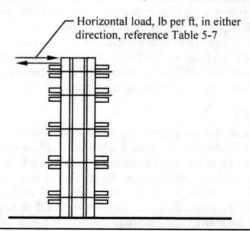

Horizontal load, lb per ft, in either direction, reference Table 5-7

FIGURE 5-16   Minimum lateral load for bracing wall form systems.

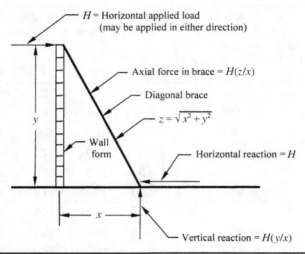

FIGURE 5-17   Bracing system for concrete formwork.

geometry of a bracing system. The brace is positioned a distance x from the base of the form and the height of the form is a distance y. The diagonal length z of the brace can be calculated from the equation, $z = \sqrt{x^2 + y^2}$.

As shown in Figure 5-17, the horizontal force H at the top of the form is applied in a direction to the right, which causes an axial compression in the brace, a horizontal reaction to the left, and an upward vertical reaction at the bottom of the brace. However, the horizontal force may be applied to the top of the form in a direction to the left, which would reverse the direction of the forces, causing axial tension in the brace, a horizontal reaction to the right, and a downward reaction at the bottom of the brace.

The horizontal load H is cause by wind pressure acting against the form. Wind pressures and resulting forces on a structure should be calculated according to local codes. The publication by the American Society of Civil Engineers, *Minimum Design Loads for Buildings and Other Structures*, SEI/ASCE-7, provides a comprehensive coverage of wind loads based upon substantial research.

Example 5-19

Diagonal braces will be placed horizontally at 9 ft on centers along the top of a 10-ft-high wall form. Each brace will be attached at the top of the wall form and the bottom of each brace will be attached at a distance of 6 ft horizontally from the bottom of the form. The local code prescribes a 25 lb per sq ft wind pressure. For this wind pressure, Table 5-7 shows a minimum lateral load of 125 lb per lin ft acting on the top of the wall form. Calculate the axial force in each brace and the reactions at the bottom of each brace.

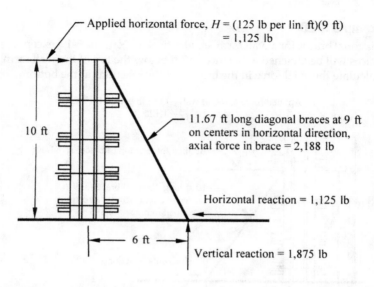

The geometric properties can be calculated as follows:

$$x = 6\text{-ft}$$
$$y = 10\text{-ft}$$
$$z = \sqrt{x^2 + y^2}$$
$$= \sqrt{(6)^2 + (10)^2}$$
$$= 11.67 \text{ ft}$$

The forces and reactions of the bracing can be calculated as follows:

Applied horizontal load = (125 lb per lin ft)(9 ft)
$$= 1,125 \text{ lb}$$
Axial force in brace = $H(z/x)$
$$= (1,125 \text{ lb})(11.67/6)$$
$$= 2,188 \text{ lb}$$
Vertical reaction at bottom of brace = $H(y/x)$
$$= (1,125 \text{ lb})(10/6)$$
$$= 1,875 \text{ lb}$$
Horizontal reaction at bottom of brace = H
$$= 1,125 \text{ lb}$$

The axial compression load in the brace is 2,188-lb and the unbraced length is 11.67 ft. Table 5-5 shows many wood members are adequate to resist this load. Each brace must be securely attached at the top and bottom.

Example 5-20

Diagonal braces for a wall form are identical to Example 5-19, except the braces will be attached a distance of 2 ft below the top of the wall form. Calculate the axial force in the brace and the reactions at the bottom.

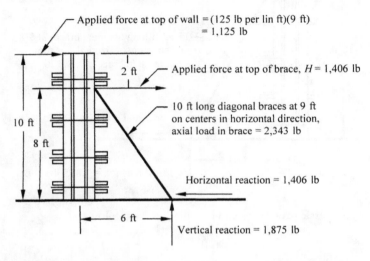

Applied force at top of wall = (125 lb per lin ft)(9 ft)
= 1,125 lb

2 ft

Applied force at top of brace, $H = 1,406$ lb

10 ft long diagonal braces at 9 ft on centers in horizontal direction, axial load in brace = 2,343 lb

10 ft

8 ft

Horizontal reaction = 1,406 lb

6 ft

Vertical reaction = 1,875 lb

Because the brace is attached 2 ft below the top of the wall, the horizontal force against the top of the brace will be higher than the horizontal force against the top of the wall. The top of the brace is attached at 8 ft above the ground, rather than 10 ft above the ground. The force acting horizontally at the top of the brace can be calculated as follows:

$$H = 1,125 \text{ lb } (10/8)$$
$$= 1,406 \text{ lb}$$

Refer to Figure 5-19 to calculate the following geometric properties:

$$x = 6 \text{ ft}$$
$$y = 8 \text{ ft}$$
$$z = \sqrt{x^2 + y^2}$$
$$= \sqrt{(6)^2 + (8)^2}$$
$$= 10\text{-ft}$$

The forces and reactions of the bracing can be calculated as follows:

Axial force in brace = $H(z/x)$
$$= (1,406 \text{ lb})(10/6)$$
$$= 2,343 \text{ lb}$$

Vertical reaction at bottom of brace = $H(y/x)$
$$= (1,406 \text{ lb})(8/6)$$
$$= 1,875 \text{ lb}$$

Horizontal reaction at bottom of brace = H

$$= 1,406 \text{ lb}$$

The axial compression load in the brace is 2,343 lb and the unbraced length is 10 ft. Table 5-5 shows that many wood members are adequate to resist this load. Each brace must be securely attached at the top and bottom.

Design of Forms for a Concrete Slab

The procedure for designing a floor slab is similar to the procedure for designing wall forms. For a slab, joists perform the same function as the studs for a wall—they support the plywood sheathing. The stringers for a slab form support the joists, just as the wales support the studs in a wall form. The shores support the stringers for a slab form, whereas the form ties provide support for the wales in a wall form.

The use of the equations and principles developed previously will be illustrated by designing the forms for a concrete slab (Figure 5-18).

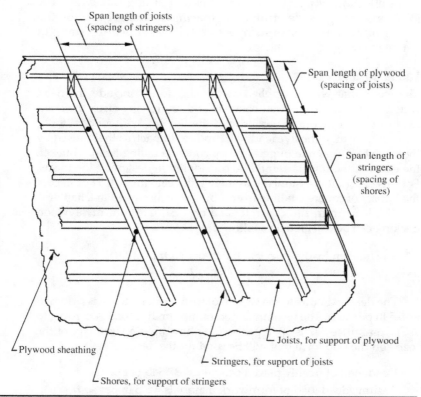

FIGURE 5-18   Allowable span lengths and components of a slab form.

The design will include the decking, joists, and stringers. Requirements for shoring of formwork are discussed in the following section. Similar to wall forms, in the design of slabs the thickness and grade of the plywood are often selected based on the availability of materials. The following design criteria will apply:

1. Thickness of concrete slab, 7 in.

2. Concrete unit weight, 150 lb per cu ft.

3. Deflection limited to $l/360$, but not greater than $\frac{1}{16}$ in.

4. All form lumber will be S4S No. 2 grade Douglas Fir-Larch with no splits, assuming a short-term load-duration and wet condition.

5. Decking for the slab form will be $\frac{7}{8}$ in. plywood sheathing, from the Group II species with a S-2 stress rating.

The design will be performed in the following sequence: determine the dead and live loads on the slab forms; determine the spacing of joists based on the strength and deflection of plywood decking; determine the spacing of stringers based on the strength and deflection of joists; check the bearing strength between the joists and stringers; determine the spacing of shores based on the strength and deflection of stringers; and check the bearing strength between stringers and shores.

Loads on Slab Forms

The loads that will be applied to the slab forms include the dead weight of the freshly placed concrete and reinforcing steel, the dead load of the formwork materials, and the live load of workers, tools, and equipment. If the live load will produce impact, as that caused by motor-driven concrete buggies or by concrete falling from a bucket, the effect of the impact should be considered in designing the forms. For this design it is assumed that there will be no impact. The effect of the impact of concrete falling from a bucket is discussed in Chapter 7.

The load from the concrete slab on a square foot of plywood decking can be calculated as follows:

$$\text{Uniform load per sq ft} = 150 \text{ lb per cu ft } (7/12 \text{ ft})$$
$$= 87.5 \text{ lb per sq ft}$$

The dead load due to the weight of formwork material is assumed as 5.5 lb per sq ft. The recommended minimum live load according to ACI Committee 347 is 50 lb per sq ft. Following is a summary of the loads on the slab form that will be used for the design:

Dead load of freshly placed concrete = 87.5 lb per sq ft

Assume dead load of formwork materials = 5.5 lb per sq ft

Live load of workers and tools = _50.0 lb per sq ft_
Design load = 143.0 lb per sq ft

Plywood Decking to Resist Vertical Load

Because the type and thickness of plywood have already been chosen, it is necessary to determine the allowable span length of the plywood, that is, the distance between joists, such that the bending stress, shear stress, and deflection of the plywood will be within allowable limits. Table 4-9 provides the section properties for the $7/8$-in.-thick plywood selected for the design. The plywood will be placed with the face grain across the supporting joists. Therefore, the properties for stress applied parallel to face grain will apply. The physical properties for a 1-ft-wide strip of the $7/8$-in.-thick plywood installed with the face grain across supports are as follows:

Approximate weight = 2.6 lb per sq ft
Effective thickness for shear, $t = 0.586$ in.
Cross-sectional area, $A = 2.942$ in.^{2}
Moment of inertia, $I = 0.278$ in.^{4}
Effective section modulus, $S_e = 0.515$ in.^{3}
Rolling shear constant, $Ib/Q = 8.05$ in.^{2}

Table 4-10 indicates the allowable stresses for the plywood of Group II species and S-2 stress rating for normal duration of load. However, values for stresses are permitted to be increased by 25% for short-duration load conditions, less than 7 days, as discussed in Chapter 4 (Table 4-4). The 25% increase does not apply to the modulus of elasticity. For this design, it is assumed that a wet condition will exist. Following are the allowable stresses that will be used for the design:

Allowable bending stress, $F_b = 1.25(820) = 1,025$ lb per sq in.
Allowable rolling shear stress, $F_s = 1.25(44) = 55$ lb per sq in.
Modulus of elasticity, $E = 1,300,000$ lb per sq in.

Bending Stress in Plywood Decking

Because the plywood will be installed over multiple supports, three or more, Eq. (5-41) can be used to determine the allowable span length of the plywood based on bending.

From Eq. (5-41), $l_b = [120F_b S_e / w_b]^{1/2}$
$$= [120(1,025)(0.515)/(143)]^{1/2}$$
$$= 20.0 \text{ in.}$$

Rolling Shear Stress in Plywood Decking

The allowable span length of the plywood based on rolling shear stress can be calculated.

$$\text{From Eq. (5-48), } l_s = 20F_s(Ib/Q)/w_s$$
$$= 20(55)(8.05)/143$$
$$= 61.9 \text{ in.}$$

Bending Deflection in Plywood Decking

For a maximum permissible deflection of $l/360$, but not greater than $\frac{1}{16}$ in., and plywood with three or more spans, Equations (5-57a) and (5-57b) can be used to calculate the allowable span lengths for bending deflection.

$$\text{From Eq. (5-57a), } l_d = [1{,}743EI/360w_d]^{1/3} \text{ for } \Delta = l/360$$
$$= [1{,}743(1{,}300{,}000)(0.278)/360(143)]^{1/3}$$
$$= 23.0 \text{ in.}$$
$$\text{From Eq. (5-57b), } l_d = [1{,}743EI/16w_d]^{1/4} \text{ for } \Delta = \frac{1}{16} \text{ in.}$$
$$= [1{,}743(1{,}300{,}000)(0.278)/16(143)]^{1/4}$$
$$= 22.9 \text{ in.}$$

Summary for the $\frac{7}{8}$-in.-Thick Plywood Decking

For bending, the maximum span length of the plywood = 20.0 in.

For shear, the maximum span length of the plywood = 61.9 in.

For deflection, the maximum span length of the plywood = 22.9 in.

For this design, bending controls the maximum span length of the plywood sheathing. The joists that support the plywood must be placed at a spacing that is no greater than 20.0 in. For uniformity, choose a joist spacing of 20.0 in.

Joists for Support of Plywood

The joists support the plywood and the stringers must support the joists. Therefore, the maximum allowable spacing of stringers will be governed by the bending, shear, and deflection in the joists.

Consider using 4×4 S4S joists whose actual size is $3\frac{1}{2}$ in. by $3\frac{1}{2}$ in. From Table 4-1, the physical properties can be obtained for 4×4 S4S lumber as follows:

Area, $A = 12.25$ in.^{2}

Section modulus, $S = 7.15$ in.^{3}

Moment of inertia, $I = 12.51$ in.^{4}

The 4 × 4 S4S lumber selected for the forms is No. 2 grade Douglas Fir-Larch with no splits. The allowable stresses can be obtained by adjusting the reference design values in Table 4-3 by the adjustment factors for size, short-duration loading, and a wet condition. Table 4-3a provides the adjustments for size, Table 4-4 for short-duration loads, and Table 4-5 for a wet condition application. Following are the allowable stresses that will be used for the design:

Allowable bending stress, $F_b = C_F \times C_D \times C_M \times$ (Bending reference value)

$$= 1.5(1.25)(0.85)(900)$$

$$= 1,434 \text{ lb per sq in.}$$

Allowable shear stress, $F_v = C_D \times C_M \times$ (shear stress reference value)

$$= 1.25(0.97)(180)$$

$$= 218 \text{ lb per sq in.}$$

Modulus of elasticity, $E = C_M \times$ (reference value of modulus of elasticity)

$$= 0.9(1,600,000)$$

$$= 1,440,000 \text{ lb per sq in.}$$

Bending in 4 × 4 S4S Joists

Each joist will support a horizontal strip of decking 20 in. wide, which will produce a uniform load of 143 lb per sq ft × 20/12 ft = 238 lb per lin ft along the entire length of the joist. The allowable span length of the joists, that is, the distance between stringers, based on bending can be determined as follows:

From Eq. (5-34), $l_b = [120F_b S/w]^{1/2}$

$$= [120(1,434)(7.15)/238]^{1/2}$$

$$= 71.9 \text{ in.}$$

Shear in 4 × 4 S4S Joists

The allowable span length based on shear in the joists can be calculated.

From Eq. (5-36), $l_v = 192F_v bd/15w + 2d$

$$= 192(218)(3.5)(3.5)/15(238) + 2(3.5)$$

$$= 150.6 \text{ in.}$$

Deflection in 4 × 4 S4S Joists

The permissible deflection is $l/360$, but not greater than $\frac{1}{16}$ in. Using Eqs. (5-37a) and (5-37b) the allowable span lengths due to deflection can be calculated as follows.

From Eq. (5-37a), $l_\Delta = [1{,}743EI/360w]^{1/3}$ for $\Delta = l/360$
$$= [1{,}743(1{,}440{,}000)(12.51)/360(238)]^{1/3}$$
$$= 71.6 \text{ in.}$$
From Eq. (5-37b), $l_\Delta = [1{,}743EI/16w]^{1/4}$ for $\Delta = \frac{1}{16}$ in.
$$= [1{,}743(1{,}440{,}000)(12.51)/16(238)]^{1/4}$$
$$= 53.6 \text{ in.}$$

Summary for 4 × 4 S4S Joists

For bending, the maximum span length of joists = 71.9 in.
For shear, the maximum span length of joists = 150.6 in.
For deflection, the maximum span length of joists = 53.6 in.

For this design, deflection governs the maximum span length of the joists. Because the stringers must support the joists, the maximum spacing of the wales must be no greater than 53.6 in. For constructability, a spacing of 48.0 in. center to center will be used for stringers.

Stringers for Support of Joists

For this design, consider using 4-in.-thick stringers. Thus, the actual size of a stringer will be 3½ in. Determine the unit stress in bearing between joists and stringers.

$$\text{Contact area in bearing, } A = 3.5 \text{ in.} \times 3.5 \text{ in.}$$
$$= 12.25 \text{ sq in.}$$

The total load in bearing between a joist and a stringer will be the unit pressure of the concrete acting on an area based on the joist and stringer spacing. For this design, the area will be 20 in. wide by 48 in. long. Therefore, the pressure acting between the joist and the stringer will be

$$P = 143 \text{ lb per sq ft } [20/12 \times 48/12]$$
$$= 953 \text{ lb}$$

The calculated applied unit stress in bearing, perpendicular to grain, between a joist and a stringer will be

$$f_{c\perp} = 953 \text{ lb}/12.25 \text{ sq in.}$$
$$= 77.8 \text{ lb per sq in.}$$

The allowable compression stress perpendicular to grain for Douglas Fir-Larch can be obtained by multiplying the value tabulated in Table 4-3 by the adjustment factor for a wet condition ($C_M = 0.67$).

There is no load-duration adjustment factor in the compression perpendicular to grain.

Allowable compression stress perpendicular to grain is

$$F_{c\perp} = (0.67)(625 \text{ lb per sq in.})$$
$$= 418.7 \text{ lb per sq in.}$$

The allowable compressive unit stress of 418.7 lb per sq in. is greater than the applied unit stress of 77.8 lb per sq in. Therefore, the unit stress is within the allowable value for this species and grade of lumber.

Size of Stringer Based on Selected 48-in. Spacing

Although the loads transmitted from the joists to the stringers are concentrated, it is generally sufficiently accurate to treat them as uniformly distributed loads having the same total value as the concentrated loads when designing forms for concrete slabs.

With a 48-in. spacing and a 143 lb per sq ft pressure, the uniform load on a stringer will be 143 lb per sq ft × 48/12 ft = 572 lb per lin ft. Consider using 4 × 6 S4S stringers.

From Table 4-1, the physical properties for 4 × 6 S4S lumber are

$$A = 19.25 \text{ in.}^2$$
$$S = 17.65 \text{ in.}^3$$
$$I = 48.53 \text{ in.}^4$$

The allowable stresses for the No. 2 grade Douglas Fir-Larch can be obtained by adjusting the reference design values in Table 4-3 by the adjustment factors for size, short-duration loading, and a wet condition. Table 4-3a provides the adjustments for size, Table 4-4 for short-duration loads, and Table 4-5 for a wet condition application. Following are the allowable stresses that will be used for the design:

Allowable bending stress,
$$F_b = C_F \times C_D \times C_M \times \text{(bending reference value)}$$
$$= 1.3(1.25)(0.85)(900)$$
$$= 1,243 \text{ lb per sq in.}$$

Allowable shear stress,
$$F_v = C_D \times C_M \times \text{(shear stress reference value)}$$
$$= 1.25(0.97)(180)$$
$$= 218 \text{ lb per sq in.}$$

Modulus of elasticity,
$$E = C_M \times \text{(reference modulus of elasticity)}$$
$$= 0.9(1,600,000)$$
$$= 1,440,000 \text{ lb per sq in.}$$

Bending in 4 × 6 S4S Stringers

Based on bending, the span length of the stringers must not exceed the value calculated.

From Eq. (5-34), $l_b = [120F_bS/w]^{1/2}$

$$= [120(1,243)(17.65)/572]^{1/2}$$

$$= 67.8 \text{ in.}$$

Shear in 4 × 6 S4S Stringers

Based on shear, the allowable span length of the stringers must not exceed the value calculated.

From Eq. (5-36), $l_v = 192F_vbd/15w + 2d$

$$= 192(218)(3.5)(5.5)/15(572) + 2(5.5)$$

$$= 104.9 \text{ in.}$$

Deflection in 4 × 6 S4S Stringers

The permissible deflection is $l/360$, but not greater than $\frac{1}{16}$ in. Using Eqs. (5-37a) and (5-37b) the allowable span lengths due to deflection can be calculated as follows:

From Eq. (5-37a), $l_\Delta = [1,743EI/360w]^{1/3}$ for $\Delta = l/360$

$$= [1,743(1,440,000)(48.53)/360(572)]^{1/3}$$

$$= 83.9 \text{ in.}$$

From Eq. (5-37b), $l_\Delta = [1,743EI/16w]^{1/4}$ for $\Delta = \frac{1}{16}$ in.

$$= [1,743(1,440,000)(48.53)/16(572)]^{1/4}$$

$$= 60.4 \text{ in.}$$

Summary for 4 × 6 S4S Stringers

For bending, the maximum span length of stringers = 67.8 in.

For shear, the maximum span length of stringers = 104.9 in.

For deflection, the maximum span length of stringers = 60.4 in.

For this design, deflection governs the maximum span length of the stringers. Because the shores must support the stringers, the maximum spacing of shores must be no greater than 60.4 in. For simplicity in layout and constructability, a spacing of 60.0 in. center to center will be used.

Shores for Support of Stringers

Based on the strength of the stringers in bending, shear, and deflection, the stringers must be supported every 60 in. by a shore. A shore must resist the concrete pressure that acts over an area consisting of the spacing of the stringers and the span length of the stringers, 48 in. by 60 in. The vertical load on a shore can be calculated as follows:

$$\text{Area} = 48 \text{ in.} \times 60 \text{ in.}$$
$$= 2{,}880 \text{ sq in.}$$
$$= 20.0 \text{ sq ft}$$
$$\text{Vertical load on a shore} = 143 \text{ lb per sq ft } (20.0 \text{ sq ft})$$
$$= 2{,}860 \text{ lb}$$

Thus, a shore is required that will withstand a vertical load of 2,860 lb. There are numerous patented shores available that can sustain this load. For this design, S4S wood shores will be installed with a 10-ft unbraced length. Using a wood shore with an allowable compression stress parallel to grain of 1,552 lb per sq in. and a modulus of elasticity for column stability of 580,000 lb per sq in., Table 5-5 indicates an allowable load of 4,672 lb for a 4×4 S4S wood shore.

Bearing between Stringer and Shore

For this design, 4×6 S4S stringers will bear on 4×4 S4S wood shores. Determine the unit stress in bearing between stringers and shores.

$$\text{Contact area, } A = 3.5 \text{ in.} \times 3.5 \text{ in.}$$
$$= 12.25 \text{ sq in.}$$

The calculated unit stress in bearing, perpendicular to grain, between a stringer and a shore will be

$$f_{c\perp} = 2{,}860 \text{ lb}/12.25 \text{ sq in.}$$
$$= 233.5 \text{ lb per sq in.}$$

From Table 4-3, the tabulated design value for unit compression stress, perpendicular to grain, for No. 2 grade Douglas Fir-Larch is 625 lb per sq in. The allowable design stress can be calculated by multiplying the reference design value by the wet condition adjustment factor ($C_M = 0.67$) (refer to Table 4-5). There is no load-duration adjustment factor in compression stress perpendicular to grain. The allowable bearing stress will be

$$F_{c\perp} = (0.67)(625 \text{ lb per sq in.})$$
$$= 418.7 \text{ lb per sq in.}$$

This allowable unit stress of 418.7 lb per sq in. is greater than the applied unit stress of 233.5 lb per sq in. for this grade and species of wood. The formwork must also be adequately braced to resist lateral forces on the slab forming system. This topic is discussed in the following sections.

Design Summary of Forms for Concrete Slab

The allowable stresses and deflection criteria used in the design are based on the following values.

For $7/8$-in.-thick plywood from Group II species with S-2 stress rating:

> Allowable bending stress = 1,025 lb per sq in.
>
> Allowable rolling shear stress = 55 lb per sq in.
>
> Modulus of elasticity = 1,300,000 lb per sq in.

For 4×4 S4S joists, No. 2 grade Douglas Fir-Larch lumber:

> Allowable bending stress = 1,434 lb per sq in.
>
> Allowable shear stress = 218 lb per sq in.
>
> Modulus of elasticity = 1,440,000 lb per sq in.
>
> Permissible deflection less than $l/360$, but not greater than $1/16$ in.

For 4×6 S4S stringers, No. 2 grade Douglas Fir-Larch lumber:

> Allowable bending stress = 1,243 lb per sq in.
>
> Allowable shear stress = 218 lb per sq in.
>
> Modulus of elasticity = 1,440,000 lb per sq in.
>
> Permissible deflection less than $l/360$, but not greater than $1/16$ in.

The forms for the concrete slab should be built to the following conditions:

Item	Nominal Size and Spacing
Decking	$7/8$-in.-thick plywood
Joists	4×4 S4S at 20 in. on centers
Stringers	4×6 S4S at 48 in. on centers
Shores	60 in. maximum spacing along stringers

This design example describes the calculations for determining the minimum size and spacing of members using 4-in.-wide S4S lumber. The purpose of the example is to illustrate the procedures and calculations for designing components of slab forms for concrete roofs and floors. The 4-in.-wide lumber was selected because workers can easily walk and work on a 4-in.-square member, compared to a 2-in.-wide member. Also, the 4-in.-wide member will not roll over, or fall sideways, as it is being placed compared to a 2-in.-wide member. Furthermore, the larger size 4-in. lumber will not require the amount of bridging or lateral bracing, compared to 2-in. size lumber. Figure 5-19 summarizes the design, including the size and spacing of decking, joists, and stringers.

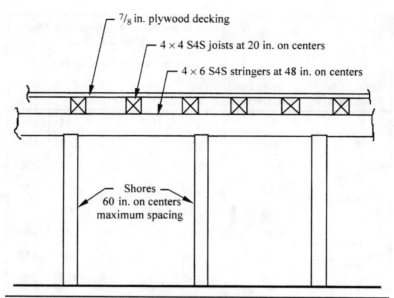

FIGURE 5-19   Summary of design for a slab form.

Minimum Lateral Force for Design of Slab Form Bracing Systems

The minimum lateral loads required for the design of slab form bracing are presented in Table 5-8. The horizontal load shown in Table 5-8 is in lb per lin ft, applied along the edge of the slab in either direction as illustrated in Figure 5-20. Where unbalanced loads from the placement of concrete or from starting or stopping of equipment may occur, a structural analysis of the bracing system should be made.

Minimum Time for Forms and Supports to Remain in Place

The minimum time for stripping forms and removal of supporting shores is a function of concrete strength, which should be specified by the engineer/architect. The preferred method of determining stripping time is using tests of job-cured cylinders or concrete in place. The American Concrete Institute ACI Committee 347 [3] provides recommendations for removal of forms and shores. Table 5-9 provides the suggested minimum times that forms and supports should remain in place under ordinary conditions. These recommendations are in accordance with ACI Committee 347.

Solid Slab Thickness, in.	Dead Load, lb per sq ft	P, lb per lin ft, Applied along Edge of Slab in Either Direction				
		Width of Slab in Direction of Force, ft				
		20	40	60	80	100
4	65	100	100	100	104	130
6	90	100	100	108	144	180
8	115	100	100	138	184	230
10	140	100	112	168	224	280
12	165	100	132	198	264	330
14	190	100	152	228	304	380
16	215	100	172	258	344	430
20	265	106	212	318	424	530

Notes:
1. Reprinted from "Formwork for Concrete," ACI Special Publication No. 4, the American Concrete Institute.
2. Slab thickness given for concrete weighing 150 lb per cu ft; allowance of 15 lb per sq ft for weight of forms. For concrete of a different weight or for joist slabs and beams and slab combinations, estimate dead load per sq ft and work from dead load column, interpolating as needed on a straight-line basis. Do not interpolate in ranges that begin with 100 lb minimum load.
3. Special conditions may require heavier bracing.

TABLE 5-8 Minimum Lateral Force for Design of Slab Form Bracing, lb per lin ft

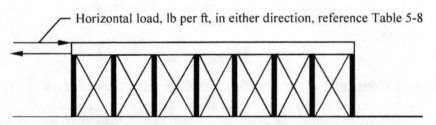

Horizontal load, lb per ft, in either direction, reference Table 5-8

FIGURE 5-20 Minimum lateral force for bracing slab forming systems.

Minimum Safety Factors for Formwork Accessories

Table 5-10 provides the minimum safety factors for formwork accessories, including form ties, anchors, hangers, and inserts used as form ties. These recommendations are in accordance with ACI Committee 347.

Item	Time	
Walls	12 hours	
Columns	12 hours	
Sides of beams and girders	12 hours	
Pan joist forms		
30 in. wide or less	3 days	
over 30 in. wide	4 days	
	Where structural live load is	
	Less than structural dead load	Greater than structural dead load
Arch centers	14 days	7 days
Joist, beam, or girder soffits		
Under 10 ft clear span between structural supports	7 days	7 days
10 to 20 ft clear span between structural supports	14 days	14 days
Over 20 ft clear span between supports	21 days	14 days
One-way floor slabs		
Under 10 ft clear span between supports	4 days	3 days
10 to 20 ft clear span between supports	7 days	4 days
Over 20 ft clear span between supports	10 days	7 days
Two-way slab systems	Removal times are contingent on reshores, where required, being placed as soon as practicable after stripping operations are complete but not later than the end of the working day in which stripping occurs. Where reshores are required to implement early stripping while minimizing sag or creep, the capacity and spacing of such reshores should be specified by the engineer/architect.	
Posttensioned slab system	As soon as full posttensioning has been applied	

Notes:

1. Reprinted from ACI Committee 347, Guide to Formwork for Concrete, American Concrete Institute.

2. Where such forms also support formwork for slab or beam soffits, the removal time of the latter should govern.

3. Distances between supports refer to structural supports and not to temporary formwork or shores.

4. Where forms may be removed without disturbing shores, use half of values shown, but not less than 3 days.

5. See Chapter 6 for special conditions affecting number of floors to remain shored or reshored.

TABLE 5-9 Suggested Minimum Time Forms and Supports Should Remain in Place under Ordinary Conditions

Accessory	Safety Factor	Type of Construction
Form tie	2.0	All applications
Form anchor	2.0	Formwork supporting form weight and concrete pressures only
	3.0	Formwork supporting weight of forms, concrete, construction live loads, and impact
Form hangers	2.0	All applications
Anchoring inserts used as form ties	2.0	Precast concrete panels when used as formwork

Notes:
1. Reprinted from ACI Committee 347, Guide to Formwork for Concrete, American Concrete Institute.
2. Safety factors are based on ultimate strength of accessory.

TABLE 5-10   Minimum Safety Factors for Formwork Accessories

References

1. "National Design Specification for Wood Construction," ANSI/AF&PA NDS-2005, American Forest & Paper Association, Washington, DC, 2005.
2. "Design Values for Wood Construction," Supplement to the National Design Specification, American Forest & Paper Association, Washington, DC, 2005.
3. "Concrete Forming," APA—The Engineered Wood Association, Tacoma, WA, 2005.
4. "Plywood Design Specification," APA—The Engineered Wood Association, Tacoma, WA, 1997.
5. "Formwork for Concrete," Special Publication No. 4, American Concrete Institute, Detroit, MI.
6. "Guide to Formwork for Concrete," ACI Committee 347-04, American Concrete Institute, Detroit, MI, 2004.
7. "Timber Construction Manual," 5th Edition, American Institute of Timber Construction, John Wiley & Sons, Hoboken, NJ, 2005.
8. "Design of Wood Structures," 6th Edition, The McGraw-Hill Companies, New York, NY, 2007
9. "Minimum Design Loads for Buildings and Other Structures," SEI/ASCE 7-02, American Society of Civil Engineers, Reston, VA, 2003.

CHAPTER **6**

Shores and Scaffolding

General Information

Vertical shores, or posts, and scaffolding are used with formwork to support concrete girders, beams, floor slabs, roof slabs, bridge decks, and other members until these members gain sufficient strength to be self-supporting. Many types and sizes of each are available from which those most suitable for a given use may be selected. They may be made from wood or steel or from a combination of the two materials. Aluminum shores and scaffolding are also available.

Shores

In general, shores are installed as single-member units that may be tied together at one or more intermediate points with horizontal and diagonal braces to give them greater stiffness and to increase their load-supporting capacities.

If shores are to provide the load capacities that they are capable of providing, at least two precautions must be taken in installing them. They should be securely fastened at the bottom and top ends to prevent movement or displacement while they are in use. Because the capacities of shores are influenced by their slenderness ratios, two-way horizontal and diagonal braces should be installed with long shores at one or more intermediate points to reduce the unsupported lengths.

Concrete is usually pumped or placed from buckets, which permits the concrete to fall rapidly onto a limited area. The use of either method may produce a temporary uplift in the forms near the area under the load. If this should happen, it is possible that a portion of the forms will be lifted off the tops of one or more shores. Unless the tops of the shores are securely fastened to the formwork which they support, their positions may shift. For the same reason, the bottom ends of shores should be securely held in position. There are reports that forms have collapsed because the positions of shores shifted

163

while concrete was being placed. The subject of failures of formwork is discussed in Chapter 7.

Table 5-5 and Figure 5-12 show the relations between the capacities of shores and their effective lengths. For example, the indicated allowable load on a 4 × 4 S4S wood shore 6 ft long is 11,076 lb for wood with an allowable unit stress in compression of 1,650 lb per sq in. and a modulus of elasticity for column stability of 580,000 lb per sq in., whereas the indicated allowable load on the same shore with an effective length of 12 ft is only 3,318 lb. The former load is approximately three times the latter load. In this instance, doubling the effective length of the shore reduces the allowable load to 30% of the allowable load for the 6-ft length.

Table 6-1 gives the allowable loads on wood shores whose unsupported lengths vary from 6 to 14 ft, expressed as percentages of the

Size of Shore, in. F^*_c, lb per sq in. E'_{min}, lb per sq in.	Percent of the Allowable Load on a Shore 6 ft Long								
	Effective Length, ft								
	6	7	8	9	10	11	12	13	14
4 × 4 $F^*_c = 1,650$ $E'_{min} = 580,000$	100	79.3	63.3	49.8	42.3	35.4	30.0	25.7	22.2
4 × 4 $F^*_c = 1,552$ $E'_{min} = 580,000$	100	79.8	63.8	52.0	42.9	35.8	30.7	26.1	22.6
4 × 4 $F^*_c = 1,495$ $E'_{min} = 470,000$	100	77.7	61.7	49.9	40.9	34.5	28.9	24.8	21.4
4 × 4 $F^*_c = 1,322$ $E'_{min} = 510,000$	100	80.1	64.4	52.3	43.3	36.2	30.7	26.8	22.8
4 × 6 $F^*_c = 1,600$ $E'_{min} = 580,000$	100	79.5	63.6	51.5	42.5	35.7	31.2	25.9	22.5
4 × 6 $F^*_c = 1,485$ $E'_{min} = 580,000$	100	81.2	66.5	54.7	45.6	38.7	33.0	28.4	24.6
4 × 6 $F^*_c = 1,430$ $E'_{min} = 470,000$	100	78.6	62.6	49.9	41.6	34.8	29.4	25.3	21.8
4 × 6 $F^*_c = 1,265$ $E'_{min} = 510,000$	100	80.7	64.9	53.0	43.7	36.7	30.9	26.6	23.2

TABLE 6-1 Relations between Allowable Loads and Unsupported Lengths of Wood Shores

allowable loads on shores that are 6 ft long. The information applies to wood with the allowable unit compressive stress parallel to grain and the modulus of elasticity as shown.

The pattern of dramatic reduction in allowable load with increase in effective length applies to wood members having other physical properties and also to materials other than wood and to patented shores.

Wood Post Shores

Although the common practice today is to use patented shores, there are some instances where wood shores are fabricated for use on a particular job. Wood shores have several advantages and several disadvantages when they are compared with patented shores.

Among the advantages are the following:

1. The initial cost is low.
2. They are usually readily available.
3. They possess high capacity in relation to their weight.
4. It is easy to attach and remove braces.

Among the disadvantages are the following:

1. It is difficult to adjust their lengths.
2. The cost of labor for installing wood shores may be higher than for installing patented shores.
3. Unless they are stored carefully, they may develop permanent bows, which will reduce their load capacities.
4. They may develop rot or permanent bows.

If wood shores are too long for a given use, they must be sawed to fit the required length, which results in additional labor costs and waste of materials. If they are too short, it is necessary to splice them, which may result in a weakening of the shores when compared to the strengths of unspliced shores. Also, the cost of the labor required to make such splices may be substantial.

Usually the final adjustment in the height of the top of a wood shore is made with two wood wedges driven under the bottom of the shores from opposite sides. Because of the higher allowable unit stress in compression perpendicular to the grain, wedges made of hardwoods are better than those made of softwoods. Both wedges should be nailed to the mud sills, or other boards on which they rest, to prevent displacement.

Field-constructed butt or lap splices of timber shoring should not be used unless the connections are made with hardware of adequate strength and stability. To prevent buckling, splices should not be placed near mid-height of unbraced shores or midway between points of lateral support.

Patented Shores

Patented shores are more commonly used than job-fabricated wood shores for supporting formwork for concrete beams and slabs. When compared with wood shores, they have several advantages and several disadvantages.

The advantages of patented shores include the following:

1. They are available in several basic lengths.
2. They are readily adjustable over a wide range of lengths.
3. For most of them, adjustments in length can be made in small increments.
4. In general, they are rugged, which ensures a long life.
5. The shore heads are usually long enough to give large bearing areas between the shores and the stringers that rest on them.

Among the disadvantages of patented shores are the following:

1. The initial cost is higher than for wood shores.
2. For some, but not all, it is more difficult to attach intermediate braces than it is for wood shores.
3. Because of their slenderness, some of them are less resistant to buckling than wood shores.

Ellis Shores

The method of shoring of this company consists of two 4 × 4 S4S wood posts, fastened by two special patented clamps. The bottom of one post rests on the supporting floor, whereas the second post is moved upward along the side of the lower one. Two metal clamps, made by Ellis Construction Specialties, are installed around the two posts, as illustrated in Figure 6-1. The top post is raised to the desired height, and the two clamps automatically grip the two posts and hold them in position.

The shore specifications include lower shore members composed of two Ellis clamps with pivotal plates, permanently attached with threaded nails near the top, 12 in. apart, center to center. The upper shore member is of sufficient length to obtain the desired height. Both the lower and upper members are No. 1 grade Douglas Fir or Yellow Pine, free of heart center, stained with ends squared.

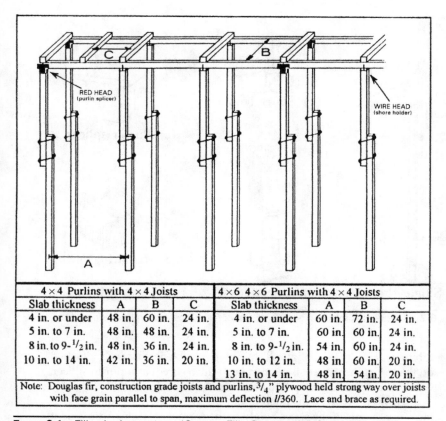

4 × 4 Purlins with 4 × 4 Joists				4 × 6 4 × 6 Purlins with 4 × 4 Joists			
Slab thickness	A	B	C	Slab thickness	A	B	C
4 in. or under	48 in.	60 in.	24 in.	4 in. or under	60 in.	72 in.	24 in.
5 in. to 7 in.	48 in.	48 in.	24 in.	5 in. to 7 in.	60 in.	60 in.	24 in.
8 in. to 9-$^1/_2$ in.	48 in.	36 in.	24 in.	8 in. to 9-$^1/_2$ in.	54 in.	60 in.	24 in.
10 in. to 14 in.	42 in.	36 in.	20 in.	10 in. to 12 in.	48 in.	60 in.	20 in.
				13 in. to 14 in.	48 in.	54 in.	20 in.

Note: Douglas fir, construction grade joists and purlins, $^3/_4$" plywood held strong way over joists with face grain parallel to span, maximum deflection $l/360$. Lace and brace as required.

FIGURE 6-1   Ellis shoring system. (*Source: Ellis Construction Specialties, Ltd.*)

The patented Ellis clamp is designed with a solid rectangular collar with two pivotal plates, which are scorated on the flat inner surface for firm gripping. Two clamps nailed 12 in. apart near the top of a 4 × 4 make a lower shore member. The upper shore member is another 4 × 4 of desired length.

Figure 6-1 illustrates the Ellis shoring system, including the recommended spacing pattern for the indicated forming condition. The manufacturer recommends a maximum load of 4,000 lb for heights up to 12 ft, provided a horizontal lace is installed in each direction at the midpoint on the shore, attached to the lower member. For lower heights, higher loads may be obtained in accordance with the manufacturer's recommendations. The lower shore member, with two clamps attached, is available for rent from the company.

Accessories are available for installation of the Ellis shoring system. Figure 6-2 shows accessories for the tops and bottoms of shores. A wire head, configured from a ½-in. rod, is used to securely hold the shore to the purlins. A metal attachment, fabricated from 4-in.-square

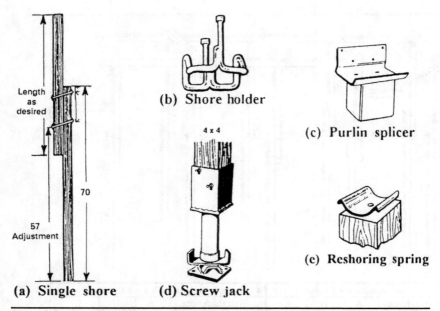

Length as desired

70

57 Adjustment

(a) Single shore

(b) Shore holder

4 x 4

(d) Screw jack

(c) Purlin splicer

(e) Reshoring spring

Figure 6-2   Accessories for wood shores. (*Source: Ellis Construction Specialties, Ltd.*)

tubing and ³⁄₁₆-in. U-configured steel plate, can be permanently attached on the top of the upper shore member, which serves as a purlin splicer. One side is open, allowing it to be stripped from below.

Screw jacks can be installed under the lower shore member, allowing adjustments of the shore height to the correct position. Adjustment handles turn easily on machined threads, allowing a 6-in. standard range of adjustment, 3 in. up and 3 in. down. The screw jack has a safe working load of 10,000 lb with a 2.5:1 safety factor.

A reshore spring, made of high-carbon spring steel, is available to hold the Ellis shore in place during reshoring. It is used to keep the shore tight against the slab during reshoring concrete slabs. The reshore spring is nailed to the top of each shore, eliminating the need for cutting and nailing lumber to the top and bottom of each shore. The spring returns to its original shape for many reuses. A 200-lb load will compress the reshore spring flat.

Symons Shores

The single-post steel shores of the Symons Corporation, illustrated in Figure 6-3, are available in three models which provide adjustable shoring heights from 5 ft 7 in. to 16 ft. Each post shore consists of two parts: a base post with a threaded collar and a staff member which fits into the base post. The assembly weight of a single-post shore varies from 67 to 80 lb, which permits a shore to be handled by one person.

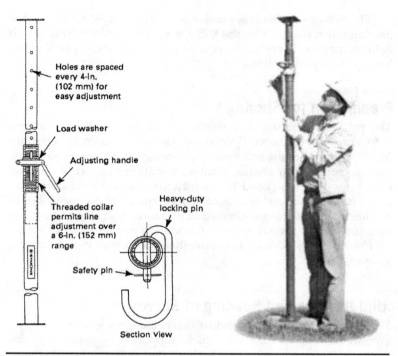

Holes are spaced every 4-in. (102 mm) for easy adjustment

Load washer

Adjusting handle

Threaded collar permits line adjustment over a 6-in. (152 mm) range

Heavy-duty locking pin

Safety pin

Section View

FIGURE 6-3   Single-post steel shore. (*Source: Symons Corporation*)

Symons post shores each carry load ratings of up to 10,000 lb, depending on the shore height.

The Symons post shores have a unique locking pin that, under normal use conditions, cannot be easily broken or lost. For approximate height adjustment, this pin is inserted into one of the holes spaced at 4-in. intervals along the length of the staff. A safety pin secures the locking pin and eliminates accidental slippage of the base and staff. After the post shore has been set in position, a threaded collar with handle permits fine adjustment of the post shore height over a 6-in. range.

To support 4-in. stringers, or 8-in. wide flange stringers, a 4-in. by 8-in. U-head is inserted into the top of the post shore staff. The U-head is inserted through the end plate hole on the shore and attached by a ½-in.-diameter attachment pin and a ⅛-in.-diameter hairpin clip. Nail holes on the side of the U-head provide a convenient means of securing lumber stringers. A steel beam clamp should be used to secure the steel stringers to the U-head.

A 5-in. by 8-in. J-head is also available for quick installment under aluminum beam stringers. The J-head must be pinned to the top of the post shore using the ½-in.-diameter attachment and hairpin clip. A steel beam clamp or aluminum attachment clip assembly should be used to secure the aluminum beam to the J-head.

The Symons timber brace nailer plate allows rapid attachment of bracing when required by the U.S. Occupational Safety and Health Administration. These single-post shores are fully compatible with Symons heavy-duty shoring.

Site Preparation for Shoring

The proposed shoring area should be cleared of all obstructions before erection of shores. Special care should be given to determining the capacity of the soil for every shoring job, to ensure adequate bearing support for shores. Weather conditions can arise that can create a poor ground condition, which may cause a hazardous situation. If fill is required in shored areas, a qualified engineer should be consulted to determine remedial measures that should be taken to ensure safety. A solid concrete foundation pad on grade or a mud sill should be installed to distribute the shoring load over a suitable ground area.

Selecting the Size and Spacing of Shores

This example illustrates a method of determining the spacing of shores. Consider a concrete slab 5 in. thick, supported by concrete beams spaced at 20 ft center to center, as illustrated in Figure 6-4. The beam

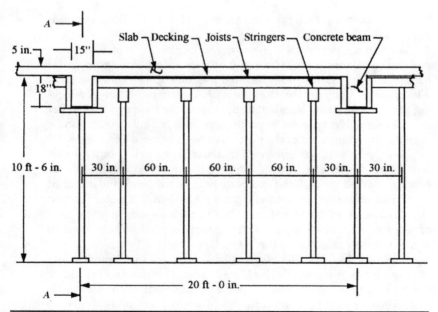

FIGURE 6-4   Formwork for slab and beam.

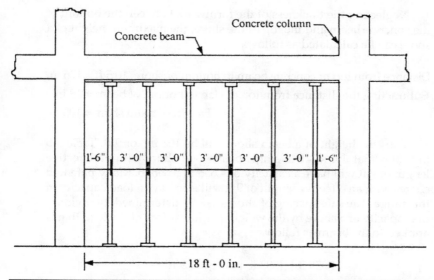

FIGURE 6-5   Spacing of shores for concrete beam (refer to section A–A in Figure 6-4).

stem will be 18 in. deep and 15 in. wide. Thus, the total depth of the beam is 23 in., 18-in. stem, plus the 5-in. slab. The concrete beam is 18 ft 0 in. long, which is the clear span between columns (see Figure 6-5). The clear height from the lower floor to the bottom of the slab will be 10 ft 6 in.

In addition to the weight of the concrete, assume a temporary construction live load of 50 lb per sq ft. Because the shoring for the slab will be placed 2.5 ft, or 30 in., from the centerline of the shoring for the beam, the load on the beam shoring must sustain the live load acting over this distance along the beam form (Figure 6-5).

Consider the 23-in.-deep by 15-in.-wide beam and that portion of the 5-in. slab adjacent to the shores that support the slab. The weight, including the live load that will be applied to the beam shores can be calculated:

Beam weight = [23 in. × 15 in.]/144 × (150 lb per cu ft)
  = 360 lb per lin ft

Slab weight = 5 in. × [30 in. − 15 in.]/144 × (150 lb per cu ft)
  = 78 lb per lin ft

Assume dead load of formwork = 8 lb sq ft × 2.5 ft = 20 lb per lin ft

Live load = 50 lb per sq ft × 2.5 f = 125 lb per lin ft

Total load = 583 lb per lin ft

Total load from beam = 583 lb per lin ft × 18 ft
  = 10,494 lb

Neglecting the thickness of the formwork between the bottom of the concrete beam and the top of the shore, the maximum height of a shore can be calculated as follows:

Distance from top of concrete beam to bottom of shore, 10.5 ft = 126 in.

Subtracting the distance from top of slab to bottom of beam = 18 in.

Height of beam shore = 108 in.

Thus the height of a beam shore will be 108 in., or 9 ft. Table 5-5 indicates that a 4 × 4 S4S single-post wood shore with an effective length of 9 ft will have a capacity of 4,655 to 5,656 lb. Many patented shores with an effective length of 9 ft will also have a load capacity in this range. The safe spacing of shores can be determined by dividing the capacity of a shore by the weight per linear foot of the load that is applied to the beam as follows:

[4,655 lb]/[583 lb per ft] = 7.9 ft

The distance of 7.9 ft is the maximum safe spacing based on the capacity of a shore with an allowable load of 4,655 lb. According to the foregoing calculations, one shore should be placed about every 8.0 ft along the beam. However, this spacing disregards the strength of the beam forms between the shores. Usually the beam bottom is limited to a deflection not to exceed $l/360$. If the beam bottom is made of 2-in. nominal thickness lumber and if any restraining support provided by nailing the beam sides to the beam bottom is neglected, the maximum safe spacing of the shores may be determined from the stress and deflection equations given in Chapter 5. Because the beam bottoms will be installed over multiple supports, the equations for beam 9 in Table 5-2 will apply.

The 15-in. wide beam bottom will be fabricated from 12-in. wide No. 2 grade Southern Pine. Assume a dry condition of lumber and no short-duration load. From Table 4-2, the reference design value is 975 lb per sq in. Because the lumber will be laid flat, the reference bending stress can be increased by 1.2 (refer to Table 4-7). The following allowable stresses for the beam bottom lumber:

Allowable bending stress, F_b = 1.2 (975) = 1,170 lb per sq in.

Allowable shear stress, F_v = 175 lb per sq in.

Modulus of elasticity, E = 1,600,000 lb per sq in.

The width of the beam bottom is 15 in. and the actual thickness is 1.5 in. for 2-in. nominal S4S lumber. The physical properties of the beam bottom will be as follows.

From Eq. (5-9), the section modulus $S = bd^2/6$

$$= (15 \text{ in.})(1.5 \text{ in.})^2/6$$
$$= 5.625 \text{ in.}^3$$

From Eq. (5-8), the moment of inertia $I = bh^3/12$

$$= (15 \text{ in.})(1.5 \text{ in.})^3/12$$
$$= 4.218 \text{ in.}^4$$

The allowable span length based on bending can be calculated as follows.

For bending:

From Eq. (5-34), $l_b = [120F_b S/w]^{1/2}$

$$= [120(1,170)(5.625)/583]^{1/2}$$
$$= 36.8 \text{ in.}$$

For shear:

From Eq. (5-36), $l_v = 192F_s bd/15w + 2d$

$$= 192(175)(15)(1.5)/15(583) + 2(1.5)$$
$$= 89.4 \text{ in.}$$

For deflection:

From Eq. (5-37a), $l_\Delta = [1{,}743EI/360w]^{1/3}$ for $\Delta = l/360$

$$= [1{,}743(1{,}600{,}000)(4.218)/360(583)]^{1/3}$$
$$= 38.2 \text{ in.}$$

Bending governs and the maximum allowable span length for the beam bottom is 36.8 in. Thus, a shore must be placed at a spacing that does not exceed 37.4 in. For constructability, limit the spacing of the shores to 36 in., locating the shores along the beam forms at 3 ft. 0 in. (see Figure 6-5).

Consider the shores to support the slab between the concrete beams. The weight of the 5-in.-thick concrete slab, including live loads, will be:

Dead load, 5/12 ft × 150 lb per cu ft =   62.5 lb per sq ft

Assume dead load of form material =   8.0 lb per sq ft

Live load due to construction = 50.0 lb per sq ft

Total load = 120.5 lb per sq ft

For calculations, use 121 lb per sq ft

The entire weight from the floor slab will be supported by shores, with no portion of this weight transmitted to the beam forms. The height of a shore will be:

Distance from top of concrete slab to bottom of shore, 10.5 ft + 5 in.
= 131.0 in.

Depth of slab, decking, joists, stringers: 5.0 + 1.0 + 7.25 + 7.25
= 20.5 in.

Height of slab shores = 110.5 in.

Thus, the height of a slab shore will be 110.5 in., or slightly over 9 ft. Consider the strength of a 10-ft-high shore. For a 4×4 S4S wood shore with an effective length of 10 ft, Table 5-5 indicates an allowable load of 3,819 to 4,688 lb. The maximum area that can be supported by a 3,819 lb shore can be calculated as follows:

[3,819 lb]/[121 lb per sq ft] = 31.5 sq ft.

The shores for the floor slab will be spaced 5 ft apart in rows which are spaced 3 ft apart (refer to Figures 6-4 and 6-5). This spacing of shores provides a total area of 3 ft × 5 ft = 15 sq ft, which is less than the allowable 31.5 sq ft. This spacing of shores provides a margin of safety of 31.5/15 = 2.1 for the shoring. The decking, joists, and stringers must have adequate strength to transmit the loads from these components of the formwork to the slab shores. The methods of determining the size and spacing of joists and stringers have been discussed in Chapter 5 and are discussed further in Chapter 11.

Tubular Steel Scaffolding Frames

Tubular steel scaffolding, as illustrated in Figure 6-6, has several advantages over single-post shores for supporting loads from concrete, including the following:

1. A two-frame section, with X-braces between frames, will stand without additional supports.

2. A choice of frame heights, with adjustable screw jacks on the legs will permit easy and fast adjustment to the desired exact height.

3. A choice of lengths of diagonal braces will permit variations in the spacing of frames over a wide range.

4. The use of screw jacks on the bottom of the legs will compensate for uneven floor conditions.

5. It provides better safety for workers.

6. It can be used for purposes other than shoring formwork.

FIGURE 6-6 Tubular steel scaffolding to support formwork. (*Source: Symons Corporation*)

Most types of steel scaffolding can be either rented or purchased. Due to the high initial investment cost of scaffolding, its purchase is not economically justifiable unless the scaffolding will be used a great many times and at frequent intervals. Many contractors buy a reasonable quantity of scaffolding and rent additional units to suit the needs of a particular job.

Commercial frames are usually available in several widths, varying from 2 to 5 ft for standard size frames, depending on the manufacturer. Each frame consists of two legs with heights that vary from 2 to 6 ft per frame, in increments of 1 ft. Up to three frames can be stacked on top of each other to obtain a three-tier shoring system.

The rated safe load capacity is generally around 10,000 lb per leg, or 20,000 lb per frame. The load capacity will increase or decrease depending on the actual number of tiers used and the total extension of the jacks. The manufacturer should be consulted to determine the safe load capacity for each job application.

Diagonal cross braces are available in lengths which are multiples of 1 ft, such as 3, 4, 5, 6, or 7 ft, to brace frames. Locking devices are used to securely attach diagonal braces. Screw jacks allow fine adjustments of the height of the frames. Figure 6-7 illustrates standard shore frames available from the Symons Corporation.

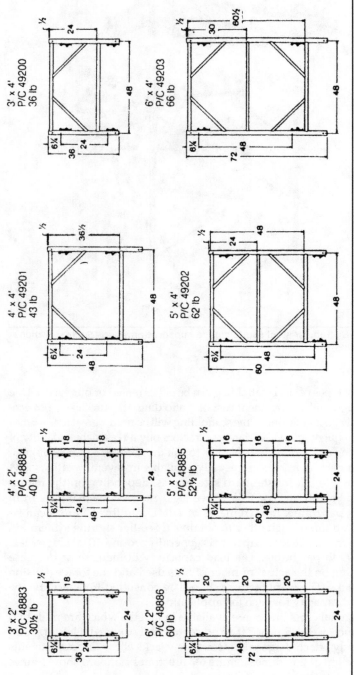

Figure 6-7 Standard size frames for shoring. (*Source: Symons Corporation*)

176

Accessory Items for Tubular Scaffolding

The following accessory items are available for use with tubular scaffolding:

1. Base plates to provide solid footing for the scaffolding
2. Adjustable bases and screw jacks for adjustment to exact height
3. Locking devices to attach tube braces between adjacent scaffolds
4. Diagonal cross braces to provide rigidity to the scaffold assembly
5. Coupling pins to connect two frames vertically
6. U-shaped shore heads for attachment of wood or metal stringers

Figure 6-8 illustrates typical types of connections that are available for scaffolding of Patent Construction Systems, and Figure 6-9 presents typical accessories available for steel scaffolding frames of the Symons Corporation. Figure 6-10 shows a jobsite with tubular steel frames in place.

Steel Tower Frames

Commercial steel tower frames are available to support high and heavy loads, such as bridge piers or heavy beams. Tower frames are erected in sections, each consisting of four legs with heights of 8 or 10 ft. Each section is box shaped, fabricated from two welded frames and two X-welded frames, with equal dimensions in the horizontal and vertical directions. Up to four additional sections may be added to obtain a maximum height of 40 ft for a four-tier tower. Each tower section is securely fastened by frame couplers at each leg. A horizontal brace, adjusted by a turnbuckle, is used to brace adjacent sections of the tower. Tower frames are available with working load capacities of up to

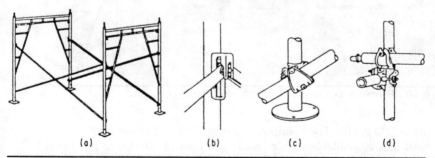

(a)   (b)   (c)   (d)

FIGURE 6-8 Illustrative connections for scaffolding. (a) Frame assembly. (b) SideLoks. (c) Coupler for diagonal connections. (d) Coupler for right-angle connection. (*Source: Patent Construction Systems*)

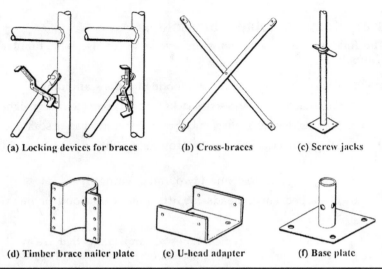

(a) Locking devices for braces   (b) Cross-braces   (c) Screw jacks

(d) Timber brace nailer plate   (e) U-head adapter   (f) Base plate

FIGURE 6-9 Typical accessories for tubular steel scaffolding frames. (*Source: Symons Corporation*)

FIGURE 6-10 Formwork supported by tubular steel frames on a job.

100,000 lb per leg. For example, a 10-ft by 10-ft by 10-ft-high tower with four legs is available with a maximum load of 400,000 lb per tower frame. An 8- by 8- by 8-ft-high tower is also available with a 50,000-lb working load per leg, or 200,000 lb per frame. Figure 6-11 illustrates a tower frame available from the Economy Forms Corporation (EFCO).

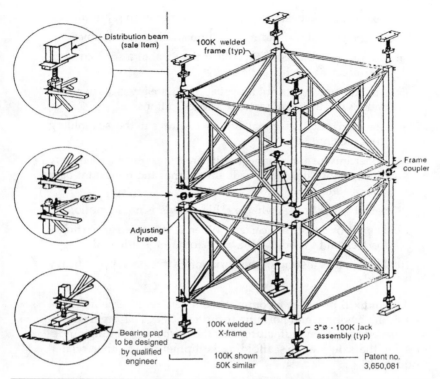

Distribution beam
(sale Item)

100K welded
frame (typ)

Frame
coupler

Adjusting
brace

Bearing pad
to be designed
by qualified
engineer

100K welded
X-frame

100K shown
50K similar

3"ø - 100K jack
assembly (typ)

Patent no.
3,650,081

Figure 6-11   Section of steel tower frame. (*Source: EFCO*)

The fine adjustment in the height of tower frames is made with screw jacks installed in the bottom or top, or both, of all legs. Also, the lower jack or adjustable base is used to compensate for uneven ground or floor conditions. In general, the bottom jacks should be used to level and plumb the scaffolding, and the top jacks should be used for exact adjustments in height.

If the scaffolding is to be moved from one location to another on a project, casters, which support the scaffolding only while it is being moved, may be attached to the bottoms of the legs. An entire tower frame can be moved as a single unit with a crane.

Safety Practices Using Tubular Scaffolding

As for all construction operations, there are certain practices which should be observed when using tubular scaffolding to support form-work. These include the following:

1. Install sufficient sills or underpinning in addition to base plates on all scaffolding to provide adequate support from the ground.

2. Use adjustable bases to compensate for uneven ground.

3. Ensure that all scaffolding is plumb and level at all times.

4. Do not force braces to fit. Adjust the level of a scaffold until the proper fit can be made easily.

5. Anchor running scaffolding at appropriate intervals in accordance with the manufacturer's specifications.

6. Use horizontal diagonal braces to prevent the scaffolding from racking.

7. Either guy or interconnect with braces to other towers, any tower whose height exceeds three times the minimum base dimension.

8. Set all caster brakes when a rolling tower is not in motion.

9. Do not overload the casters. When a tower is supporting a load, extend the adjustable bases to take the full load.

10. Use the legs instead of the horizontal struts to support heavy loads.

Appendix B provides recommended safety requirements of the Scaffolding, Shoring, and Forming Institute (SSFI). This organization was formed by U.S. manufacturers for the purpose of setting guidelines for the safe use of scaffolding equipment and other related products.

Horizontal Shores

Adjustable horizontal steel shores, consisting of two members per unit, are frequently used to support the forms for concrete slabs, beams, and bridge decking. The members are designed to permit one member to telescope inside the other, thereby providing stepless adjustments in the lengths within the overall limits for a given shore. Horizontal flanges or prongs projecting from the ends of the shores rest on the supporting beams, stringers, or walls. They are available for spans varying from about 5 to 20 ft.

Horizontal shores can be used as joists directly under the plywood sheathing, which eliminates the requirement for lumber. The units are manufactured with adjustable built-in camber setting devices to compensate for deflection under loading.

Among the advantages of horizontal shores are the following:

1. They are relatively light in weight.

2. The adjustable lengths, which permit a wide range of uses, reduce the need for excessive inventory.

3. The built-in camber permits adjustment for deflection under loading.

4. They may be installed and removed rapidly.

5. Their use reduces the need for vertical shores, thus leaving the lower floor area free for other construction work or for the storage of materials.

Figure 6-12 shows adjustable-length horizontal shores that are available from the Symons Corporation. Additional information on horizontal shores is given in Chapter 12.

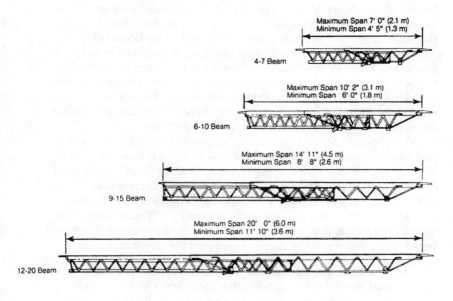

FIGURE 6-12 Adjustable-length horizontal steel shores for forming concrete slabs. (*Source: Symons Corporation*)

Shoring Formwork for Multistory Structures

When concrete is placed for the beams or slabs for a multistory building, the concrete is supported by decking, joists, stringers, and shores, which in turn must be supported by the concrete previously placed for the lower floors, some of which has not gained sufficient strength to support its floors plus the additional load from the shores resting on it. Thus, it is necessary to provide a safe method of support that will not endanger the lower concrete floors. If vertical posts or steel frames are used to support the decking system, they usually rest on a lower floor, and that floor should be supported by shores correctly located on the floors below it.

When constructing a multistory building, it is desirable to know how many lower floors should be supported by shores that are left in place or are placed there by a reshoring operation. Frequently, this decision is left to the superintendent, who may, in his or her desire to economize by reusing the lower forms before it is safe to remove them, cause a failure in one or more of the floor slabs.

The question of how long the original shores should be left in place, how many reshores are required, and how long the reshores should be left in place, varies with each project. Each structure must be evaluated by the form designer to ensure safety of workers and compliance with code requirements. This topic is discussed in "Formwork for Concrete," published by the American Concrete Institute [1]. The following are some of the factors that should be considered in determining the number of floors that should be shored to support the construction loads:

1. Structural design load of slab or members, including live load, partition loads, and other loads for which the engineer designed the slab. Also, any variations in live load that the engineer may have used in design

2. Dead load weight of concrete and formwork

3. Construction live loads such as placing crews, equipment, and materials storage

4. Design strength of concrete specified

5. Cycle time between placements of successive floors

6. Strength of concrete at time it is required to support shoring loads from above

7. Distribution of loads among floor slabs, shores, and reshores or backshores at the time of placing concrete, stripping formwork, and removal of reshoring and backshoring

8. Span of slab or structural member between permanent supports

9. Type of formwork systems; that is, span of horizontal formwork components, individual shore loads, etc.

10. Minimum age where appropriate

The form designer must perform an analysis of all loads, including construction loads, form loads, and dead loads of slabs based on the number of stories of shoring and reshoring. The results of the analysis must be compared to the allowable load on the slabs based on the ultimate design load capacity, multiplied by the percentage of strength and divided by a safety factor. The comparison of the load analysis to the allowable load must be performed at any stage during construction. The local punching shear in the slab under heaviest shore load must also be checked.

Numerous repetitive calculations are required to analyze shoring. The Scaffolding, Shoring, and Forming Institute offers a computer program for execution on a personal computer for performing such calculations.

Thus, shoring and reshoring of multistory structure represents special conditions, particularly in relation to removal of forms and shores. The decision regarding the safe condition for removal of shores should be made by a competent engineer-architect in cooperation with the structural designer of the concrete structure.

References

1. Dayton Superior Corporation, Parsons, KS.
2. EFCO, Economy Forms Corporation, Des Moines, IA.
3. Ellis Construction Specialties, Ltd, Oklahoma City, OK.
4. Patent Construction Systems, Paramus, NJ.
5. Safway Steel Products, Milwaukee, WI.
6. Scaffolding, Shoring, and Forming Institute, Inc., Cleveland, OH.
7. Symons Corporation, Des Plaines, IL.
8. Scaffolding & Shoring Institute, Cleveland, OH.
9. X. L. Liu, H. M. Lee, and W. F. Chen, "Analysis of Construction Loads on Slabs and Shores by Personal Computer," Concrete International: Design and Construction, Vol. 10, No. 6, Detroit, MI, 1988.
10. "Formwork for Concrete," Special Publication No. 4, 7th ed., American Concrete Institute, Detroit, MI, 2004.

CHAPTER 7

Failures of Formwork

General Information

The failure of formwork is always embarrassing, expensive, and a sad situation for everyone involved in a project. Safety must be a concern of everyone, workers and supervisors. Accidents not only affect the workers but also their families. Everyone in a project must be alert to unsafe conditions, and all work must be performed in accordance with safety regulations and the requirements specified in the form design.

A failure may result in a collapse of part or all of the forms. Also it may result in a distortion or movement of the forms that will require the removal and replacement of a section of concrete. Sometimes repairs, such as expensive chipping and grinding operations, may be required to bring the section within the specified dimension limitations. Such failures should not and will not occur if the formwork is constructed with adequate strength and rigidity.

Formwork should be designed by an engineer or by someone who has sufficient knowledge of forces and resistance of form materials. Some states and cities require that forms, other than the simplest concrete structures, be designed by registered professional engineers and that following their erection they be inspected for adequacy by an engineer.

Causes of Failures of Formwork

There are numerous causes of failures of formwork. Following is a partial list:

1. Improper or inadequate shoring
2. Inadequate bracing of members
3. Lack of control of rate of concrete placement

4. Improper vibration or consolidation of concrete

5. Improper or inadequate connections

6. Improper or inadequate bearing details

7. Premature stripping of formwork

8. Errors in placement of reshoring

9. Improper, or lack of, design of formwork

10. Inadequate strength of form material

11. Failure to follow codes and standards

12. Modifications of vendor-supplied equipment

13. Negligence of workers or supervisors

In many instances the failure of formwork is a result of improper or inadequate shoring for slabs and roofs. Because shoring systems constitute the greatest danger of form failures, special care should be exercised in designing and inspecting the shoring system for a concrete structure.

Designers and constructors of formwork are responsible for compliance with the local, state, and federal codes applicable to formwork. The Occupational Safety and Health Act (OSHA), Subpart Q, provides safety regulations for formwork [1]. The "Building Code Requirements for Reinforced Concrete," produced by ACI Committee 318 of the American Concrete Institute, also provides requirements applicable to formwork [2]. ACI Committee 347 publishes "Guide to Formwork for Concrete," which contains formwork standards [3].

Traditionally, the responsibility for formwork has been with the construction contractor. Although codes and standards are developed to ensure safety related to construction, it is the duty of all parties involved in a project to achieve safety. This includes form designers, engineer-architects, workers, supervisors, and inspectors.

Forces Acting on Vertical Shores

The forces acting on vertical shores include dead and live loads plus impact. Dead loads include the weight of the concrete plus the weight of any construction materials that are stored on the concrete slab while the shores are in place. Live loads include the weights of workers and construction equipment that move around over the decking or the concrete as it is being placed, or soon thereafter. Although the magnitudes of these loads may be determined or estimated with reasonable accuracy, it appears that the effect of the impact of moving equipment or falling concrete may have been underestimated in some instances. This refers to concrete that is discharged from crane-handled buckets, pumping of concrete, and motor-driven concrete buggies.

Force Produced by Concrete Falling on a Deck

Deck forms that support fresh concrete are designed to withstand the weight of the concrete plus an additional load, frequently referred to as a live load, resulting from workers, concrete buggies, and materials that may be stored on the concrete slab before it attains full strength. Where concrete is placed with a bucket or by pumping from a concrete pump truck, consideration should be given to the additional pressure or force resulting from the sudden reduction in velocity as the concrete strikes the deck.

Figure 7-1 represents a condition where concrete flows from a bucket suspended above a deck and strikes either the deck or the top surface of fresh concrete already on the deck. Assume that the velocity of the concrete is V_2 at point 2 and the velocity V_3 is zero at point 3, a short distance below point 2.

For this condition, the following symbols will apply:

W = original weight of concrete in bucket, lb
h = height of fall from point 1 to point 2, ft
y = distance between points 2 and 3, ft
T = time required to empty bucket at uniform rate of flow, sec
w = weight of falling concrete, lb per sec, $= W/T$
$m = w/g = w/32.2 = W/32.2T$
a = acceleration or deceleration of concrete between points 2 and 3, ft per sec^2
t = time required for a given particle of concrete to travel from point 2 to point 3, sec

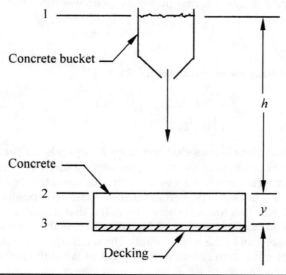

FIGURE 7-1   Force produced by dropping concrete on decking.

g = acceleration of gravity = 32.2 ft per sec^2
V_2 = velocity of concrete at point 2
V_3 = velocity of concrete at point 3
$V_2 = \sqrt{2gh}$
$V_3 = 0$

The average velocity between points 2 and 3 is $(V_2 + V_3)/2 = V_2/2$.

$$
\begin{aligned}
t &= y/[(V_2)/2] \\
&= 2y/V_2 \\
a &= (V_2 - V_3)/t \\
&= (V_2 - 0)/t \\
&= V_2/t \\
&= \left(\sqrt{2gh}\right)/t
\end{aligned}
$$

Only that mass of concrete between points 2 and 3 will be undergoing a deceleration at a given time. Only this mass need be considered in determining the force resulting from deceleration. The weight of this concrete will be:

$$
\begin{aligned}
w' &= wt \\
m &= wt/g \\
&= Wt/32.2T
\end{aligned}
$$

When a given mass of concrete undergoes a change in velocity between points 2 and 3, designated as an acceleration, the resulting force F is equal to:

$$F = ma$$

Substituting for m and a, we get: 2×8

$$
\begin{aligned}
F &= (Wt/32.2T) \cdot \left(\sqrt{2gh}\right)/t \\
&= \left[W\sqrt{2gh}\right]/32.2T \text{ lb}
\end{aligned}
\tag{7-1}
$$

Equation (7-1) shows that the force F depends on the original weight of the concrete in the bucket, the vertical distance from the top surface of the concrete in the bucket to the deck, and the time required to empty the bucket. The vertical distance between points 2 and 3 does not affect this force. It should be noted that the force is directly related to the rate of emptying a bucket, whereas the force is related to the square root of the height of fall. Thus, reducing the rate of flow is more effective than reducing the height of fall if the force is to be reduced.

Example 7-1

Determine the resulting force when a bucket containing 3,000 lb of concrete is emptied in 5 sec with a maximum height of fall of 4 ft. Applying Eq. (7-1), we get:

$$F = \left[W\sqrt{2gh} \right]/32.2T$$
$$= \left[3,000 \text{ lb}\sqrt{2(32.2 \text{ ft per sec}^2)(4 \text{ ft})} \right]/(32.2 \text{ ft per sec}^2)(5 \text{ sec})$$
$$= 300 \text{ lb}$$

If this force acts on an area of 2 sq ft of deck, the resulting increase in pressure, equivalent to an increase in weight, will average 150 lb per sq ft. Assume that the decking is 1-in. nominal thickness lumber supported by 2×8 S4S joists spaced 24 in. apart. The effect of this force on the decking lumber will be a temporary increase in the deflection and in the bending moment in the decking. Assume that the entire force is applied to a strip of decking 12 in. wide and 24 in. long, all of which is located between two joists. Consider this as a simple beam 12 in. wide, ¾ in. deep, and 24 in. long, subjected to a uniform load of 150 lb per lin ft.

The bending moment for a simple span beam with a uniform load will be:

$$\text{From Eq. (5-5), } M = wl^2/96 \text{ in.-lb}$$
$$= 150 \text{ lb per ft } (24 \text{ in.})^2/96$$
$$= 900 \text{ in.-lb} \tag{a}$$

Rearranging the terms in Eq. (5-10), the resisting moment will be:

$$M = f_b bd^2/6$$
$$= f_b(12 \text{ in.})(3/4 \text{ in.})^2/6$$
$$= 1.125 f_b \tag{b}$$

Equating (a) and (b),

$$1.125 f_b = 900$$
$$f_b = 800 \text{ lb per sq in.}$$

Thus, for the stated conditions the decking will be subjected to a temporary additional fiber stress in bending equal to 800 lb per sq in. This is a substantial increase in stress for lumber. However, at the time this impact force acts, the assumed live load, in excess of the weight of the concrete slab, is not acting on the slab, which provides some factor of safety, or relief.

Consider the effect this force may have in increasing the unit stress in horizontal shear in a joist. Assume that the full force acts on the deck at one end of a single joist. This will produce a temporary increase of approximately 300 lb in the vertical shear in that end of the joist.

$$V = 300 \text{ lb} \tag{a}$$

Rearranging the terms in Eq. (5-13), the shear force will be:

$$V = 2f_v bd/3 \tag{b}$$

Equating (a) and (b), we get:

$$2f_v bd/3 = 300$$
$$f_v = 3 \times 300/2bd$$
$$= 3(300 \text{ lb})/[2 \ (1\frac{1}{2} \text{ in.}) \times (7\frac{1}{4} \text{ in.})]$$
$$= 41.4 \text{ lb per sq in.}$$

Motor-Driven Concrete Buggies

The current method of placing concrete is with buckets from a crane or pumping from a concrete pump truck. However, in the past, motor-driven concrete buggies were frequently used for placing concrete. Although concrete buggies are not commonly used today, the material contained in this section is presented for those isolated situations where buggies may be used.

The weight of a loaded buggy can be as much as 3,000 lb. Some states limited the speed of buggies to a maximum 12 mi/hr. If the buggies are stopped quickly, substantial horizontal forces can be produced. It is possible that the thrusts resulting from these horizontal forces can exceed the resisting strength of horizontal and diagonal bracing systems.

Consider the horizontal force that may be caused by a loaded motor-driven buggy that is stopped quickly, based on the following assumptions and conditions:

Weight of loaded buggy = 3,000 lb

Maximum speed, 10 mi/hr = 14.7 ft per sec

Assume buggy is stopped in 5 sec

The resulting force is given by the following equation:

$$F = Ma$$
$$= Wa/g \qquad (7\text{-}2)$$

where F = average horizontal force, lb
M = mass of loaded buggy, W/g
g = acceleration of gravity, 32.2 ft per sec^2
a = average acceleration or deceleration of buggy, ft per sec^2

a = (14.7 ft per sec)/(5 sec)
= 2.94 ft per sec^2

$F = Wa/g$
= [(3,000 lb)(2.94 ft per sec^2]/(32.2 ft per sec^2)
= 274 lb

If this buggy is stopped in 3 sec, the horizontal force will be 457 lb. The forces produced by various speeds and stopping times are tabulated in Table 7-1.

The horizontal forces listed in Table 7-1 are for one buggy. If more than one buggy will be stopped at the same time, the force listed should be multiplied by the number of buggies in order to obtain the probable total force on the formwork.

If motor-driven concrete buggies are to be used in placing the concrete for slabs, the formwork, especially the vertical shoring system, must be constructed to resist the horizontal forces that will act on the forms. It will be necessary to install two-way horizontal braces to reduce the unsupported lengths of the shores. In addition, two-way diagonal braces should be installed, extending from the tops to the bottoms of shores to resist the horizontal forces resulting from the operation of the buggies. Also, the tops of the shores should be securely fastened to the stringers which they support.

Impact Produced by Motor-Driven Concrete Buggies

Although it is difficult or impossible to calculate accurately the effect of the impact on formwork caused by motor-driven concrete buggies, it is known that the effect may be substantial, especially when the buggies are traveling at relatively high speeds. It is possible that the combined weight and impact of a buggy on a given span may lift a stringer in an adjacent span off a shore, thereby permitting the shore to shift its position from under the stringer unless the two members are securely connected together.

If the shores consist of two tiers, it is possible that the bottoms of the top tier of shores may have moved with respect to the lower system of supporting shores, thereby producing unstable supports for the top shores. If the forces resulting from the movement of equipment, such as motor-driven concrete buggies, or the forces resulting from any other kind of load or impact are likely to produce an uplift

Maximum Speed, mi/hr	Time to Start or Stop a Buggy, sec	Rate of Acceleration or Deceleration, ft/sec²	Horizontal Force Produced, lb
4	2	2.94	274
	4	1.47	137
	6	0.98	91
	8	0.74	69
	10	0.59	55
5	2	3.66	341
	4	1.83	171
	6	1.22	114
	8	0.92	86
	10	0.73	68
6	2	4.40	410
	4	2.20	205
	6	1.47	137
	8	1.10	102
	10	0.88	82
8	2	5.88	548
	4	2.94	274
	6	1.96	183
	8	1.47	137
	10	1.17	109
10	2	7.34	683
	4	3.67	342
	6	2.45	228
	8	1.83	171
	10	1.47	137
12	2	8.82	822
	4	4.41	411
	6	2.94	274
	8	2.20	205
	10	1.76	164
	12	1.47	137

Note:
The specified horizontal forces are based on a buggy whose total loaded weight is 3,000 lb. For any other weight, the force may be obtained by the quotient of its weight divided by 3,000 lb. For example, if the total weight of a loaded buggy is 2,000 lb, its maximum speed is 8 mi/hr, and it is stopped in 4 sec, the horizontal force will be $2,000/3,000 \times 274 = 183$ lb.

TABLE 7-1   Horizontal Forces Caused by Starting and Stopping Loaded Motor-Driven Concrete Buggies

on the decking or shoring system, all form members should be connected together in a manner that will prevent the members from separating or moving with respect to each other.

Design of Formwork to Withstand Dynamic Forces

The analyses presented in the previous sections of this chapter show that formwork may be subjected to dynamic forces in addition to those resulting from static loads. The magnitudes of dynamic forces on formwork will depend on the methods of placing the concrete. When these types of forces are anticipated, the strength and rigidity of the formwork should be increased adequately by additional bracing, increased strength of the form material, and stronger connections between adjoining members.

Examples of Failure of Formwork and Falsework

The following examples briefly describe several failures of formwork and falsework during the erection of concrete bridges. This information is presented to emphasize the need to exercise care in designing, erecting, using, and removing formwork or falsework for concrete structures.

Canada Larga Road Bridge: During the construction of this 88-ft-long structure, a falsework frame collapsed as it was being adjusted to grade, causing several injuries and a substantial delay in the completion of the project (see Figure 7-2).

Oceanside Bridge: The span for this bridge, a 127-ft-long box girder, had been poured, and the forms had been removed with only the

Figure 7-2 This falsework collapsed while being adjusted to grade.

Figure 7-3 Steel-pipe scaffolding fell while concrete was being placed at a span over a canyon.

steel beams and the falsework bents, which supported the forms, to be removed. During the removal of one of the steel beams, something slipped, and the other steel beams moved out of position, causing the failure. Several persons were injured.

San Bruno Bridge: This failure occurred during the construction of a concrete bridge over a railroad. The bridge was designed to use prestressed concrete beams 95 ft long to span the railroad. The ends of the beams were supported temporarily during erection by two falsework bents, one on each side of the railroad. The permanent supports for the beams were to be concrete girders cast later on top of permanent bents and to enclose the ends of the precast beams in the concrete for the girders. During the placing of the precast beams, they rolled over on their sides and, because they were not designed to be self-supporting in this position, they collapsed and fell onto the railroad tracks. Because prestressed concrete beams are used extensively in concrete structures, it is imperative that constructors use effective procedures to prevent the beams from rolling over during erection.

Arroyo Seco Bridge: During the construction of this bridge, the 230-ft-long center span was supported on demountable towers made of steel pipe. Some of the main beams that rested on the towers and supported the forms for the concrete structure were previously used steel sections that may have been damaged and weakened prior to use on this project. This failure resulted in six fatalities and several others were injured. Figure 7-3 shows the project after the failure.

Prevention of Formwork Failures

The safety of workers is a concern of all parties: owners, designers, and contractors. Safety is everyone's responsibility, including workers in the field, supervisors, and top management. There are many risks in the process of erecting and dismantling forming systems. Every precaution should be taken to ensure a safe working environment. Below is a partial list of rules that can be used to reduce the potential of formwork failures.

1. Prepare a formwork plan that includes detailed drawings and written specifications for fabricating, erecting, and dismantling of the formwork. The plan should be prepared by a person who is competent in the design of formwork.

2. Follow all state, local, and federal codes, ordinances, and regulations pertaining to formwork, shoring, and scaffolding.

3. Post guidelines for shoring and scaffolding in a conspicuous place and ensure that all persons who erect, dismantle, or use shoring are aware of them (refer to Appendix B).

4. Ensure compliance of all OSHA rules and regulations (refer to Appendix C).

5. Follow all instructions, procedures, and recommendations from manufacturers of formwork components used in the formwork.

6. Survey the jobsite for hazards, such as loose earth fills, ditches, debris, overhead wires, and unguarded openings.

7. Ensure adequate fall protection for workers during erection of formwork, pouring of concrete, and dismantling of formwork.

8. Inspect all shoring and scaffolding before using it, to ensure it is in proper working condition and to ensure workers are using the equipment properly.

9. Make a thorough check of the formwork system after it is erected and immediately before a pour, in particular connections between formwork components.

10. Never take chances. If in doubt regarding the safety, contact a safety officer and management. It is best to prevent an accident.

References

1. Occupational Safety and Health Standards for the Construction Industry, Part 1926, Subpart Q: Concrete and Masonry Construction, U.S. Department of Labor, Washington, DC.
2. Building Code Requirements for Reinforced Concrete, ACI Committee 318, American Concrete Institute, Detroit, MI.

CHAPTER 8

Forms for Footings

General Information

In this book, concrete footings and foundations are defined as those components of structures, relatively low in height, whose primary functions are to support structures and equipment. In general, they include:

1. Wall footings and low foundation walls
2. Column footings
3. Footings for bridge piers
4. Foundations for equipment

Footings usually are of considerable length, including those that extend around a building, plus those that extend across a building. They are constructed in excavated ground trenches. If the earth adjacent to the footing is sufficiently firm, a trench can be excavated to the desired width, and the earth can serve as the side form.

The pressure, which will vary from zero at the top of the forms to a maximum value at the bottom, may be determined as discussed in Chapter 3. Because the forms are filled rapidly, within an hour or less, the maximum pressure will be 150 lb per sq ft times the depth in feet.

The forces which the forms must resist are the lateral pressure from the concrete plus any uplift pressure that may result where the footings are constructed with sloping or battered sides and tops. Because the depth of concrete usually is small, the pressure on the forms will be relatively low. Materials used for forms include lumber, plywood, hardboard, steel, and fiber tubes.

Forms for Foundation Walls

Foundation walls are constructed on and along wall footings, as illustrated in Figure 8-1. Because of their low heights, usually varying from 2 to 6 ft, the pressure on the forms will be less than that on forms

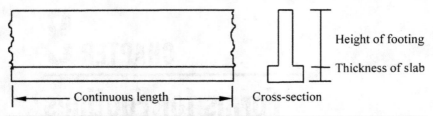

FIGURE 8-1   Concrete footing and foundation wall.

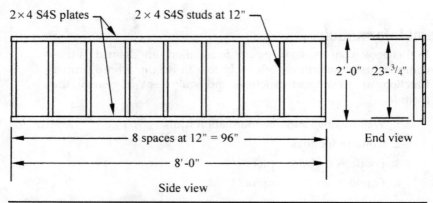

FIGURE 8-2   All-wood form panels for concrete footings.

for major walls. Job-built or patented form panels may be used. Panels 2 ft wide, in lengths that are multiples of 2 ft, up to 8 ft, will permit the use of 4- by 8-ft sheets of Plyform without waste. It may be desirable to provide some odd-length panels for use on walls whose lengths are not multiples of 2 ft.

Figure 8-2 shows views of a job-built panel that uses ¾ 1 × 4-in. Plyform for the facing and 2 × 4 S4S lumber for the plates and studs. It will be noted that the edges of the Plyform are flush with the exterior surfaces of the two end studs. However, the top and bottom edges of the plywood extend ⅛ in. above and below the exterior top and bottom surfaces of the 2 × 4 plates. This will permit fast and easy placing of spreader-type form ties through the wall and the forms.

Example 8-1

Design the forms for a 4-ft-high foundation wall using the panel forms shown in Figure 8-2. The foundation wall will be formed by stacking two of the form panels in the vertical direction, shown as follows.

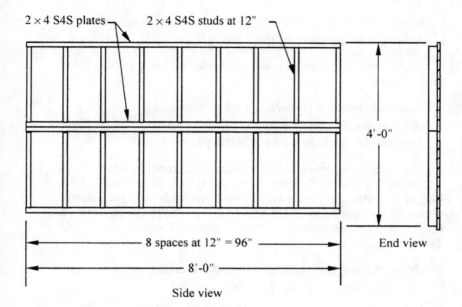

The following conditions will apply:

Maximum depth of concrete, 4 ft
Concrete weight, 150 lb per cu ft
Sheathing, ¾-in. Plyform Class I
Lumber, 2 × 4 No. 2 grade Douglas Fir-Larch
Assume dry condition of lumber
Assume short-duration load, less than 7 days
Permissible deflection < $l/360$, not to exceed ⅛ in.
Tie strength, 4,000 lb

The type of cement, temperature, and rate of placement are not stipulated; therefore, use Eq. (3-1) to determine the concrete pressure on the bottom form panel.

From Eq. (3-1), $P_m = wh$
$$= (150 \text{ lb per cu ft})(4 \text{ ft})$$
$$= 600 \text{ lb per sq ft}$$

Determine the maximum safe spacing of studs, based on using ¾-in. Class I Plyform. From Table 4-11, the physical properties per foot of width for the face grain placed across support studs will be:

Approximate weight = 2.2 lb per sq ft
Effective section modules, $S_e = 0.455$ in.[3]

Moment of inertia, $I = 0.199$ in.^{4}

Rolling shear constant, $Ib/Q = 7.187$ in.^{2}

From Table 4-12 the indicated allowable stresses for Class I Plyform will be:

Allowable bending stress, $F_b = 1,930$ lb per sq in.

Allowable rolling shear stress, $F_s = 72$ lb per sq in.

Modulus of elasticity, $E = 1,650,000$ lb per sq in.

The maximum pressure of the freshly placed concrete on the forms will be 600 lb per sq ft. Consider a strip of Plyform 12 in. wide and 8 ft long at the bottom of the forms. For the 12-in.-wide strip the value of uniformly distributed load will be $w = 600$ lb per lin ft. The pressure on the Plyform will occur over multiple supports; therefore, Beam 9 in Table 5-2 will apply.

Determine the spacing of studs based on bending of sheathing.

$$\text{From Eq. (5-41), } l_b = [120F_b S_e / w_b]^{1/2}$$
$$= [120(1,930)(0.455)/600]^{1/2}$$
$$= 13.2 \text{ in.}$$

Determine the spacing of studs based on rolling shear of sheathing.

$$\text{From Eq. (5-48), } l_s = 20F_s(Ib/Q)/w_s$$
$$= 20(72)(7.187)/600$$
$$= 17.2 \text{ in.}$$

Determine the spacing of studs based on a maximum deflection of $l/360$, but not greater than $\frac{1}{16}$ in.

$$\text{From Eq. (5-57a), } l_d = [1,743EI/360w_d]^{1/3} \text{ for } \Delta = l/360$$
$$= [1,743(1,650,000)(0.199)/360(600)]^{1/3}$$
$$= 13.8 \text{ in.}$$

$$\text{From Eq. (5-57b), } l_d = [1,743EI/16w_d]^{1/4} \text{ for } \Delta = \frac{1}{16} \text{ in.}$$
$$= [1,743(1,650,000)(0.199)/16(600)]^{1/4}$$
$$= 15.6 \text{ in.}$$

Summary for ¾-in. Plyform Sheathing

For bending, the maximum spacing of studs $= 13.2$ in.

For shear, the maximum spacing of studs $= 17.2$ in.

For deflection, the maximum spacing of studs $= 13.8$ in.

The maximum span length of the ¾-in. Plyform sheathing is governed by bending at 13.2 in. For constructability, choose 12 in.

Therefore, the maximum spacing of studs will be 12 in. center to center, or 1 ft.

Determine the maximum spacing of form ties based on the horizontal concrete pressure where the two panels are joined: the 2×4 top plate of the bottom panel and the bottom plate of the top panel. Thus, two plates will resist the pressure along this row of form ties. When the forms are filled with concrete, the depth at this section will be 2 ft from the top of the form. The maximum pressure from Eq. (3-1) will be 2 ft $\times$ 150 lb per cu ft = 300 lb per sq ft. Consider a horizontal strip 12 in. above and 12 in. below this section. The lateral pressure on the form will be 2 ft $\times$ 300 lb per sq ft = 600 lb per lin ft.

The uniformly distributed load of 600 lb per lin ft is resisted by 2 plates. Check the adequacy of plates for bending, shear and deflection. The physical properties of 2×4 lumber can be obtained from Table 4-1 as follows:

Area, $A = 2(5.25) = 10.5$ in.^{2}
Section modulus, $S = 2(3.06) = 6.12$ in.^{3}
Moment of inertia, $I = 2(5.36) = 10.72$ in.^{4}

Referring to Tables 4-3, 4-3a, and 4-4 for 2×4 S4S No. 2 grade Douglas Fir-Larch lumber adjusted for size and short-load duration the allowable stresses will be:

Bending stress, $F_b = C_F \times C_D \times$ (reference design value for bending)
$$= 1.5(1.25)(900 \text{ lb per sq in.})$$
$$= 1,687 \text{ lb per sq in.}$$

Shear stress, $F_v = C_D \times$ (reference design value for shear)
$$= 1.25(180 \text{ lb per sq in.})$$
$$= 225 \text{ lb per sq in.}$$

Modulus of elasticity, $E = 1,600,000$ lb per sq in.

The allowable span lengths for bending, shear, and deflection in the double 2×4 plates can be calculated as follows:

Bending in the double plates:

Eq. (5-34), $l_b = [120 F_b S / w]^{1/2}$
$$= [120(1,687)(6.12)/600]^{1/2}$$
$$= 45.4 \text{ in.}$$

Shear in the double plates:

Eq. (5-36), $l_v = 192 F_v b d / 15w + 2d$
$$= 192(225)(10.5)/15(600) + 2(3.5)$$
$$= 57.4 \text{ in.}$$

Deflection in the double plates for $\Delta = l/360$:

Equation (5-37a), $l_\Delta = [1{,}743EI/360w]^{1/3}$

$$= [1{,}743(1{,}600{,}000)(10.72)/360(600)]^{1/3}$$

$$= 51.7 \text{ in.}$$

Deflection in the double plates for $\Delta = \frac{1}{16}$ in.:

Equation (5-37b), $l_\Delta = [1{,}743EI/80w]^{1/4}$

$$= [1{,}743(1{,}600{,}000)(10.72)/16(600)]^{1/4}$$

$$= 42.0 \text{ in.}$$

Summary for Double 2 × 4 Plates

For bending, maximum span length = 45.4 in.

For shear, maximum span length = 57.4 in.

For deflection, maximum span length = 42.0 in.

The maximum permissible span length of the double plates is governed by deflection at 42.0 in. For constructability, choose a spacing for the form ties at 36 in.

Check the strength of the tie based on the concrete pressure area of 24 in. by 36 in. The load on a tie can be calculated as follows:

$$\text{Area} = 2.0 \text{ ft} \times 3.0 \text{ ft}$$

$$= 6.0 \text{ sq ft}$$

$$\text{Tie tension load} = 600 \text{ lb per sq ft (6.0 sq ft)}$$

$$= 3{,}600 \text{ lb}$$

The ties for the wall must have a safe tension load of 3,300 lb. Therefore, use a 4,000-lb working load tie. This panel will weigh approximately 100 lb. Although this panel was designed by using equations to illustrate the methods, it could have been designed more quickly by using the information given in the tables for the design of plywood or Plyform.

Procedure for Erection of Forms for Footings

Figure 8-3 will be used to illustrate and describe the step-by-step procedure for erecting wall forms using these panels. At the time the wall footing was poured, short pieces of lumber were set in the concrete flush with the top surface of the footing at about 4-ft intervals to serve as nailer inserts. The procedure in erecting the forms is as follows:

1. Place two rows of plywood-faced 4 × 4 S4S lumber, with the inner faces separated by the thickness of the wall, along the

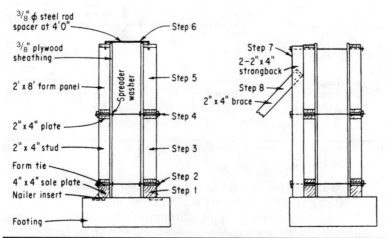

3/8" φ steel rod spacer at 4'0" — Step 6

3/8" plywood sheathing

Step 7
2–2" x 4" strongback

2' x 8' form panel — Spreader washer — Step 5

Step 8
2" x 4" brace

2" x 4" plate — Step 4

2" x 4" stud — Step 3

Form tie

4" x 4" sole plate — Step 2

Nailer insert — Step 1

Footing

FIGURE 8-3 Procedure in erecting forms for footings.

footing as sole plates. The top edge of the plywood facing extends ⅛ in. above the top of each plate. Toe-nail the plates to the nailer inserts in the footing.

2. Place through-wall button-end spreader ties at the desired spacing across the sole plates. Either hit each tie with a hammer to embed it in the projecting edge of the plywood or pre-cut notches in the edge of the plywood to receive the ties.

3. Set the lower tier of form panels on the sole plates extending along the footing, with the ends of the panels butting against each other. Fasten the end studs of adjacent panels together, using bolts, short form ties, or double-headed nails. Then nail the lower plate of each form panel to the 4- by 4-in. sill plate, and wedge-clamp the form ties.

4. Space form ties along the top surface of the tier panels.

5. Set another tier of form panels on top of the lower tier. Fasten the ends of the panels together, as described previously, and then nail the bottom plate of the second tier to the top plate of the first tier. Then wedge-clamp the second row of form ties.

6. Install form spacers across the top tier of forms, using 1×4 wood boards, metal straps, U-shaped steel rods set through prebored holes in the top plate, or some other suitable method. The U-shaped rods can be inserted and removed quickly and should have a very long life.

7. Install vertical double 2×4 wood strongbacks along the inside wall, at 8-ft spacings at the end junctures of panels, using through-wall form ties to hold them in place.

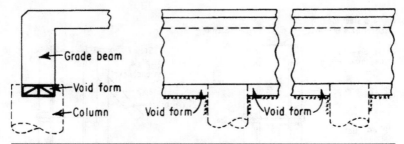

FIGURE 8-4 Use of void forms under a grade beam.

8. As the forms are aligned, attach a brace near the top of each strongback and to a stake set in the ground.

Erecting and stripping the forms can proceed at a very rapid rate after the form crew becomes familiar with this procedure.

Forms for Grade Beams

The side forms for grade beams can be fabricated and erected in a manner similar to those previously described for foundation walls. The bottom of a beam may rest on the ground with no required bottom form. However, designs frequently require a void space between the bottom of a beam and the ground under it. The void space is required for clay-type soils that are sensitive to expanding when the moisture of the soil increases, causing extensive uplift pressures on the bottom of concrete grade beams or slabs on grade. For this purpose fiber void units, such as carton forms made from cardboard, are installed under the bottom of a beam between its supporting piers, columns, or footings, as illustrated in Figure 8-4. The units are initially strong enough to support the fresh concrete, but they will soon soften and decay to leave a void space under the beam.

Forms for Concrete Footings

The selection of forms for concrete footings will depend on the size and shape of the footings and the number of times the forms can be reused without modification. Rectangular footings with constant cross sections are easily formed, but stepped footings are more complicated.

Figure 8-5 illustrates a form for a rectangular footing that can be erected and stripped rapidly and reused many times if other footings have the same dimensions.

The sheathing may be made from plywood attached to wood frames made of 2 × 4 S4S or larger lumber. The use of a steel rod at each corner permits quick assembly and removal, and it also ensures adequate anchorage. When the four rods are removed, the panels can be reassembled in another location within a few minutes.

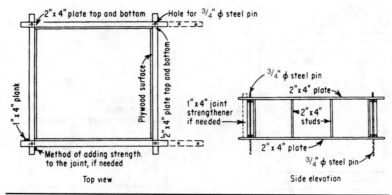

Figure 8-5 Forms for concrete footings.

The studs and plates can be made from 2 × 4, or 2 × 6 lumber, depending on the size and species of lumber used for the forms and the pressure from the concrete. If the footing is 18 in. deep, the maximum footing width will be less than 5 ft, using 2 × 4 plates for the species of lumber commonly used in formwork.

Where the size and depth of a footing produce excessive compression stresses between the pin and the wood in the bottom plate, 1- by 4-in. vertical planks can be nailed to the plates, as shown in Figure 8-5, to increase the strength of the joints, or nails can be used to supplement the strength of the pins.

The top view of Figure 8-5 shows that, by starting with long top and bottom plates on opposite sides of the forms, the length of a footing may be increased while the width remains constant. By adding strips of plywood, adequately attached to the plates, the forms can quickly be converted to larger sizes. A similar method can be used to increase both dimensions of a footing. In general, it is better to use the forms for the smallest footings first, and then go to the next larger sizes. However, the forms may also be used in the reverse order.

Additional Forms for Concrete Footings

There are many types of forms that can be used for concrete footings, depending on the footing shape, as described in the following sections and illustrated in Figures 8-6 through 8-10.

Figure 8-6 shows a plan and elevation view that is sometimes used for square or rectangular footings whose depths vary from 12 to 36 in. For footing depths up to 24 in., ¾-in. Plyform sheathing is supported by 2 × 4 studs, spaced at 12 in. center to center. For 36-in.-deep footings, ⅞-in. Plyform with 2 × 4 size lumber is used. The procedure for determining the required size of lumber for a particular job application is illustrated in previous sections of this chapter and in Chapter 5 of this book. The forms should be adequately braced.

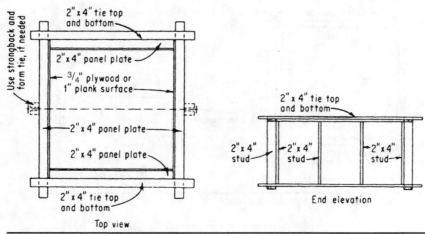

Use strongback and form tie, if needed

2" x 4" tie top and bottom

2" x 4" panel plate

³/₄" plywood or 1" plank surface

2" x 4" panel plate

2" x 4" panel plate

2" x 4" tie top and bottom

Top view

2" x 4" tie top and bottom

2" x 4" stud   2" x 4" stud   2" x 4" stud

End elevation

FIGURE 8-6   Forms for large size concrete footings.

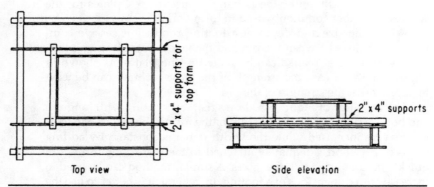

2" x 4" supports for top form

Top view

2" x 4" supports

Side elevation

FIGURE 8-7   Forms for stepped footings.

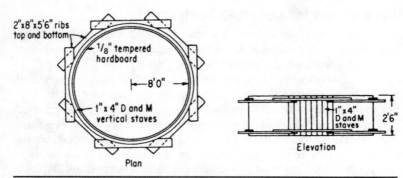

2"x8"x5'6" ribs top and bottom

¹/₈" tempered hardboard

8'0"

1" x 4" D and M vertical staves

Plan

1" x 4" D and M staves

2'6"

Elevation

FIGURE 8-8   Forms for a large-diameter job-built round footing.

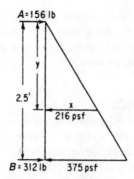

FIGURE 8-9 Concrete pressure on footing form.

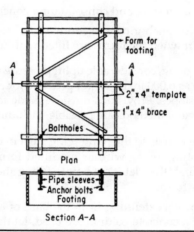

FIGURE 8-10 Template for anchor bolts

Additional strength can be obtained by using vertical strong-backs, as illustrated in Figure 8-6. If strongbacks are used, form ties should be installed between strongbacks on opposite sides of the footing, as shown in the top view of this plan.

Forms for Stepped Footings

Figure 8-7 illustrates a set of forms for constructing a stepped footing. The forms for the lower and upper sections are made in a conventional manner, with the upper form supported in the proper location on the lower form, using two pieces of 2×4 lumber attached to the bottom of the upper form, with the 4-in. face vertical. Form ties should be used if and where they are needed.

The lower form is filled, and the concrete is allowed to set until it is stiff enough to resist the hydrostatic pressure from the concrete in the upper form, after which time the upper form is filled. Caution must be taken to prevent an unwanted cold joint.

Forms for Sloped Footings

Footings with sloped surfaces sometimes are used instead of stepped footings, primarily to save concrete. Considering the higher cost of forms, it is questionable whether an overall economy is realized. Among the disadvantages are the following:

1. Possible increase in the cost of material
2. Increase in the cost of labor for fabricating forms
3. Difficulty in placing and consolidating concrete
4. Necessity of anchoring the forms down to prevent hydrostatic uplift when the forms are filled with concrete

For a sloped footing, considerable uplift pressure is exerted against the form, as much as 6,000 lb of vertical pressure for a 6-ft-square footing with 1:1 side slopes. Provisions must be made to resist this pressure, for which several methods are possible, including the following:

1. Attach the panels to the reinforcing steel in the lower 11½ in. of the footing, using wire or modified form ties. Place this concrete first, then let it set before filling the remaining portion of the form.
2. If steel dowels extending from the base of a footing into the bottom of a concrete column are used, let the lower concrete set, and then attach the top panels to the dowels.
3. Install a horizontal platform around the top panels and load it with ballast.
4. Drive stakes or steel pins well into the ground on the outside of the footing, and anchor the forms to the stakes.

Forms for Round Footings

For relatively small diameters, less than 48 in., laminated fiber tubes may be used for forms for cylindrical-type round footings. Several manufacturers provide circular forming tubes. These tubes can also be used for stepped footings, provided the diameters of the footings correspond to the diameters of the available tubes.

Forms for round footings whose diameters do not correspond to the available diameters of fiber tubes or whose diameters exceed the maximum size fiber tubes may be made from sheets of steel or from lumber that is lined with plywood or tempered hardboard. Table 4-21 gives the minimum bending radii for tempered hardboard. Figure 8-8 illustrates a form for a round footing.

For the job-built round form in Figure 8-8, determine the maximum safe spacing of the ribs, based on filling the footing with concrete in one operation. The pressure on the vertical surface will vary uniformly from zero at the top to 375 lb per sq ft as shown in Figure 8-9.

If the vertical staves are assumed to be supported by the ribs at the top and bottom ends, which is an exaggerated span on the safe side, and if a vertical strip 12 in. wide is considered, the reaction at A will be 156 lb, and at B it will be 312 lb. The depth y to the point of maximum moment, which is the point of zero shear, is obtained from the triangle whose area is

$$xy/2 = 156 \text{ lb} \tag{a}$$

Also, from similar triangles,

$$x/375 = y/2.5$$
$$x = 375y/2.5 \tag{b}$$

Combining Eqs. (a) and (b) and solving for y:

$$y = 1.443 \text{ ft}$$

This is the depth to the point of maximum moment. Considering all forces acting above line x, determine the applied bending moment at this depth:

$$M = 156 \text{ lb}(1.443 \text{ ft}) - [(216/2)(1.443)^2/3)]$$
$$= 225.1 - 74.9$$
$$= 150.2 \text{ ft-lb}$$
$$= 1{,}802.4 \text{ in.-lb}$$

Consider No. 1 grade 1×4 S4S lumber for sheathing that has an allowable bending stress of 1,625 lb per sq in. The resisting moment of a strip of sheathing 12 in. wide will be:

$$M = F_b S$$
$$= F_b(bd^2/6)$$
$$= 1{,}625[(12)(0.75)^2/6]$$
$$= 1{,}828 \text{ in.-lb}$$

Thus, the resisting moment is greater than the applied bending moment. In order to limit the deflection, the ribs will be spaced not more than 24 in. apart by placing them 3 in. below the top and 3 in. above the bottom of the staves.

Determine the number of nails required to fasten adjacent rib planks together. The maximum force, which tends to burst the form, will act on the bottom rib. It will be a force of 312 lb on each foot of diameter for a total value of 2,496 lb as calculated as follows:

$$2P = 312(8) = 2,496 \text{ lb}$$

Each of the two ribs, on opposite sides of the form, must resist one-half of this force. Thus, each rib must resist 1,248 lb, which the joint between adjacent ribs must withstand. Using 20d nails with a lateral resistance of 185 lb per nail, the number of nails required at the joints can be calculated as follows:

$$\text{Number of nails} = (1,248 \text{ lb})/(185 \text{ lb/nail})$$
$$= 6.7$$

Thus, 7 nails are required at each joint.

Placing Anchor Bolts in Concrete Foundations

When concrete foundations support machines or other equipment, it usually is necessary to install anchor bolts at the time the concrete is placed. A wood template, with bolt holes at the exact locations, permits the bolts to be placed correctly, as illustrated in Figure 8-10. The template may be made from planks or a sheet of plywood. Usually, a short section of steel pipe or metal cylinder is installed around each bolt to permit a slight lateral movement of the bolt and ensure that it will fit the hole spacing in the base of the machine. The space inside the pipe may be filled with a nonshrinkage grout.

CHAPTER 9

Forms for Walls

General Information

In general, forms for walls may be divided into one of three categories:

1. Those that are built in place, using plywood for sheathing and lumber for studs and wales

2. Prefabricated, job-built panels, using plywood sheathing, attached semipermanently to forms made from lumber, such as 2 × 4 or 2 × 6 S4S lumber

3. Patented form panels, using plywood facings attached to, and backed up by, steel or wood, or a combination of steel and wood forms $\frac{1}{16}$

For a single use, built-in-place forms are usually the most economical. For multiple uses, where standard-size panels can be reused without modification, it may be desirable to use either job-built or patented-type panels. However, if there are enough reuses, patented panels may be cheaper than job-built panels because of the longer life and possible lower labor costs for erecting and dismantling them. Also, because of the better dimension control in making them, patented panels usually will give a better fit-up than job-built panels when they are assembled into wall forms.

The common thicknesses of 4-ft-wide by 8-ft-long sheets of plywood used for wall forms are $\frac{3}{4}$, $\frac{7}{8}$, and 1 in. Smaller thicknesses may be used as form liners. Thicknesses of 1 in. and $1\frac{1}{8}$ in. are available for heavy formwork. However, the larger thicknesses make them more difficult to handle by workers, and the cost of the thicker panels is generally considerably higher than the small thicknesses. Special treatment of the surfaces of the plywood will increase the life of plywood considerably, permitting more than 200 uses per face in some reported instances.

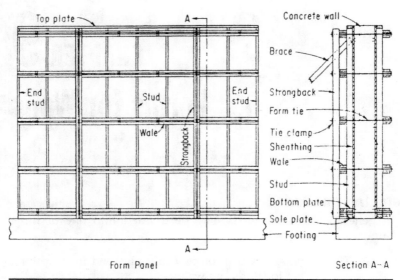

FIGURE 9-1   Forms for concrete wall.

Definition of Terms

Sometimes there are variations in the terminology used in the construction industry. Figure 9-1 illustrates the components of a set of
wall forms. Walls with pilasters, counterforts, setbacks, corbels, and
other structural components require modified forms, which will be
discussed later in this chapter.

The following terms related to job-built formwork for wall are
used in this book:

1. *Sheathing* is the plywood on each side of the wall against which
the fresh concrete is placed. The sheathing provides resistance to the
pressure of the freshly placed concrete.

2. *Studs* are the members to which the sheathing is attached. They
are normally installed vertically. Studs are usually single members of
2×4, 2×6, or larger lumber, depending on the pressure on the forms.
Studs provide support for the sheathing.

3. *Wales*, are usually double 2×4, 2×6, or larger lumber with
separators, are installed on opposite sides of wall forms, perpendicular to the studs, to hold the studs in position, to ensure good
alignment for the forms, and to receive the form ties. Other names by
which this member is called are walers or waling. The wales provide
support for the studs.

4. *Strongbacks* sometimes are installed perpendicular to wales to
provide additional strength and rigidity to high forms.

5. *Top plates* are installed and fastened to the tops of studs as parts of the panel frames.

6. *Bottom plates* are installed and fastened to the bottoms of studs as parts of the panel frames.

7. *Sole plates* are placed and fastened along each side of wall footings to provide initial alignment and support for wall forms. Also, the use of sole plates permits easy use of form ties near the bottom of wall forms.

8. *Braces*, fastened to one side of the forms and to stakes set in the ground about 8 to 10 ft apart, prevent the forms from shifting when the concrete is placed. If the sole plates are not securely fastened to the footing, braces should also be attached to the bottoms of the forms on one side of the wall. Screw-adjusting braces are commercially available which allow fast and accurate aligning of forms.

9. *Form ties*, with a clamping device on each end, are installed through the forms to resist the bursting pressure exerted by the concrete. Some are equipped with devices which enable them to serve as form spreaders or spacers. Many types and sizes are available, with allowable working strengths varying from 1,500 to 50,000 lb or more. The common strengths of form ties are 3,000 and 4,000 lb. Form ties provide support for the wales.

Designing Forms for Concrete Walls

Forms for concrete walls should provide sufficient strength and rigidity at the lowest practical cost, considering materials, labor, and any construction equipment used in making, erecting, and removing them. The designer should know the magnitudes of the forces that act on the component parts and the strengths of these parts in resisting the forces. Chapter 3 presented pressures and loads that act on formwork.

The strength of lumber varies with the species and grade used for formwork. The allowable unit stresses in bending, horizontal shear, compression perpendicular to the grain, and compression parallel to the grain vary with the species and grade of lumber (see Chapter 4). Each design should be based on information that applies to the allowable unit stresses for the species and grade of lumber used for the formwork.

As presented in Chapter 5, many calculations are required to determine the adequacy of the strength of formwork members. It requires considerable time to analyze each beam element of formwork to determine the applied and allowable stresses for a given load condition, and to compare the calculated deflection with the permissible deflection. For each beam element, a check must be made for bending, shear, and deflection. For simplicity in the design of formwork it is common practice to rewrite the equations for stresses and deflection in order to calculate the permissible span length of a member in terms of the member size, allowable stress, and the loads on the member. Table 5-3 summarizes the equations for calculating allowable span lengths of wood members, and Table 5-4 summarizes the equations for calculating the allowable span lengths of plywood members. These two tables enable the designer to design formwork in an organized manner.

There are two general approaches to the design for wall forms. One approach is to select first the material to be used for formwork, then use the equations in Tables 5-3 and 5-4 to determine the spacing of the materials; including studs, wales, and form ties. The other approach is to use tables that have been developed by a company that supplies formwork systems and accessories. In the latter approach, the materials and recommendations must be followed by the company that supplies the formwork and accessories.

Usually the steps in designing forms for concrete walls follow the sequence shown below:

1. Determine the maximum pressure exerted by the concrete on the wall form, considering the depth of the forms, rate of placing concrete, temperature of the concrete, type of cement and additivies, density of concrete, and method of consolidating the concrete. See Chapter 3 for the equations and tables for pressures on wall forms. Note the maximum pressure of concrete on formwork is limited by wh, where w is the density of concrete and h is the height of concrete placement.

2. Select the kind, grade, and thickness of the material to be used for sheathing (see Tables 4-9 and 4-10 for plywood and Tables 4-11 and 4-12 for Plyform, the plywood manufactured especially for formwork). Then determine the maximum safe spacing of the studs, considering the allowable unit fiber stress in bending, the horizontal shear stress, and the permissible deflection in the sheathing. See Table 5-4 for the equations for calculating the allowable span lengths of plywood based on bending and shear stress and the deflection criteria.

3. Select the grade and size of lumber to be used for studs. See Table 4-1 for physical properties of lumber and Tables 4-2 and 4-3 give reference design values of wood members. Table 4-4 through 4-8 provides adjustment factors that are applied to the reference design values to determine allowable stresses. Determine the maximum safe spacing of wales, based on the permissible span length of the studs, checking the bending and shear stresses and deflection of studs. Table 5-3 provides the equations for calculating the allowable span length of lumber based on stresses and deflection.

4. Select the grade and size of lumber to be used for wales (see Tables 4-1 and 4-2). Determine the maximum safe spacing of form

ties, considering the allowable unit stresses in bending and shear, and the permissible deflection for the wales, whichever limits the spacing. See Table 5-3 for the equations to calculate the allowable span length. A check must also be made for the bearing stress between the stud and the wale.

5. Select form ties that will safely withstand the forces transmitted to them by the wales. An alternate method is to select the form ties, then to determine the maximum safe spacing along the wales, checking to be certain that this spacing does not exceed that deter-mined in step 4.

Physical Properties and Allowable Stresses for Lumber

For clarity in illustrating the design of wall forms, the physical properties and allowable stresses of S4S dimension lumber and Plyform are reproduced here from Chapter 4. These values are shown in Tables 9-1 through 9-4. Also, the definition of terms and the summary of equations from Chapter 5 for the design of formwork are repeated for clarity (see Tables 9-5 and 9-6).

Table 9-2 gives the tabulated design values of Southern Pine lumber for normal load duration and dry service conditions (<19% moisture content). The values tabulated in Table 9-2 are only for Southern Pine. If other species and grades are used for formwork, the tabulated stresses should be adjusted to conform to the kind of lumber used.

Physical Properties and Allowable Stresses for Plyform

Table 9-3 provides the physical properties of Plyform. These properties are for a 12-in. width of plywood. The allowable stresses for Plyform are shown in Table 9-4.

Table of Equations for Calculating Allowable Span Lengths for Wood Beams and Plywood Sheathing

Table 9-5 provides a summary of the equations used for calculating the allowable span lengths for wood beams with uniformly distributed loads, and Table 9-6 provides the equations for calculating the allowable span lengths of Plyform sheathing for uniformly distributed concrete pressure. The allowable stresses for bending F_b and horizontal shear F_v are in lb per sq in. The unit for allowable span length is inches. The units for physical properties are defined in the previous pages of this chapter.

Nominal Size, in.	Actual Size, Thickness × Width, in.	Net Area in.^{2}	X-X axis x —[ ]— x		Y-Y axis x ═══ x	
			I_x, in.^{4}	S_x, in.^{3}	I_y, in.^{4}	S_y, in.^{3}
1 × 4	¾ × 3½	2.62	2.68	1.53	0.123	0.328
1 × 6	¾ × 5½	4.12	10.40	3.78	0.193	0.516
1 × 8	¾ × 7¼	5.43	23.82	6.57	0.255	0.680
1 × 10	¾ × 9¼	6.93	49.47	10.70	0.325	0.867
1 × 12	¾ × 11¼	8.43	88.99	15.82	0.396	1.055
2 × 4	1½ × 3½	5.25	5.36	3.06	0.984	1.313
2 × 6	1½ × 5½	8.25	20.80	7.56	1.547	2.063
2 × 8	1½ × 7¼	10.87	47.63	13.14	2.039	2.719
2 × 10	1½ × 9¼	13.87	98.93	21.39	2.603	3.469
2 × 12	1½ × 11¼	16.87	177.97	31.64	3.164	4.219
3 × 4	2½ × 3½	8.75	8.93	5.10	4.56	3.65
3 × 6	2½ × 5½	13.75	34.66	12.60	7.16	5.73
3 × 8	2½ × 7¼	18.12	79.39	21.90	9.44	7.55
3 × 10	2½ × 9¼	23.12	164.88	35.65	12.04	9.64
3 × 12	2½ × 11¼	28.12	296.63	52.73	14.65	11.72
4 × 4	3½ × 3½	12.25	12.50	7.15	12.51	7.15
4 × 6	3½ × 5½	19.25	48.53	17.65	19.65	11.23
4 × 8	3½ × 7¼	25.38	111.14	30.66	25.90	14.80
4 × 10	3½ × 9¼	32.37	230.84	49.91	33.05	18.89
4 × 12	3½ × 11¼	39.37	415.28	73.83	40.20	22.97
5 × 5	4½ × 4½	20.25	34.17	15.19	34.17	15.19
6 × 6	5½ × 5½	30.25	76.25	27.73	76.26	27.73
6 × 8	5½ × 7½	41.25	193.35	51.56	103.98	37.81
6 × 10	5½ × 9½	52.25	393.96	82.73	131.71	47.90
6 × 12	5½ × 11½	63.25	697.06	121.2	159.44	57.98
8 × 8	7½ × 7½	56.25	263.7	70.31	263.67	70.31
8 × 10	7½ × 9½	71.25	535.9	112.8	333.98	89.06
8 × 12	7½ × 11½	86.25	950.5	165.3	404.29	107.81
10 × 10	9½ × 9½	90.25	678.8	142.9	678.75	142.90
10 × 12	9½ × 11½	109.25	1204.0	209.4	821.65	173.98
12 × 12	11½ × 11½	132.25	1457.5	253.5	1,457.50	253.48

TABLE 9-1 Properties of S4S Dry Lumber, Moisture < 19%

Reference Design Values for Southern Pine, lb per sq in.

Nominal Size	Grade	Extreme Fiber in Bending F_b	Shear Parallel to Grain F_v	Compression Perpendicular to Grain $F_{c\perp}$	Compression Parallel to Grain $F_{c//}$	Modulus of Elasticity	
						E	E_{min}
2"–4" thick and 2"–4" wide	No. 1	1,850	175	565	1,850	1,700,000	620,000
	No. 2	1,500	175	565	1,650	1,600,000	580,000
2"–4" thick and 6" wide	No. 1	1,650	175	565	1,750	1,700,000	620,000
	No. 2	1,250	175	565	1,600	1,600,000	580,000
2"–4" thick and 8" wide	No. 1	1,500	175	565	1,650	1,700,000	620,000
	No. 2	1,200	175	565	1,550	1,600,000	580,000
2"–4" thick and 10" wide	No. 1	1,300	175	565	1,600	1,700,000	620,000
	No. 2	1,050	175	565	1,500	1,600,000	580,000
2"–4" thick and 12" wide	No. 1	1,250	175	565	1,600	1,700,000	620,000
	No. 2	975	175	565	1,450	1,600,000	580,000
Multiplier[3] of values for loads <7 days		1.25	1.25	1.0	1.25	1.0	1.0
Multiplier[4] of values for moisture >19%		0.85*	0.97	0.67	0.8**	0.9	0.9

Notes:
1. Values shown are for visually graded Southern Pine, ref. [2].
2. Values for Southern Pine are already adjusted for size, $C_F = 1.0$.
3. See Table 4-4 through 4-8 for other adjustment factors that may be applicable.
4. *denotes when $(F_b)(C_F) \leq 1,150$ lb per sq in., $C_M = 1.0$
5. **denotes when $(F_b)(C_F) \leq 750$ lb per sq in., $C_M = 1.0$
6. See the National Design Specification for Wood Construction for other grades and species of wood and for additional adjustments that may be appropriate for conditions where wood is used.

TABLE 9-2 Reference Design Values of Southern Pine with < 19% Moisture

Thick-ness, in.	Approx Weight, lb/ft²	Properties for Stress Applied Parallel to Face Grain (face grain across supports)			Properties for Stress Applied Perpendicular to Face Grain (face grain along supports)		
		I Moment of Inertia, in.⁴/ft	S_e Effective Section Modulus, in.³/ft	Ib/Q Rolling Shear Constant, in.²/ft	I Moment of Inertia, in.⁴/ft	S_e Effective Section Modulus, in.³/ft	Ib/Q Rolling Shear Constant, in.²/ft
Class I							
15/32	1.4	0.066	0.244	4.743	0.018	0.107	2.419
1/2	1.5	0.077	0.268	5.153	0.024	0.130	2.739
19/32	1.7	0.115	0.335	5.438	0.029	0.146	2.834
5/8	1.8	0.130	0.358	5.717	0.038	0.175	3.094
23/32	2.1	0.180	0.430	7.009	0.072	0.247	3.798
3/4	2.2	0.199	0.455	7.187	0.092	0.306	4.063
7/8	2.6	0.296	0.584	8.555	0.151	0.422	6.028
1	3.0	0.427	0.737	9.374	0.270	0.634	7.014
1 1/8	3.3	0.554	0.849	10.430	0.398	0.799	8.419
Class II							
15/32	1.4	0.063	0.243	4.499	0.015	0.138	2.434
1/2	1.5	0.075	0.267	4.891	0.020	0.167	2.727
19/32	1.7	0.115	0.334	5.326	0.025	0.188	2.812
5/8	1.8	0.130	0.357	5.593	0.032	0.225	3.074
23/32	2.1	0.180	0.430	6.504	0.060	0.317	3.781
3/4	2.2	0.198	0.454	6.631	0.075	0.392	4.049
7/8	2.6	0.300	0.591	7.990	0.123	0.542	5.997
1	3.0	0.421	0.754	8.614	0.220	0.812	6.987
1 1/8	3.3	0.566	0.869	9.571	0.323	1.023	8.388
Structural I							
15/32	1.4	0.067	0.246	4.503	0.021	0.147	2.405
1/2	1.5	0.078	0.271	4.908	0.029	0.178	2.725
19/32	1.7	0.116	0.338	5.018	0.034	0.199	2.811
5/8	1.8	0.131	0.361	5.258	0.045	0.238	3.073
23/32	2.1	0.183	0.439	6.109	0.085	0.338	3.780
3/4	2.2	0.202	0.464	6.189	0.108	0.418	4.047
7/8	2.6	0.317	0.626	7.539	0.179	0.579	5.991
1	3.0	0.479	0.827	7.978	0.321	0.870	6.981
1 1/8	3.3	0.623	0.955	8.841	0.474	1.098	8.377

Notes:
1. Courtesy, APA—The Engineered Wood Association, "Concrete Forming", 2004.
2. The section properties presented here are specifically for Plyform, with its special layup restrictions.

TABLE 9-3 Effective Section Properties of Plyform with 1-ft Widths

Item		Plyform Class I	Plyform Class II	Structural I Plyform
Bending stress, lb/in.^{2}	F_b	1,930	1,330	1,930
Rolling shear stress, lb/in.^{2}	F_s	72	72	102
Modulus of Elasticity—(psi, adjusted, use for bending deflection calculation)	E	1,650,000	1,430,000	1,650,000
Modulus of Elasticity—(lb/in.^{2}, unadjusted, use for shear deflection calculation)	E_e	1,500,000	1,300,000	1,500,000

Note:
1. Courtesy, APA—The Engineered Wood Association.

TABLE 9-4   Allowable Stresses for Plyform, lb per sq in.

	Equations for Calculating Allowable Span Lengths, in Inches, of Wood Beams			
			Bending Deflection	
Span Condition	Bending l_b, in.	Shear l_v, in.	l_Δ, in. for $\Delta = l/360$	l_Δ, in. for $\Delta = 1/16$-in.
Single spans	$[96F_bS/w]^{1/2}$ Eq. (5-30)	$16F_vbd/w + 2d$ Eq. (5-32)	$[4,608EI/1,800/w]^{1/3}$ Eq. (5-33a)	$[4,608EI/80w]^{1/4}$ Eq. (5-33b)
Multiple spans	$[120F_bS/w]^{1/2}$ Eq. (5-34)	$192F_vbd/15w + 2d$ Eq. (5-36)	$[1,743EI/360w]^{1/3}$ Eq. (5-37a)	$[1,743EI/16w]^{1/4}$ Eq. (5-37b)

Notes:
1. Physical properties of wood beams can be obtained from Table 9-1.
2. See Tables 4-2 and 4-3 for allowable stresses with adjustments from Tables 4-3a through 4-8.
3. Units for stresses and physical properties are pounds and inches.
4. Units of w are lb per lineal foot along the entire length of beam.

TABLE 9-5   Equations for Calculating Allowable Span Lengths, in Inches, for Wood Beams Based on Bending, Shear, and Deflection for Beams with Uniformly Distributed Loads

	Equations for Calculating Allowable Span Lengths, in Inches, for Plywood and Plyform				
			Bending Deflection		Shear Deflection
Span Condition	Bending l_b, in.	Shear l_s, in.	l_d, in. for $\Delta = l/360$	l_d, in. for $\Delta = 1/16$ in.	l_s, in. for $\Delta = l/360$
Single Spans	$[96F_bS_e/w_b]^{1/2}$ Eq. (5-40)	$24F_s(lb/Q)/w_s$ Eq. (5-46)	$\left[\dfrac{921.6EI}{360w_d}\right]^{1/3}$ Eq. (5-55a)	$\left[\dfrac{921.6EI}{16w_d}\right]^{1/4}$ Eq. (5-55b)	$\dfrac{1{,}270E_eI}{360w_sCt^2}$ Eq. (5-60a)
Two Spans	$[96F_bS_e/w_b]^{1/2}$ Eq. (5-40)	$19.2F_s(lb/Q)/w_s$ Eq. (5-47)	$\left[\dfrac{2{,}220EI}{360w_d}\right]^{1/3}$ Eq. (5-56a)	$\left[\dfrac{2{,}220EI}{16w_d}\right]^{1/4}$ Eq. (5-56b)	$\dfrac{1{,}270E_eI}{360w_sCt^2}$ Eq. (5-60a)
Three or more Spans	$[120F_bS_e/w_b]^{1/2}$ Eq. (5-41)	$20F_s(lb/Q)/w_s$ Eq. (5-48)	$\left[\dfrac{1{,}743EI}{360w_d}\right]^{1/3}$ Eq. (5-57a)	$\left[\dfrac{1{,}743EI}{16w_d}\right]^{1/4}$ Eq. (5-57b)	$\dfrac{1{,}270E_eI}{360w_sCt^2}$ Eq. (5-60a)

Notes:
1. See Table 9-3 for physical properties of Plyform.
2. Refer Table 9-4 for allowable stresses of Plyform.
3. Units for stresses and physical properties are in pounds and inches.
4. Units for w are lb per lineal foot for a 12-in.-wide strip of Plyform.

TABLE 9-6 Equations for Calculating Allowable Span Lengths, in Inches, for Plywood and Plyform Based on Bending, Shear, and Deflection due to Uniformly Distributed Pressure

Design of Forms for a Concrete Wall

The use of the equations and principles developed in Chapter 5 will be illustrated by designing the forms for a concrete wall. The design will include the sheathing, studs, wales, and form ties.

In the design of wall forms, the thickness and grade of the plywood is often selected based on availability of materials. For this design example, a ¾-in.-thick Plyform Class I has been selected. The lumber will be No. 2 grade S4S Southern Pine with no splits. The following design criteria will apply:

1. Height of wall, 10 ft 0 in.

2. Rate of filling forms, 5 ft per hr.

3. Temperature, 80°F.

4. Concrete unit weight, 150 lb per cu ft.

5. Type I cement will be used without retarders.

6. A dry condition for wood, less than 19% moisture content.

7. Short load duration, less than 7 days.

8. Limit deflection to $l/360$, but no greater than ⅟₁₆ in.

9. Studs and wales will be 2-in. nominal framing.

10. Lumber will be No. 2 grade Southern Pine (refer to Table 9-2).

11. Sheathing will be 3/4-in. Plyform (refer to Tables 9-3 and 9-4).

Figure 9-2 provides an illustration of the span lengths and spacing of components for the wall form. Tables 9-5 and 9-6 provide the equations that are necessary to design the wall forms. The design will be performed in the following sequence:

1. Determine the maximum lateral pressure of the concrete on the wall forms based on temperature and rate of filling the forms.

2. Determine the spacing of studs based on the strength and deflection of plywood sheathing.

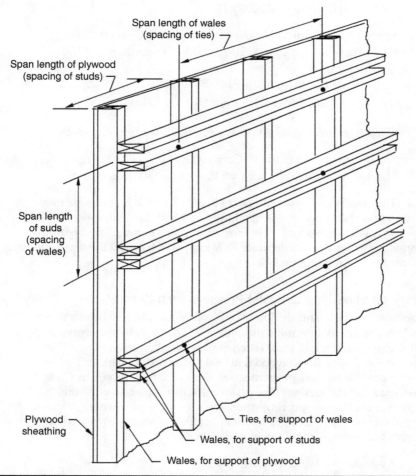

Span length of wales (spacing of ties)

Span length of plywood (spacing of studs)

Span length of suds (spacing of wales)

Plywood sheathing

Ties, for support of wales

Wales, for support of studs

Wales, for support of plywood

FIGURE 9-2 Span length and spacing of components of wall form.

3. Determine the spacing of wales based on the strength and deflection of studs.

4. Check the bearing strength between the studs and wales.

5. Determine the span length of wales between ties based on the strength and deflection of wales.

6. Determine the required strength of form ties.

Lateral Pressure of Concrete on Forms

The lateral concrete pressure against the wall forms can be obtained from Table 3-3 as 713 lb per sq ft. Or the pressure can be calculated using Eq. (3-2), which applies to wall placement heights up to 14 ft and rates of placement less than 7 ft per hr. Using Eq. (3-2) to calculate the pressure:

$$P_m = C_w C_c [150 + 9,000R/T]$$

From Table 3-1, for 150 lb per cu yd concrete, $C_w = 1.0$
From Table 3-2, for Type I cement without retarders, $C_c = 1.0$

$$P_m = 1.0 \times 1.0 \times [150 + 9,000(5/80)]$$
$$= 712.5 \text{ lb per sq ft, round up to 713 lb per sq ft}$$

Checks on limitations on pressures calculated from Eq. (3-2):

Limited to greater than 600 $Cw = 600 \times (1.0) = 600$ lb per sq ft
Limited to less than $P_m = wh = 150 \times (10) = 1500$ lb per sq ft

The calculated value from Eq. (3-2) is 713, which is above the minimum of 600 and below the maximum of 1,500. Therefore, use 713 lb per sq ft lateral pressure on the forms. The 713 lb per sq ft maximum pressure will occur at a depth of $713/150 = 4.75$ ft below the top of the form as shown in Figure 9-3.

Plyform Sheathing to Resist Pressure from Concrete

Because the type and thickness of Plyform has already been chosen, it is necessary to determine the allowable span length of the plywood (the distance between studs) such that the bending stress, shear stress, and deflection of the plywood will be within allowable limits. Table 9-3 provides the section properties for the ¾-in.-thick Plyform Class I selected for the design. The plywood will be placed with the face grain across the supporting studs; therefore, the properties for stress applied parallel to face grain will apply. Below are the physical properties for a one foot wide strip of the Plyform.

Approximate weight = 2.2 lb per sq ft
Moment of inertia, $I = 0.199$ in.[4]

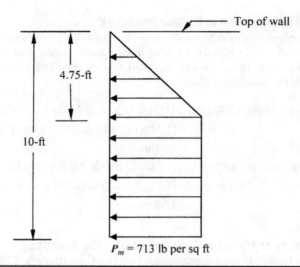

FIGURE 9-3   Distribution of concrete pressure for Example 9-1.

Effective section modulus, $S_e = 0.455$ in.[3]
Rolling shear constant, $Ib/Q = 7.187$ in.[2]

Table 9-4 gives values of bending stress, shear stress, and modulus of elasticity for Plyform Class I as follows:

Allowable bending stress, $F_b = 1,930$ lb per sq in.
Allowable rolling shear stress, $F_s = 72$ lb per sq in.
Modulus of elasticity for shear deflection, $E_e = 1,500,000$ lb per sq in.
Modulus of elasticity for bending deflection, $E = 1,650,000$ lb per sq in.

Bending Stress in Plywood Sheathing
Because the Plyform will be installed over multiple supports, three or more, Eq. (5-41) from Table 9-6 can be used to determine the allowable span length of the plywood based on bending:

$$\text{From Eq. (5-41): } l_b = [120 F_b S_e / w_b]^{1/2}$$
$$= [120(1,930)(0.455)/(713)]^{1/2}$$
$$= 12.1 \text{ in.}$$

Rolling Shear Stress in Plywood Sheathing
The allowable span length of the plywood based on rolling shear stress can be calculated using Eq. (5-48) from Table 9-6:

$$\text{From Eq. (5-48): } l_s = 20 F_s (Ib/Q) / w_s$$
$$= 20(72)(7.187)/713$$
$$= 14.5 \text{ in.}$$

Bending Deflection in Plyform Sheathing

For a maximum permissible deflection of $l/360$, but no greater than 2×4 $\frac{1}{16}$ in., and plywood with three or more spans, Eqs. (5-57a) and (5-57b) from Table 9-6 can be used to calculate the allowable span length for bending deflection.

From Eq. (5-57a): $l_d = [1{,}743EI/360w_d]^{1/3}$ for $\Delta = l/360$

$$= [1{,}743(1{,}6500{,}000)(0.199)/360(713)]^{1/3}$$

$$= 13.1 \text{ in.}$$

From Eq. (5-57b): $l_d = [1{,}743EI/16w_d]^{1/4}$ for $\Delta = \frac{1}{16}$ in.

$$= [1{,}743(1{,}650{,}000)(0.199)/16(713)]^{1/4}$$

$$= 14.9 \text{ in.}$$

Summary of Allowable Span Lengths for the Sheathing

For bending, the maximum span length of sheathing = 12.1 in.

For shear, the maximum span length of sheathing = 14.5 in.

For bending deflection, the maximum span length of sheathing = 13.1 in.

For this design, bending stress controls the maximum span length of the Plyform sheathing. The studs that support the Plyform must be placed at a spacing that is no greater than 12.1 in. For uniformity and constructability, choose a maximum stud spacing of 12.0 in., center to center.

Check Total Deflection in Plyform

Calculate the total deflection, including bending deflection and shear deflection and compare it to the maximum permissible deflection of $l/360$ and $\frac{1}{16}$ in. For bending deflection with 2-in. nominal studs for support the effective span length is the clear span + $\frac{1}{4}$ in. (refer to Figure 5-10):

$$l_d = 12 \text{ in.} - 1.5 \text{ in.} + \frac{1}{4} \text{ in.}$$

$$= 10.75\text{-in.}$$

Calculate bending deflection from Eq. (5-51) as follows:

$$\Delta_d = w_d l_d^4/1{,}743EI$$

$$= 713(10.75)^4/(1{,}743)(1{,}650{,}000)(0.199)$$

$$= 0.017 \text{ in.}$$

For shear deflection with 2-in. nominal studs to support the effective span length is the clear span (refer to Figure 5-10).

$$l_s = 12 \text{ in.} - 1.5 \text{ in.}$$

$$= 10.5 \text{ in.}$$

Calculate the shear deflection from Eq. (5-60a) as follows:

$$\Delta_s = w_s C l^2 l_s^2 / 1{,}270 E_e I$$
$$= 713(120)(3/4)^2(10.5)^2 / (1{,}270)(1{,}500{,}000)(0.199)$$
$$= 0.014 \text{ in.}$$

Total deflection = bending deflection + shear deflection:
$$= 0.017 \text{ in.} + 0.014 \text{ in.}$$
$$= 0.031 \text{ in.}$$

Compare the calculated total deflection with the permissible deflection criteria of $l/360$, not to exceed $\frac{1}{16}$ in.:

Permissible deflection for $l/360 = 12$ in./$360 = 0.033 > 0.031$, therefore okay

Permissible deflection for $\frac{1}{16}$ in. $= 0.0625$ in. > 0.031, therefore okay

Therefore, the 12-in. spacing of studs is adequate for bending stress, shear stress, and deflection in the ¾-in. Class I Plyform sheathing.

Studs for Support of Plyform

The studs support the Plyform sheathing and the wales must support the studs. Therefore, the maximum allowable spacing of wales will be governed by the bending, shear, and deflection in the studs.

Consider using 2 × 4 S4S studs, whose actual size is 1½ by 3½ in. From Table 9-1. the physical properties can be obtained for 2 × 4 S4S lumber:

Area, $A = 5.25$ in.^{2}
Section modulus, $S = 3.06$ in.^{3}
Moment of inertia, $I = 5.36$ in.^{4}

The lumber selected for the forms is No. 2 grade Southern Pine. From Table 9-2, using the values with appropriate adjustments for short-duration loading, the allowable stresses are as follows:

Allowable bending stress, $F_b = 1.25(1{,}500) = 1{,}875$ lb per sq in.
Allowable shear stress, $F_v = 1.25(175) = 218$ lb per sq in.
Modulus of elasticity, $E = 1.0(1{,}600{,}000) = 1{,}600{,}000$ lb per sq in.

Bending in 2 × 4 S4S Studs

Each stud will support a vertical strip of sheathing 12 in. wide, which will produce a uniform load of 713 lb per sq ft × 1.0 ft = 713 lb per lin ft on the lower portion of the stud. Because the studs will extend to the full height of the wall, the design should be based on continuous beam action for the studs. The allowable span length of studs, the distance between wales, based on bending can be calculated from Eq. (5-34) in Table 9-5 as follows:

$$\text{From Eq. (5-34): } l_b = [120F_bS/w]^{1/2}$$
$$= [120(1{,}875)(3.06)/713]^{1/2}$$
$$= 31.1 \text{ in.}$$

Shear in 2 × 4 S4S Studs

The allowable span length based on shear in the studs can be calculated from Eq. (5-36) in Table 9-5 as follows:

$$\text{From Eq. (5-36) } l_v = 192F_vbd/15w + 2d$$
$$= 192(218)(1.5)(3.5)/15(713) + 2(3.5)$$
$$= 27.5 \text{ in.}$$

Deflection in 2 × 4 S4S Studs

The permissible deflection is $l/360$, but not greater than $\frac{1}{16}$ in. Using Eqs. (5-37a) and (5-37b) from Table 9-5, the allowable span lengths due to deflection can be calculated as follows:

$$\text{From Eq. (5-37a): } l_\Delta = [1{,}743EI/360w]^{1/3} \text{ for } \Delta = l/360$$
$$= 1{,}743(1{,}600{,}000)(5.36)/360(713)]^{1/3}$$
$$= 38.8 \text{ in.}$$
$$\text{From Eq. (5-37b): } l_\Delta = [1{,}743EI/16\tfrac{5}{8}\,w]^{1/4} \text{ for } \Delta = \tfrac{1}{16} \text{ in.}$$
$$= [1{,}743(1{,}600{,}000)(5.36)/16(713)]^{1/4}$$
$$= 33.8 \text{ in.}$$

Summary for 2 × 4 S4S Studs

For bending, the maximum span length of studs = 31.1 in.

For shear, the maximum span length of studs = 27.5 in.

For deflection, the maximum span length of studs = 33.8 in.

For this design, bending governs the maximum span length of the studs. Because the wales must support the studs, the maximum spacing of the wales must be no greater than 27.5 in. For constructability and ease of fabrication, it is desirable to space the studs 24 in. apart, center to center. This will allow easier fabrication of the forms because the 24 in. spacing will easily match the 8-ft-length of the plywood panels.

Bearing Strength between Studs and Wale

For this design, double wales will be used. Wales for wall forms of this size are frequently made of two-member lumber whose nominal thickness is 2 in., separated by short pieces of 1-in.-thick blocks.

Determine the unit stress in bearing between the studs and the wales. If a wale consists of two 2-in. nominal thick members, the contact area between a stud and a wale will be:

$$A = 2 \times [1.5 \text{ in.} \times 1.5 \text{ in.}]$$
$$= 4.5 \text{ sq in.}$$

The total load in bearing between a stud and a wale will be the unit pressure of the concrete acting on an area based on the stud and wale spacing. For this design, the area will be 12 in. long by 24 in. high. Therefore, the pressure acting between the stud and wale will be:

$$P = 713 \text{ lb per sq ft}[1.0 \text{ ft} \times 2.0 \text{ ft}]$$
$$= 1,426 \text{ lb}$$

The calculated unit stress in bearing, perpendicular to grain, between a stud and a wale will be:

$$f_c\perp = 1,426 \text{ lb}/4.5 \text{ sq in.}$$
$$= 317 \text{ lb per sq. in.}$$

From Table 9-2, the indicated allowable unit compression stress perpendicular to the grain for mixed southern pine is 565 lb per sq in. The allowable bearing stress of 565 lb per sq in. is greater than the applied bearing stress of 317 lb per sq in. Therefore, the unit bearing stress is within the allowable value for this species and grade of lumber. If another species of lumber, having a lower allowable unit stress, is used, it may be necessary to reduce the spacing of the wales, or to use wales whose members are thicker than 2 in. The wales will be spaced 24 in. apart, center to center.

Size of Wale Based on Selected 24 in. Spacing of Studs

Although the loads transmitted from the studs to the wales are concentrated, it is generally sufficiently accurate to treat them as uniformly distributed loads, having the same total values as the concentrated loads, when designing formwork for concrete walls. However, in some critical situations, it may be desirable to design the forms using concentrated loads, as they actually exist.

With a 24 in. spacing and a 713 lb per sq ft pressure, the load on a wale will be 713 lb per sq ft $\times$ 2.0 ft = 1,426 lb per ft. Consider double 2 $\times$ 4 S4S wales with the following physical properties:

From Table 9-1, for a double 2 $\times$ 4: $A = 2(5.25 \text{ in.}^2) = 10.5 \text{ in.}^2$
$$S = 2(3.06 \text{ in.}^3) = 6.12 \text{ in.}^3$$
$$I = 2(5.36 \text{ in.}^4) = 10.72 \text{ in.}^4$$

From Table 9-2, for a 2 × 4: $F_b = 1.25(1,500) = 1,875$ lb per sq in.

$$F_v = 1.25(175) = 218 \text{ lb per sq in.}$$

$$E = (1.0)(1,600,000) = 1,600,000 \text{ lb per sq in.}$$

Bending in Double 2 × 4 Wales

Based on bending, the span length of the wales must not exceed the value calculated from Eq. (5-34) in Table 9-5:

$$l_b = [120F_b S/w]^{1/2}$$
$$= [120(1,875)(6.12)/1,426]^{1/2}$$
$$= 31.1 \text{ in.}$$

Shear in Double 2 × 4 Wales

Based on shear, the span length of the wales must not exceed the value calculated from Eq. (5-36) in Table 9-5:

$$l_v = 192F_v bd/15w + 2d$$
$$= 192(218)(10.5)/15(1,426) + 2(3.5)$$
$$= 27.5 \text{ in.}$$

Deflection in Double 2 × 4 Wales

Based on the deflection criteria of less than $l/360$, the span length of the wales must not exceed the value calculated from Eq. (5-37a) in Table 9-5:

$$l_\Delta = [1,743EI/360w]^{1/3}$$
$$= [1,743(1,600,000)(10.72)/360(1,426)]^{1/3}$$
$$= 38.8 \text{ in.}$$

Based on the deflection criteria of less than $\frac{1}{16}$ in., the span length of the wales must not exceed the value calculated form Eq. (5-37b) in Table 9-5:

$$l_\Delta = [1,743EI/16w]^{1/4}$$
$$= [1,743(1,600,000)(10.72)/16(1,426)]^{1/4}$$
$$= 33.8 \text{ in.}$$

Summary for Double 2 × 4 Wales

For bending, the maximum span length of wales = 31.1 in.

For shear, the maximum span length of wales = 27.5 in.

For deflection, the maximum span length of wales = 33.8 in.

For this design, shear governs the maximum span length of the wales. Because the wales are supported by the ties, the spacing of the ties must not exceed 27.5 in. However, the ties must have adequate strength to support the load from the wales. For uniformity and constructability, choose a maximum span length of 24 in. for wales and determine the required strength of ties to support the wales.

Strength Required of Ties

Based on the strength of the wales in bending, shear, and deflection, the wales must be supported every 24 in. by a tie. A tie must resist the concrete pressure that acts over an area consisting of the spacing of the wales and the span length of the wales, 24 in. by 24 in. The load on a tie can be calculated as follows:

$$\text{Area} = 2.0 \text{ ft} \times 2.0 \text{ ft}$$
$$= 4.0 \text{ sq ft.}$$

$$\text{Load on a tie} = 713 \text{ lb per sq ft (4.0 sq ft)}$$
$$= 2,852 \text{ lb}$$

To support a wale at every 24 in., the tie must have an allowable load of 2,852 lb. Thus, a standard strength 3,000 lb working load tie is adequate. The lateral pressure of freshly placed concrete on wall forms is carried by the ties. However, the wall forms must be adequately braced to resist any eccentric loads or lateral forces due to wind. The minimum lateral force applied at the top of wall forms are presented in Chapter 5 (refer to Table 5-7).

Results of the Design of the Forms for the Concrete Wall

The allowable stresses and deflection criteria used in the design are based lumber on the following values:

For the 2 × 4 S4S No. 2 grade Southern Pine lumber:
Allowable bending stress = 1,875 lb per sq in.
Allowable shear stress = 218 lb per sq in.
Modulus of elasticity = 1,600,000 lb per sq in.
Permissible deflection less than $l/360$, but not greater than $\frac{1}{16}$ in.

For the ¾-in.-thick Plyform Class I sheathing:
Allowable bending stress = 1,930 lb per sq in.
Allowable shear stress = 72 lb per sq in.
Modulus of elasticity = 1,650,000 lb per sq in.
Permissible deflection less than $l/360$, but not greater than $\frac{1}{16}$ in.

Figure 9-4 shows the design of the forms that should be built to the following conditions:

Item	Nominal Size and Spacing
Sheathing	¾-in. thick Plyform Class I
Studs	2 × 4 at 12 in. on centers
Wales	double 2 × 4 at 24 in. on centers
Form ties	3,000 lb capacity at 24 in.

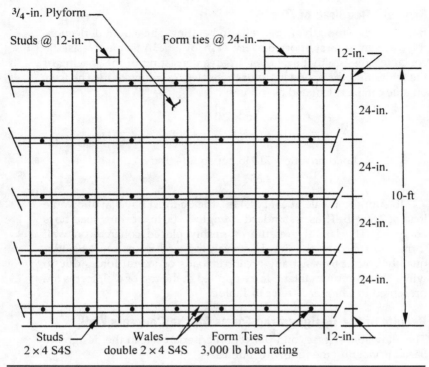

Figure 9-4 Summary of design for a concrete wall.

Tables to Design Wall Forms

An alternate method for designing wall forms is to use tables, such as those that have been developed by the APA—The Engineered Wood Association or by companies in the construction industry that supply formwork products. Tables are available that provide the allowable span length of plywood for a given concrete pressure. The design simply involves selecting the allowable concrete pressure based on the species and stress rating of the plywood, or selecting the allowable span length of framing lumber to support plywood sheathing.

Examples of these types of tables are presented in Chapter 4 for Plyform, which is manufactured by the plywood industry specifically for use in formwork for concrete. Tables 9-7 through 9-10 are reproduced in the following sections from Chapter 4 for application to the design of wall forms. When the design of formwork is based on design tables of manufacturers or associations in the construction industry, it is important to follow their recommendations.

Plywood Thickness, in.	Support Spacing, in.						
	4	**8**	**12**	**16**	**20**	**24**	**32**
15/32	2,715 (2,715)	885 (885)	355 (395)	150 (200)	— 115	— —	— —
½	2,945 (2,945)	970 (970)	405 (430)	175 (230)	100 (135)	— —	— —
19/32	3,110 (3,110)	1,195 (1,195)	540 (540)	245 (305)	145 (190)	— (100)	— —
⅝	3,270 (3,270)	1,260 (1,260)	575 (575)	265 (325)	160 (210)	— (100)	— —
23/32	4,010 (4,010)	1,540 (1,540)	695 (695)	345 (325)	210 (270)	110 (145)	— —
¾	4,110 (4,110)	1,580 (1,580)	730 (730)	370 (410)	225 (285)	120 (160)	— —
1⅛	5,965 (5,965)	2,295 (2,295)	1,370 (1,370)	740 (770)	485 (535)	275 (340)	130 (170)

Notes:
1. Courtesy APA—The Engineered Wood Association, "Concrete Forming," 2004.
2. Deflection limited to $l/360$th of the span, $l/270$th for values in parentheses.
3. Plywood continuous across two or more spans.

TABLE 9-7 Recommended Maximum Pressure on Plyform Class I—Values in Pounds per Square Foot with Face Grain Across Supports

Calculating the Allowable Concrete Pressure on Plyform

When computing the allowable pressure of concrete on plywood or Plyform, the center-to-center distances between supports should be used to determine the pressure based on the allowable bending stress in the fibers.

When computing the allowable pressure of concrete on plywood or Plyform as limited by the permissible deflection of the plywood, it is proper to use the clear span between the supports plus ¼ in. for supports whose nominal thicknesses are 2 in., and the clear span between the supports plus ⅝ in. for the supports whose nominal thicknesses are 4 in.

The recommended concrete pressures are influenced by the number of continuous spans. For face grain across supports, assume three continuous spans up to 32-in. support spacing and two spans for greater spacing. For face grain parallel to supports, assume three

| Plywood | Support Spacing, in. | | | | | |
Thickness, in.	4	8	12	16	20	24
15/32	1,385	390	110	—	—	—
	(1,385)	(390)	(150)	—	—	—
1/2	1,565	470	145	—	—	—
	(1,565)	(470)	(195)	—	—	—
19/32	1,620	530	165	—	—	—
	(1,620)	(530)	(225)	—	—	—
5/8	1,770	635	210	—	—	—
	(1,770)	(635)	(280)	120	—	—
23/32	2,170	835	375	160	115	—
	(2,170)	(835)	(400)	(215)	(125)	—
3/4	2,325	895	460	200	145	—
	(2,325)	(895)	(490)	(270)	(155)	(100)
1 1/8	4,815	1,850	1,145	710	400	255
	(4,815)	(1,850)	(1,145)	(725)	(400)	(255)

Notes:
1. Courtesy APA—The Engineered Wood Association, "Concrete Forming," 2004.
2. Deflection limited to $l/360$th of the span, $l/270$th for values in parentheses.
3. Plywood continuous across two or more spans.

TABLE 9-8   Recommended Maximum Pressure on Plyform Class I—Values in Pounds per Square Foot with Face Grain Parallel to Supports

spans up to 16 in. and two spans for 20 and 24 in. These are general rules as recommended by the APA.

There are many combinations of frame spacings and Plyform thicknesses that may meet the structural requirements for a particular job. However, it is recommended that only one thickness of plywood be used, then the frame spacing varied for the different pressures. Plyform can be manufactured in various thicknesses, but it is good practice to base design on 19/32-, 5/8-, 23/32-, and 3/4-in. Plyform Class I, because they are the most commonly available thicknesses.

Tables 9-7 through 9-10 give the recommended maximum pressures of concrete on plywood decking for Class I and Structural I Plyform. Calculations for these pressures were based on deflection limitations of $l/360$ and $l/270$ of the span, or on shear or bending strength, whichever provided the most conservative (lowest load) value. When computing the allowable pressure of concrete on plywood as limited by the allowable unit shearing stress and shearing deflection of the plywood, use the clear span between the supports.

Plywood Thickness, in.	Support Spacing, in.						
	4	8	12	16	20	24	32
15/32	3,560 (3,560)	890 (890)	360 (395)	155 (205)	115 115	— —	— —
½	3,925 (3,925)	980 (980)	410 (435)	175 (235)	100 (135)	— —	— —
19/32	4,110 (4,110)	1,225 (1,225)	545 (545)	245 (305)	145 (190)	— (100)	— —
5/8	4,305 (4,305)	1,310 (1,310)	580 (580)	270 (330)	160 (215)	— 100	— —
23/32	5,005 (5,005)	1,590 (1,590)	705 (705)	350 (400)	210 (275)	110 (150)	— —
¾	5,070 (5,070)	1,680 (1,680)	745 (745)	375 (420)	230 (290)	120 (160)	— —
1⅛	7,240 (7,240)	2,785 (2,785)	1,540 (1,540)	835 (865)	545 (600)	310 (385)	145 (190)

Notes:
1. Courtesy APA—The Engineered Wood Association, "Concrete Forming," 2004.
2. Deflection limited to $l/360$th of the span, $l/270$th for values in parentheses.
3. Plywood continuous across two or more spans.

TABLE 9-9   Recommended Maximum Pressures on Structural I Plyform—Values in Pounds per Square Foot with Face Grain Across Supports

Equations (9-1) through (9-7) were used in determining the recommended maximum pressures of concrete on plywood used for decking or sheathing, as given in Tables 9-7 through 9-10.

Allowable Pressure Based on Fiber Stress in Bending

For pressure controlled by bending stress, use Eqs. (9-1) and (9-2),

$$w_b = 96F_bS_e/(l_b)^2 \quad \text{for two spans} \tag{9-1}$$
$$w_b = 120F_bS_e/(l_b)^2 \quad \text{for three spans} \tag{9-2}$$

where w_b = uniform pressure of concrete, lb per sq ft
   F_b = allowable bending stress in plywood, lb per sq in.
   S_e = effective section modulus of a plywood strip 12 in. wide, in.^{3}/ft
   l_b = length of span, center-to-center of supports, in.

Plywood Thickness, in.	Support Spacing, in.					
	4	8	12	16	20	24
15/32	1,970 (1,970)	470 (530)	130 (175)	— —	— —	— —
1/2	2,230 (2,230)	605 (645)	175 (230)	— —	— —	— —
19/32	2,300 (2,300)	640 (720)	195 (260)	— (110)	— —	— —
5/8	2,515 (2,515)	800 (865)	250 (330)	105 (140)	— (100)	— —
23/32	3,095 (3,095)	1,190 (1,190)	440 (545)	190 (255)	135 (170)	— —
3/4	3,315 (3,315)	1,275 (1,275)	545 (675)	240 (315)	170 (210)	— (115)
1 1/8	6,860 (6,860)	2,640 (2,640)	1,635 (1,635)	850 (995)	555 (555)	340 (355)

Notes:
1. Courtesy APA—The Engineered Wood Association, "Concrete Forming," 2004.
2. Deflection limited to $l/360$th of the span, $l/270$th for values in parentheses.
3. Plywood continuous across two or more spans.

TABLE 9-10   Recommended Maximum Pressures on Structural I Plyform—Values in Pounds per Square Foot with Face Grain Parallel to Supports

Allowable Pressure Based on Bending Deflection

For pressure controlled by bending deflection, Eqs. (9-3) and (9-4),

$$w_d = 2220EI\Delta_b/(l_s)^4 \quad \text{for two spans} \qquad (9\text{-}3)$$

$$w_d = 1743EI\Delta_b/(l_s)^4 \quad \text{for three or more spans} \qquad (9\text{-}4)$$

where Δ_b = permissible deflection of plywood, in.
w_d = uniform pressure of concrete, lb per sq ft
l_s = clear span of plywood plus 1/4 in. for 2-in. supports, and clear span plus 5/8 in. for 4-in. supports, in.
E = modulus of elasticity of plywood, lb per sq in.
I = moment of inertia of a plywood strip 12 in. wide, in.^{4}/ft

Allowable Pressure Based on Shear Stress

For pressure controlled by shear stress, use Eqs. (9-5) and (9-6),

$$w_s = 19.2F_s(Ib/Q)/l_s \quad \text{for two spans} \tag{9-5}$$

$$w_s = 20F_s(Ib/Q)/l_s \quad \text{for three spans} \tag{9-6}$$

where w_s = uniform load on the plywood, lb per sq ft
F_s = allowable rolling shear stress, lb per sq in.
Ib/Q = rolling shear constant, in.^{2} per ft of width
l_s = clear span between supports, in.

Allowable Pressure Based on Shear Deflection

For pressure controlled by shear deflection, use Eq. (9-7),

$$\Delta_s = w_s Ct^2 (l_s)^2/1{,}270E_e I \tag{9-7}$$

where Δ_s = permissible deflection of the plywood, in.
w_s = uniform load, lb per sq ft
C = a constant, equal to 120 for the face grain of plywood perpendicular to the supports and equal to 60 for the face grain of the plywood parallel to the supports
t = thickness of plywood, in.
E_e = modulus of elasticity of the plywood, unadjusted for shear deflection, lb per sq in.
I = moment of inertia of a plywood strip 12 in. wide, in.^{4}/ft

Maximum Spans for Lumber Framing Used to Support Plywood

Tables 9-11 and 9-12 give the maximum spans for lumber framing members, such as studs and joists that are used to support plywood subjected to pressure from concrete. The spans listed in Table 9-11 are based on using grade No. 2 Douglas Fir or grade No. 2 Southern Pine. The spans listed in Table 9-12 are based on using grade No. 2 Hem-Fir. The allowable stresses are based on a load-duration less than 7 days and moisture content less than 19%. The deflections are limited to $l/360$ with maxima not to exceed $1/4$ in.

Although Table 9-11 and 9-12 are for single members; these tables can be adapted for use with multiple member, such as double wales. For example, suppose the total load on double 2 × 4 wales is 1,200 lb per lin ft. Because the wales are doubled, each 2 × 4 wale carries $1{,}200/2 = 600$ lb per lin ft. If the wales are No. 2 Southern Pine over more than four supports, Table 9-11 shows a 32-in. span for 600 lb per lin ft.

Equivalent Uniform Load, lb per ft	Douglas Fir #2 or Southern Pine #2 Continuous (1 or 2 Spans), 3 Supports, Nominal Size Lumber							Douglas Fir #2 or Southern Pine #2 Continuous over 4 or more Supports (3 or more Spans), Nominal Size of Lumber						
	2 × 4	2 × 6	2 × 8	2 × 10	4 × 4	4 × 6	4 × 8	2 × 4	2 × 6	2 × 8	2 × 10	4 × 4	4 × 6	4 × 8
200	48	73	92	113	64	97	120	56	81	103	126	78	114	140
400	35	52	65	80	50	79	101	39	58	73	89	60	88	116
600	29	42	53	65	44	64	85	32	47	60	73	49	72	95
800	25	36	46	56	38	56	72	26	41	52	63	43	62	82
1000	22	33	41	50	34	50	66	22	35	46	56	38	56	73
1200	19	30	38	46	31	45	60	20	31	41	51	35	51	67
1400	18	28	35	43	29	42	55	18	28	37	47	32	47	62
1600	16	25	33	40	27	39	52	17	26	34	44	29	44	58
1800	15	24	31	38	25	37	49	16	24	32	41	27	42	55
2000	14	23	29	36	24	35	46	15	23	30	39	25	39	52
2200	14	22	28	34	23	34	44	14	22	29	37	23	37	48
2400	13	21	27	33	21	32	42	13	21	28	35	22	34	45
2600	13	20	26	31	20	31	41	13	20	27	34	21	33	43
2800	12	19	25	30	19	30	39	12	20	26	33	20	31	41
3000	12	19	24	29	18	29	38	12	19	25	32	19	30	39

3200	12	18	23	28	18	28	37	12	19	24	31	18	29	38
3400	11	18	22	27	17	27	35	11	18	23	30	18	28	36
3600	11	17	22	27	17	26	34	11	18	23	30	17	27	35
3800	11	17	21	26	16	25	33	11	17	22	29	16	26	34
4000	11	16	21	25	16	24	32	11	17	22	28	16	25	33
4200	11	16	20	25	15	24	31	11	17	22	28	16	24	32
4400	10	16	20	24	15	23	31	10	16	22	27	15	24	31
4600	10	15	19	24	14	23	30	10	16	21	26	15	23	31
4800	10	15	19	23	14	22	29	10	16	21	26	14	23	30
5000	10	15	18	23	14	22	29	10	16	21	25	14	22	29

Notes:

1. Courtesy APA—The Engineered Wood Association, "Concrete Forming," 2004.
2. Spans are based on the 2001 NDS allowable stress values, $C_D = 1.25$, $C_r = 1.0$, $C_M = 1.0$.
3. Spans are based on dry, single-member allowable stresses multiplied by a 1.25 duration-of-load factor for 7-day loads.
4. Deflection is limited to $l/360$th of the span with $l/4$ in. maximum.
5. Spans are measured center-to-center on the supports.

TABLE 9-11   Maximum Spans, in Inches, for Lumber Framing Using Douglas-Fir No. 2 or Southern Pine No. 2

Equivalent Uniform Load, lb per ft	Hem-Fir #2 Continuous over 2 or 3 Supports (1 or 2 Spans), Nominal Size Lumber							Hem-Fir #2 Continuous over 4 or more Supports (3 or more Spans), Nominal Size of Lumber						
	2 × 4	2 × 6	2 × 8	2 × 10	4 × 4	4 × 6	4 × 8	2 × 4	2 × 6	2 × 8	2 × 10	4 × 4	4 × 6	4 × 8
200	45	70	90	110	59	92	114	54	79	100	122	73	108	133
400	34	50	63	77	47	74	96	38	56	71	87	58	86	112
600	28	41	52	63	41	62	82	29	45	58	71	48	70	92
800	23	35	45	55	37	54	71	23	37	48	61	41	60	80
1000	20	31	40	49	33	48	64	20	32	42	53	37	54	71
1200	18	28	36	45	30	44	58	18	28	37	47	33	49	65
1400	16	25	33	41	28	41	54	16	26	34	43	29	45	60
1600	15	23	31	39	25	38	50	15	24	31	40	26	41	54
1800	14	22	29	37	23	36	48	14	22	30	38	24	38	50
2000	13	21	28	35	22	34	45	14	23	28	36	22	35	46
2200	13	20	26	33	20	32	42	13	21	27	34	21	33	43
2400	12	19	25	32	19	30	40	12	20	26	33	20	31	41
2600	12	19	25	30	18	29	38	12	20	25	32	19	30	39
2800	12	18	24	29	18	28	36	12	19	24	31	18	28	37
3000	11	18	23	28	17	26	35	11	18	24	30	17	27	36

3200	11	17	22	27	16	25	34	11	18	23	29	17	26	34
3400	11	17	22	27	16	25	32	11	17	22	29	16	25	33
3600	11	17	21	26	15	24	31	11	17	22	28	16	24	32
3800	10	16	21	25	15	23	31	10	17	22	28	15	24	31
4000	10	16	20	24	14	23	30	10	16	21	27	15	23	30
4200	10	15	10	24	14	22	29	10	16	21	27	14	22	30
4400	10	15	19	24	14	22	28	10	16	21	26	14	22	29
4600	10	15	19	23	13	21	28	10	15	20	26	14	21	28
4800	10	14	19	22	13	21	27	10	15	20	25	13	21	28
5000	10	14	18	22	13	20	27	10	15	20	24	13	21	27

Notes:

1. Courtesy APA—The Engineered Wood Association, "Concrete Forming," 2004.
2. Spans are based on the 2001 NDS allowable stress values, $C_D = 1.25$, $C_r = 1.0$, $C_M = 1.0$.
3. Spans are based on dry, single-member allowable stresses multiplied by a 1.25 duration-of-load factor for 7-day loads.
4. Deflection is limited to $l/360$th of the span with $l/4$ in. maximum.
5. Spans are measured center-to-center on the supports.

TABLE 9-12 Maximum Spans, in Inches, for Lumber Framing Using Hem-Fir No. 2

239

Using Tables to Design Forms

Use the information in Tables 9-7 through 9-12 to design the forms for a concrete wall 12 in. thick and 11 ft high. The forms will be filled at a rate of 4 ft per hr at a temperature of 80°F. The 150 lb per cu ft concrete with Type I cement and no retarders will be consolidated with an internal vibrator at a depth not to exceed 4 ft.

Table 3-3 indicates a maximum pressure of 600 lb per sq ft. There are several possible designs. Generally, the most satisfactory design is the one that will furnish the necessary services at the lowest cost. The forms will be built at the job using Plyform Class I sheathing, No. 2 grade Hem-Fir for 2 × 4 studs and two 2 × 4 members for wales, all S4S lumber.

The Plyform will be placed with face grain across supports. Table 9-7 indicates two possible choices for thicknesses of Plyform Class I sheathing: ¾-in.-thick Plyform with spacing of supports at 12 in. on centers, or 1⅛-in.-thick Plyform with spacing of supports at 16 in. on centers. For this design, choose the ¾-in.-thick Plyform Class I.

For 12-in. spacing of support studs, the uniform load on a stud will be (600 lb per sq ft)(1 ft) = 600 lb per lin ft. Table 9-12 indicates the maximum span of 28 in. for 2 × 4 No. 2 grade Hem-Fir lumber. Therefore, the spacing of wales must not exceed 28 in.

For this spacing of wales, the load per foot on a wale will be (28/12 ft)(600 lb per sq ft) = 1,400 lb per ft. Consider using standard 3,000-lb load capacity form ties. The spacing of the ties must not exceed 3,000 lb/1,400 lb per ft = 2.14 ft, or 12(2.14) = 25.7 in. For constructability, the form ties will be placed at 24 in. center to center along the wales.

The bottom wale should be placed not more than 8 in. above the bottom of the form, and other wales at vertical intervals of 28 in. thereafter, with the top wale at 12 in. below the top of the form.

As illustrated in Figure 9-5, the pressure on the form will increase at a uniform rate from zero at the top of the wall to a maximum value of 600 lb per sq ft at a depth of 4 ft below the top of the wall. The 4 ft is calculated as (300 lb per sq ft)/(150 lb per cu ft) = 4.0 ft. The spacing of the wales in this top 4 ft of wall may be increased because of the reduced pressure, but the permissible increase will not be enough to eliminate a row of wales. Therefore, five rows of wales will be required.

Forms for Walls with Batters

Figure 9-6 illustrates a set of forms for a wall whose thickness varies. Variable-length ties are available for use with forms of this type. The procedures presented in earlier sections can be used to design the forms for this type of wall.

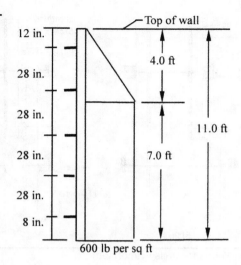

FIGURE 9-5
Wale spacing for
11-ft-high wall.

12 in.

28 in.

28 in.

28 in.

28 in.

8 in.

Top of wall

4.0 ft

11.0 ft

7.0 ft

600 lb per sq ft

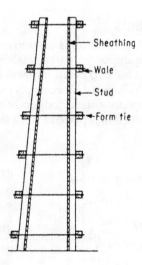

FIGURE 9-6 Forms
for walls with
batters.

Sheathing

Wale

Stud

Form tie

Forms for Walls with Offsets

Figure 9-7 illustrates two methods of erecting forms for walls with offsets. Each form tie is equipped with two washers, which are properly spaced to bear against the two inside surfaces of the sheathing to serve as form spreaders. Plan A is suitable for use where low pressures will occur, but Plan B should be used where high pressures will occur.

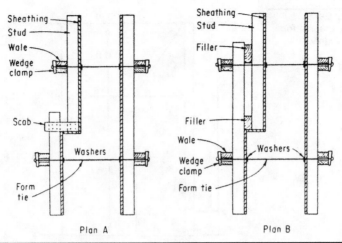

FIGURE 9-7   Forms for walls with offsets.

Forms for Walls with Corbels

Figure 9-8 illustrates the details of forms used to construct walls with corbels. Each form tie is equipped with two washers, which are properly spaced to bear against the two inside surfaces of the sheathing to serve as form spreaders. A diagonal brace or plywood gusset may be placed across the form components at the top of the corbel to resist uplift pressure from concrete that is placed from a large distance above the corbel.

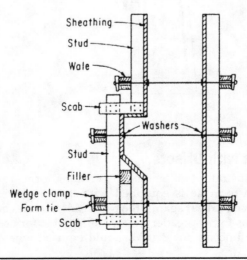

FIGURE 9-8   Forms for walls with corbels.

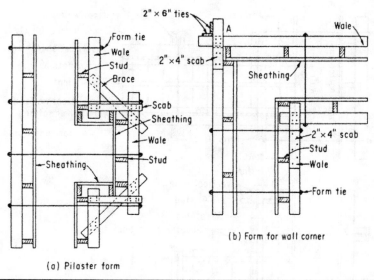

FIGURE 9-9 Details of wall form.

Forms for Walls with Pilasters and Wall Corners

Figure 9-9(a) illustrates a method of constructing forms for a wall with a pilaster. The forms for the wall are erected first, followed by the forms for the pilaster. This order is reversed in stripping the forms.

Figure 9-9(b) illustrates a method of constructing forms for a wall corner. Because nails alone may not provide sufficient strength to join the two wales at Point A, two 2 × 6 ties, extending the full height of the form, are attached as shown to provide additional strength.

Forms for Walls with Counterforts

Figure 9-10 illustrates a method of constructing forms for a wall with a counterfort. The wall forms are erected first, and then the two sides and top form for the counterfort are prefabricated and assembled in position, as illustrated.

Because of the substantial pressure exerted by the fresh concrete on the back sheathing of the counterfort, this form must be securely fastened to the wall forms to prevent it from pulling away from the wall forms. Also, the counterfort forms must be securely anchored to prevent vertical uplift from the pressure on the underside of the back sheathing.

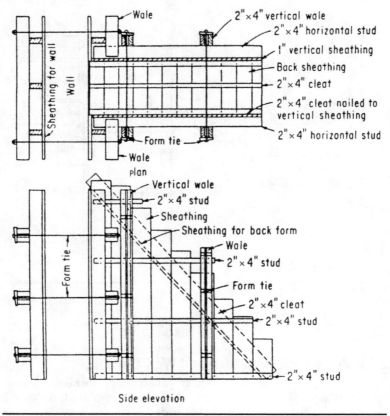

FIGURE 9-10   Forms for wall with counterforts.

Forms for Walls of Circular Tanks

The forms for walls of circular tanks must be modified because of the curvature of the walls. For tanks whose inside diameters are 30 ft or more, the sheathing usually consists of 1-in. lumber, with the planks extending around the wall, or of ⅝- or ¾-in.-thick plywood, depending on the rate of placement and the species and grade of the material used for formwork. The equations and procedures presented in previous sections and in Chapter 5 may be used to determine the required size of sheathing. See Table 4-19 for the minimum bending radii for plywood.

The size and spacing of studs and wales may be designed following the procedures described in Chapter 5. However, the wales must be installed flat against the studs, as illustrated in Figure 9-11. Template rings of 2-in.-thick lumber, sawed to the correct curvature, are assembled and, when in position, help bear against the edges of the studs on the inside form. Template rings should be spaced from 6 to 10 ft apart

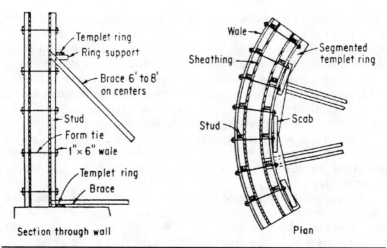

Figure 9-11   Forms for wall of circular tank, 30 ft or more in diameter.

vertically. Horizontal and inclined radial braces, spaced 6 to 8 ft apart, should be installed to secure the inside form in position.

The inside form should be erected first, complete with wales and braces. Form ties are installed along each wale adjacent to each stud. Because of the low strength of the flat wales, ties must be installed adjacent to, instead of between, the studs. The outside form is then erected, and the tie holders are secured against the wales.

Figure 9-12 illustrates the method of constructing forms for a circular tank whose diameter is less than 30 ft. Vertical planks, usually

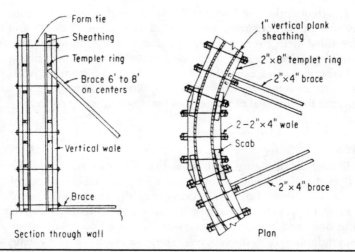

Figure 9-12   Forms for wall of circular tank, less than 30 ft in diameter.

1 in. thick, are used for sheathing. The template rings are sawed to the correct curvatures from 2-in.-thick lumber. If planks as wide as 12 in. in nominal width are used, it is possible to obtain an inside and outside ring segment from each plank. The outside segments are sawed to the correct curvature. The inside segments may be resawed to the correct inside curvature, or if a slight deviation from a true circle is not objectionable, they need not be resawed. The larger the radius, the less there will be a need to resaw the inside segment.

The vertical wales are usually double 2 × 4 planks. The inside form should be erected first and securely braced in the correct position. Then the outside form is erected, and the form ties are secured on both sides.

The equations and procedures presented in Chapter 5 may be used to determine the maximum spacing of the template rings, wales, and form ties. As indicated in Figure 9-12, the spacing of the rings may be increased for the upper portion of the wall, corresponding to the reduction in the pressure from the concrete.

Form Ties

As illustrated in Figure 9-1, form ties are used with forms for walls to hold the two sides against the pressure of the concrete. In addition to resisting the pressure from the concrete, some ties are designed to serve as form spreaders. Many types and sizes are available.

The maximum spacing of ties may be limited by the strength of the ties, the maximum safe span for the wales, or the maximum safe span of the studs in the event wales are not used. The manufacturers of ties specify the safe working load for each size and type manufactured.

It is frequently specified that ties must be removed from a concrete wall or that portions must be removed to a specified depth and the holes filled with cement mortar to eliminate the possibility of rust stains appearing later, or to prevent water from seeping through a wall around the ties. Smooth-rod ties can be pulled from the wall after the forms are removed. Other methods, illustrated hereafter, are used to permit the removal of portions of the ties within the concrete.

Snap Ties

As illustrated in Figure 9-13, these ties are made from a single rod equipped with an enlarged button or a loop at each end to permit the use of suitable tie holders. A portion of the rod in the concrete is crimped to prevent it from turning when the ends are removed by being bent and twisted. Ties can be furnished to break off at any desired depth within the concrete. They are available with washers or with tapered plastic cones attached to permit them to serve as form

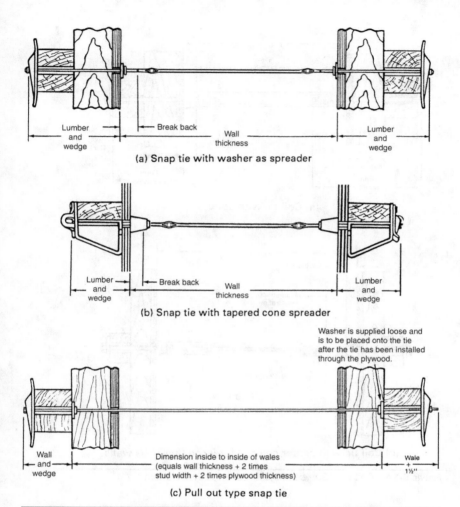

(a) Snap tie with washer as spreader

Lumber and wedge — Break back — Wall thickness — Lumber and wedge

(b) Snap tie with tapered cone spreader

Lumber and wedge — Break back — Wall thickness — Lumber and wedge

Washer is supplied loose and is to be placed onto the tie after the tie has been installed through the plywood.

Wall and wedge — Dimension inside to inside of wales (equals wall thickness + 2 times stud width + 2 times plywood thickness) — Wale + 1½"

(c) Pull out type snap tie

Figure 9-13 Snap tie assembly. (*Source: Dayton Superior Corporation*)

spreaders. Safe working-load capacities of 2,000 and 3,000 lb are available from most of the manufacturers of this type of form tie.

When ordering ties, it is necessary to specify the type and strength desired, the thickness of the concrete wall, and the dimensions of the form lumber, sheathing, studs, and wales. The desired breakback within the wall should also be specified.

Coil Ties

These ties are illustrated in Figures 9-14 and 9-15. The internal member of the assembly, which remains in the concrete, consists of two helical coils welded to two or four steel rods designated as struts.

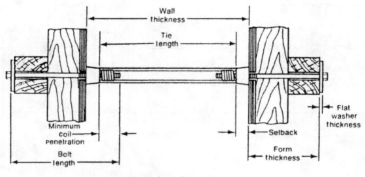

(a) Coil tie with cone

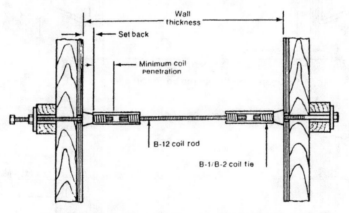

(b) Coil tie with center coil rod for large thickness walls

FIGURE 9-14   Coil tie. (*Source: Dayton Superior Corporation*)

The external members of the assembly consist of two coil bolts whose threaded ends are screwed into the coils to transmit the loads to the wales. A flat steel washer is used under the head of each bolt to provide adequate bearing area against the wale.

A wood or plastic cone may be installed at each end of the inner member to serve as a form spreader and to hold the metal back from the surface of the concrete wall. Wood cones absorb moisture from the fresh concrete and swell slightly. After the forms have been removed, the wooden cones dry out and shrink, allowing them to be removed easily. Plastic cones may be loose or screwed onto the helical coil. After cones are removed, the holes are filled with cement mortar.

For walls of great thickness, two inner members connected with a continuous threaded coil rod of the proper length may be used to produce any desired length.

FIGURE 9-15
Form panel
showing screw-on
coil ties in place
prior to installation
of reinforcing steel
and erection of
opposite form.
(*Source: Dayton
Superior
Corporation*)

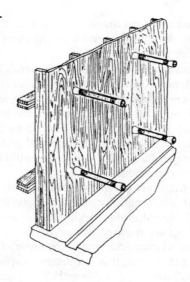

Strut coil ties are available in the range of 4,000 to 25,000 lb safe work-ing load capacity. When ordering these ties, it is necessary to give the quantity, type, safe working load, bolt diameter, tie length, wall thickness, and setback. For example, 3,000 pieces, heavy-duty two-strut, 9,000-lb capacity, ¾-in. diameter bolt, 12 in. long for a 14-in. wall, 1-in. setback.

Taper Ties

As illustrated in Figure 9-16, these ties consist of a tapered rod that is threaded at each end. A feed-through installation permits forms on both sides to be in place before ties are installed. The tapered tie system

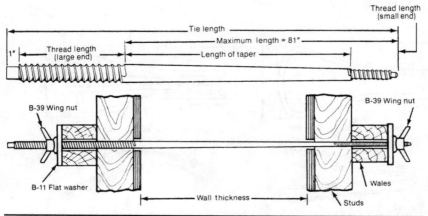

FIGURE 9-16   Taper tie. (*Source: Dayton Superior Corporation*)

can be quickly assembled and disassembled, which makes it a versatile forming system. Wing nuts or coil nuts with washers are attached at each end to secure the tie.

The ties are completely reusable. No expendable parts are left behind in the hardened concrete. However, taper ties will wear and must be inspected and replaced continually when wear or damage is noted. To facilitate removal of the taper tie from the hardened concrete, a coating with waterproof grease should be applied to the tie before installation. For best results in removal of the tie, the initial bond between the taper tie and the concrete should be broken within 24 hours of concrete placement.

Standard tie lengths are available from 34 to 60 in. Thus, the ties provide an adjustable tie length for a wide variety of wall thicknesses. These ties are available with safe working loads varying from 7,500 to 50,000 lb, depending on the coil thread diameter. The coil thread diameter is different at each end, due to the taper of the tie.

When ordering these ties, it is necessary to specify the quantity, type, large and small diameter, and length. This type of tie is ideal for walls with batters.

Coil Loop Inserts for Bolt Anchors

Coil loops are set in concrete as anchors, and then used later to support higher lifts of forms, to lift precast concrete members, and for other anchorage purposes. As illustrated in Figure 9-17, several types and sizes are available. Anchorage is obtained with coil bolts and eye bolts whose threads engage the helical coils.

The safe working load capacity of inserts depends on the diameter and length of the bolt, the edge distance, and the concrete strength. Working loads range from 9,500 lb for single flared coil loop inserts to 32,000 lb for double flared coil loops, up to 48,000 lb for triple-flared coil loop inserts.

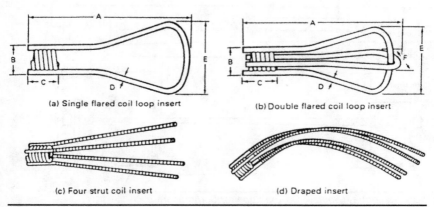

(a) Single flared coil loop insert

(b) Double flared coil loop insert

(c) Four strut coil insert

(d) Draped insert

FIGURE 9-17 Coil loop inserts for bolt anchors. (*Source: Dayton Superior Corporation*)

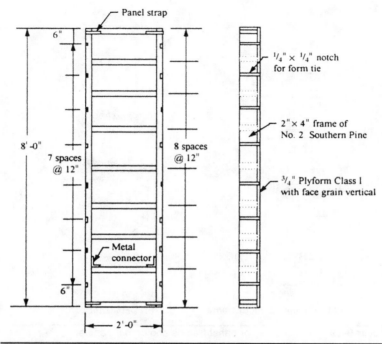

Figure 9-18 Prefabricated panel for a 600 lb per sq ft pressure.

Prefabricated Wood Form Panels

Prefabricated wood panels of the types illustrated in Figure 9-18 have several advantages when compared with wall forms built in place. They can be assembled quite rapidly once a pattern is fabricated for each part. They can be erected and removed quickly. If they are constructed solidly, they can be used many times.

They are erected side by side to produce a wall form of any desired length. Filler panels less than 2 ft wide will enable them to be used for walls whose lengths are not multiples of 2 ft. Adjacent panels should be bolted or nailed together while they are in use. Forms on opposite sides of a wall are held in position with form ties, placed in the indicated notches in the frames as the panels are erected. Two rows of horizontal wales, one along the bottom form ties and one along the top form ties, should be used to ensure good alignment.

Figure 9-18 shows a top view of an inside corner panel whose dimensions can be selected to fit any wall thickness. Figure 9-19 illustrates a method of connecting outside panels together at a corner. Steel corner clips are used to secure the corners of the panel system.

The studs must be nailed securely to the 2 × 4 frame members with 20d nails. A metal connector, of the size and type illustrated, should be installed at each end of each stud and nailed to the stud and frame. One

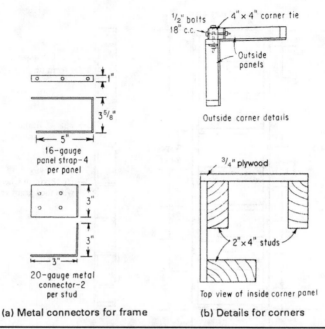

(a) Metal connectors for frame   (b) Details for corners

Figure 9-19 Connections for panel frames and corner panels.

steel panel strap applied at each corner of a frame will give the frame added strength, rigidity, and durability. Because the concrete pressure is less at the top than at the bottom, the spacing of the studs may be increased in the upper portion of a panel. However, as shown in Figure 9-18, it is better to keep the spacing constant to prevent erection of the panels in a reversed position, which could produce an unsafe installation.

Table 9-13 gives the maximum spacing of studs and form ties for a 2- by 8-ft panel for various pressures using ¾-in.-thick Plyform Class I sheathing, No. 2 grade Southern Pine 2 × 4 lumber for studs and frames, and 3,000-lb form ties. The indicated number of nails is

Pressure, lb per sq ft in.	Stud Spacing, in.	Tie Spacing, in.	Number of 20d nails per joint
300	18	48	3
350	17	48	3
400	16	45	3
450	15	40	4
500	14	36	4
550	13	32	4
600	12	30	4

Table 9-13 Maximum Spacing of Studs and Form Ties for Figure 9-17

based on using 20d nails only to fasten the studs to the frames. The use of metal connectors will permit a reduction in the number of 20d nails to a maximum of three per joint.

Commercial, or Proprietary, Form Panels

Proprietary panels, which are available in several types, are used frequently to construct forms, especially for concrete walls. Among the advantages of using these panels are many reuses, a reduction in the labor required to erect and dismantle the forms, a good fit, and a reduction in the quantity of additional lumber required for studs and wales.

Most manufacturers furnish panels of several sizes to provide flexibility in the dimensions of forms. However, where the exact dimensions of forms are not attainable with these panels, job-built filler panels may be used to obtain the desired dimensions.

A wide variety of accessories is available from manufacturers in the concrete construction industry. These accessories allow quick assembly and removal of forms and enhance safety by providing secure attachment of the form materials by patented clamping and locking devices.

Gates Single-Waler Cam-Lock System

A forming system developed by Gates & Sons, Inc. uses plywood panels, wales, strongbacks, form ties, and cam-locks to secure the tension in the ties. The panel thickness, size of frame lumber, and tie strength are designed corresponding to the pressure appropriate for each particular job.

As illustrated in Figure 9-20, the cam-lock brackets on each side of the plywood sheathing secure the loop end form tie and the horizontal

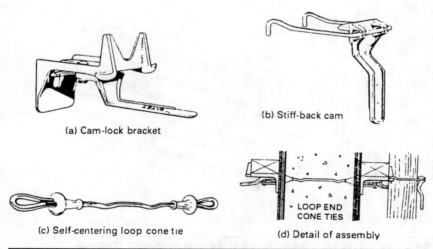

(a) Cam-lock bracket

(b) Stiff-back cam

(c) Self-centering loop cone tie

(d) Detail of assembly

LOOP END CONE TIES

Figure 9-20   Patented cam-lock hardware for wall formwork. (*Source: Gates & Sons, Inc.*)

wood walers to the plywood sheathing. The loop cone form tie is placed between two cam-lock brackets, and a 2 × 4 or 2 × 6 horizontal waler is placed on top of each cam-lock bracket. The vertical strongbacks are locked in place against the backs of the cam-lock brackets with the stiff-back cams. The stiff-back cam connects to ears on the back of the cam-lock bracket to lock the forming system in place. The self-centering plastic cone form tie is crimped to prevent turning in the concrete wall. It has high-density polyethylene cones for easy breakback of 1 or 1½ in.

This cam-lock forming system uses readily available materials, ¾- in.-thick plywood sheathing and 2 × 4 or 2 × 6 S4S lumber for horizontal wales or vertical strongbacks. Plywood panels are predrilled with ¹³⁄₁₆-in. holes to receive the ends of the form ties. As shown in Figure 9-21,

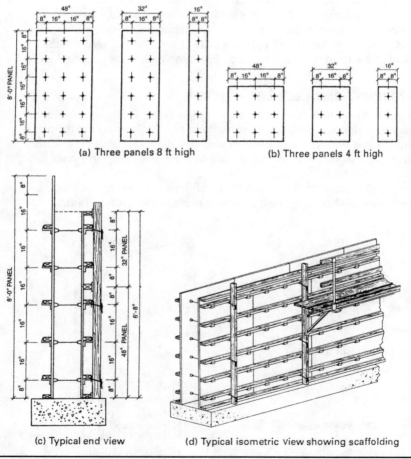

(a) Three panels 8 ft high

(b) Three panels 4 ft high

(c) Typical end view

(d) Typical isometric view showing scaffolding

Figure 9-21   Suggested form details for 16-in. tie spacing in each direction. (*Source: Gates & Sons, Inc.*)

the six basic panels suggested for tie spacing of 16 in. in each direction provide flexibility in forming for a large percentage of most jobs. By using combinations of the six basic panels, placing them in either the horizontal or vertical positions, walls may be formed with heights ranging from 16 in. to 8 ft.

A floating waler with a ¼-in. shim between the horizontal waler and the vertical strongback may be used at the top and bottom of the panels, and at the edges of adjoining panels where a form tie is not present to secure the plywood sheathing. Plyclips may also be used to stabilize the vertical and horizontal joints of plywood.

Forms for Pilasters and Corners

When a pilaster projects only a small distance from the main wall, dimension lumber may be used to form the pilaster sides. The walers should be butted against each pilaster side to give additional support. A regular plywood panel or a specially cut panel can then be nailed into the pilaster uprights with double-headed nails. Short 2 × 4s can then be locked firmly into place with cam-lock brackets. Figure 9-22(a) shows this forming arrangement using the patented single-waler cam-lock system of Gates & Sons. For wide pilasters, two ties should be used to prevent shifting or deflection. If the projection is greater than 8 in., a cross tie should be added as shown in Figure 9-22(b).

Suggested forming for a corner and a "T" wall using the Gates & Sons cam-lock system is illustrated in Figure 9-23. The outside corner on high walls may be locked securely by running two vertical 2 × 4s, as shown in Figure 9-23(b). These vertical 2 × 4s should be securely nailed into the horizontal 2 × 4s with double-headed nails.

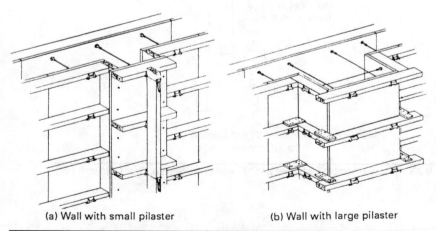

(a) Wall with small pilaster   (b) Wall with large pilaster

FIGURE 9-22 Form assembly for small and large wall pilaster. (*Source: Gates & Sons, Inc.*)

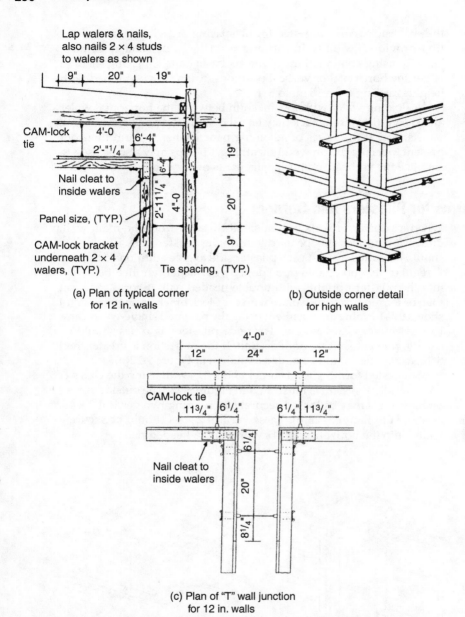

Lap walers & nails, also nails 2 × 4 studs to walers as shown

CAM-lock tie

Nail cleat to inside walers

Panel size, (TYP.)

CAM-lock bracket underneath 2 × 4 walers, (TYP.)

Tie spacing, (TYP.)

(a) Plan of typical corner for 12 in. walls

(b) Outside corner detail for high walls

CAM-lock tie

Nail cleat to inside walers

(c) Plan of "T" wall junction for 12 in. walls

FIGURE 9-23   Form assembly for corner and "T" wall end. (*Source: Gates & Sons, Inc.*)

Ellis Quick-Lock Forming System

This forming system uses a quick-lock bracket, a strongback clamp, and loop form ties to secure S4S dimension lumber and plywood panels for wall forms. Suggested details for forming low, medium, and high walls are shown in Figure 9-24.

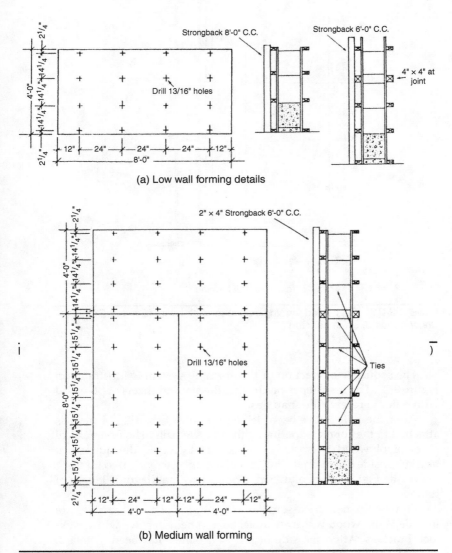

(a) Low wall forming details

(b) Medium wall forming

Figure 9-24(a) & (b) Suggested forming details for low and medium walls. (a) Low wall forming. (b) Medium wall forming. (*Source: Ellis Construction Specialties, Ltd.*)

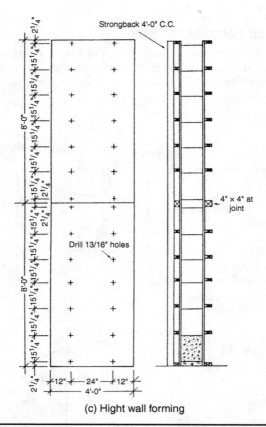

Strongback 4'-0" C.C.

8'-0"

8'-0"

Drill 13/16" holes

4" × 4" at joint

2¼" ┤12"├ 24" ┤12"├
4'-0"

(c) Hight wall forming

Figure 9-24(c) Suggested forming details for high walls. (*Source: Ellis Construction Specialties, Ltd.*)

The sequence of erection of this forming system is illustrated in Figure 9-25. The first step is predrilling the stacked sheets of plywood with holes to receive the snap ties.

After the holes have been drilled for form ties, a 2 × 4 plate is attached to the concrete footing with concrete nails. The first vertical sheet of plywood is erected and secured by using the quick-lock bracket upside down on the bottom plate, or by tacking it to the 2 × 4 plate. The first sheet of plywood is braced with a diagonal length of lumber.

Plywood panels are assembled for one side of a wall, the ties are installed, and wood wales are attached and held in place with quick-lock brackets. After one side of the wall panels has been installed, aligned, and plumbed, the loop end ties are slipped into the predrilled panels. The quick-lock brackets are then attached on the quick-lock ties to allow installation of the horizontal wood wales.

(1) Stacked sheets of 1/4-inch plywood are predrilled in preparation for wall erection.

(2) A 2 X 4 plate is then set to chalk line on concrete footing and secured with concrete nails.

(3) Vertical sheets of plywood are erected and secured by using the Quick-Lock bracket upside down on bottom plate or by tacking it to the 2 X 4 plate. Brace the first sheet with a diagonal length of lumber.

(4) Loop end ties are then slipped into the predrilled panels.

(5) Attach Quick-Lock brackets on all Quick-Lock ties and install lengths of 2 X 4 walers (see next picture).

(6) To install walers, put each one loosely across a row of Quick-Lock tilted at an angle, then strike it with a hammer to seat it firmly into the Quick-Lock Brackets.

(7) A 4 X 4 or double 2 X 4 waler should be used at the top if forms are to be stacked. Put up strong backs every 6 to 8 feet for form rigidity.

(8) Panels for subsequent lifts are placed by a man on a platform supported by scaffold brackets. These panels can be attached to the strong back with a Strong Back clamp.

(9) All in place and READY TO POUR! You know the result will be good because the job has been done right without waste, and the brackets and lumber can be reused.

FIGURE 9-25 Sequence of panel assemblies for the Ellis Quick-Lock Wall Forming System. (*Source: Ellis Construction Specialties, Ltd.*)

To install walers, each waler is loosely placed across a row of quick-locks at a tilted angle. The wale is then struck with a hammer to set it firmly into the quick-lock brackets.

A 4×4 or double 2×4 waler should be used at the top of forms, if the forms are to be stacked. Strongbacks should be placed every 6 to 8 ft for form rigidity. Panels for subsequent lifts are placed by a worker on a platform supported by scaffold brackets. These panels can be attached to the strongback with a strongback clamp.

Jahn System for Wall Forms

This system of forming walls involves the use of plywood panels, wales, strongbacks, form ties, and the patented Jahn brackets for attaching form members. The panel thickness, size, and spacing of frame lumber, and tie strength are designed corresponding to the pressure for each particular job. Figure 9-26 illustrates typical accessories for attaching studs, wales, ties, and strongbacks.

(a) Jahn "A" bracket for securing studs or wales in wall forms

(b) Jahn "C" bracket for securing strongbacks for wall forms

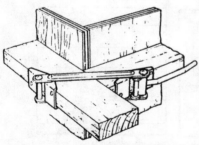

(c) Jahn Cornerlock for securing wales on outside corners of wall forms

(d) Jahn foot clip for securing plywood sheathing to footings for stem walls

Panel No. 1   Panel No. 2

(e) Jahn plywood holder for securing two adjacent panels of plywood sheathing

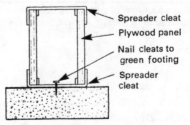

Spreader cleat
Plywood panel
Nail cleats to green footing
Spreader cleat

(f) Spreader cleats for forming footers or grade beams for short height forms

Figure 9-26   Clamping and locking devices for wall forms. (*Source: Dayton Superior Corporation*)

The commercially available Jahn system of patented hardware attachments is illustrated in Figure 9-26. The Jahn "A" bracket, patented in Canada, is shown in Figure 9-26(a). It can be installed either before or after walers are in place. The slots in the bracket allow it to slip easily over the snap tie end, eliminating laborious threading through a hole. Pressure of the bracket body is against the 2×4 instead of the plywood. The bracket can be used for any type of wall forms: round, curved, battered, beams, or columns.

Figure 9-26(b) illustrates the U.S. patented Jahn "C" bracket, which is used to attach vertical strongbacks for formwork alignment. It is designed for use with single 2×4 studs, double 2×4 wales, and snap ties. The eccentric bracket securely holds formwork while compensating for minor variations in lumber sizes. The bracket and double wales can also be used to support a horizontal plywood joint.

Figure 9-26(c) illustrates the Jahn cornerlock for securing 2×4 walers. Only two nails are needed for attachment, whereas barbed plates grip the side of the 2×4 lumber for positive nonslip action. The locking handle has a cam-action, which draws the wales together at true right angles. No special tools are needed for either installation or stripping.

Figure 9-26(d) illustrates the Jahn footing-clip, which may be used in place of a 2×4 plate for securing plywood sheathing to the top of grade beams. Figure 9-26(e) shows the Jahn plywood-holder for attachment of two adjacent plywood panels.

Figure 9-26(f) shows metal spreader cleats for securing ¾-in. plywood panels for walls up to 18 in. high without snap ties. The spreader cleats are spaced at 2-ft maximum centers.

Preparation for the forming involves gang drilling the plywood panels with holes drilled ⅛ in. larger than the snap tie head. Normally a ⅝- in.-diameter drill bit will be adequate. The ⅝-in. take-up on the Jahn "A" bracket allows a snap tie to be used with ⅝-in. or ¾-in. plywood. The ⅝-in. eccentric take up will also allow the Jahn "C" bracket and snap ties to be used with ⅝-in. or ¾-in. plywood.

The most common snap tie spacings used with the Jahn forming system are shown in Figure 9-27 for two placement rates of concrete. The spacings shown are based on using ¾-in. Plyform Class I placed in the strong direction, face grain parallel to the spacing, and 2×4 S4S studs of Douglas Fir-Larch or Southern Pine with a minimum allowable fiber stress of 1,200 lb per sq in. For different rates of pour, the manufacturer should be consulted for technical assistance.

Footing plates or clips are attached to a level footing surface as a starting point for the forms. The Jahn foot clips should be attached at 24-in. maximum spacing, with a minimum of two clips for any piece of plywood.

The first panel that is installed in the wall form is erected in a plumb position and braced temporarily. Each additional sheet of

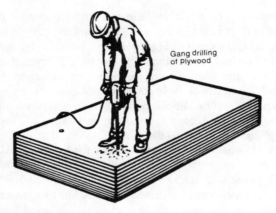

(a) Gang drilling holes in plywood for snap tie insertion

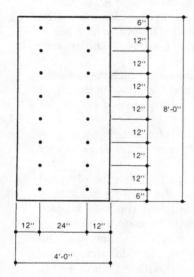

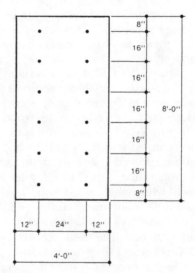

12" Vertical × 24" horizonal snap
tie spacing

Recommended rate of placement
4.5 Ft./hour at 70°F.

(b)

16" Vertical × 24" Horizontal snap
tie spacing

Recommended rate of placement
2.0 Ft./hour at 70°F.

(c)

FIGURE 9-27   Snap tie spacing and rate of placement for Jahn forming system.
(*Source: Dayton Superior Corporation*)

plywood is erected and nailed to the 2 × 4 plates and temporarily held in position by the Jahn ply holder. The vertical joints should be tight. Figure 9-28 shows the initial assembly of the forming system. Figure 9-29 illustrates the proper installation of the Jahn bracket.

After one face of plywood is installed, two workers can easily install the snap ties: one installs the ties and the other puts on the Jahn "A" bracket. Working from top to bottom, wales are installed in the bracket and securely tightened. Wale joints should occur at the bracket, or a scab should be installed at the wale joint with Jahn "C" brackets. A check should be made to ensure that plywood panels are plumb and aligned.

Erection of the inside wall panel, or second wall panel, is the same as for the outside panel, except that these plywood panels are placed over the snap tie ends. As illustrated in Figure 9-30, this can best be

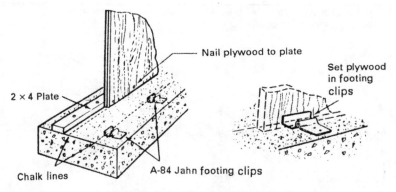

(a) Installation of footing plates or clips

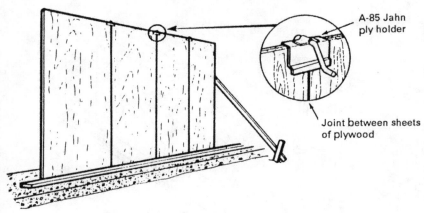

(b) Erection of plywood panels

Figure 9-28 Initial assembly of Jahn forming system. (*Source: Dayton Superior Corporation*)

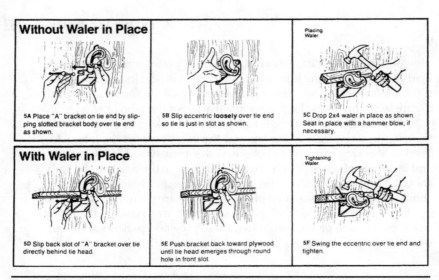

FIGURE 9-29 Installation of Jahn "A" bracket with and without waler in place. (*Source: Dayton Superior Corporation*)

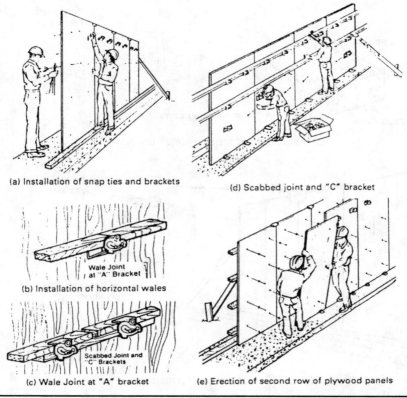

FIGURE 9-30 Erection of plywood panels for the Jahn forming system. (*Source: Dayton Superior Corporation*)

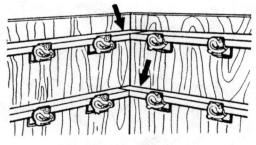

(a) Inside corner forming

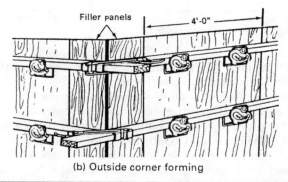

(b) Outside corner forming

FIGURE 9-31 Inside and outside corner forming. (*Source: Dayton Superior Corporation*)

accomplished by two workers slipping the sheet of plywood over the snap tie ends, starting at the bottom and moving the panel from side to side or up and down to align the holes with the snap tie ends.

Figure 9-31 illustrates inside and outside corner forming. There is no special treatment required for inside corners. The wales are simply alternated. For outside corners, cutting of full-width plywood panels can be reduced by starting on the inside wall first, using a full 4-ft 0-in. plywood panel. When the outside wall is erected, full panels are installed in line with the inside panel; and special filler panels, the same width as the wall thickness, plus the plywood thickness, are used to fill out the exterior corner.

Expensive overlapping, blocking, and nailing are eliminated by using the Jahn cornerlock. It is placed over one wale, flush at the corner, and the other wale may be run free, as shown in Figure 9-32. The cornerlock slips into place with its handle perpendicular to the wale. Nails are driven through the holes on the clamp, and the handle is pulled around 90°. A snug, tight outside corner is easily accomplished.

Strongbacks are used for form alignment, and they also act to tie stacked panels together. Loose double 2 × 4s are used for the strongbacks

FIGURE 9-32   Installation of Jahn cornerlock for forming outside corner. (*Source: Dayton Superior Corporation*)

along with the Jahn "C" brackets and snap ties. Normal spacing of the strongbacks is 8 ft on centers (Figure 9-33).

Figure 9-34 illustrates three alternatives for the attachment of adjoining upper and lower panels for large wall heights.

Installation of the second lift of plywood is illustrated in Figure 9-35. The lower plywood is raised into position. The sheet is nailed to the joint cover wale while the plywood panel is held in place with a short 2 × 4 spacer block, snap tie, and Jahn "C" bracket. As additional panels are set, they are nailed to the joint cover wale, securing them to the previously installed panel with a Jahn plywood holder. The snap ties, brackets, and wales are installed, working from the bottom to the top of the panel.

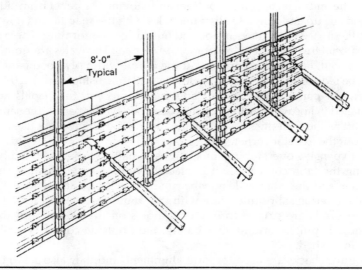

FIGURE 9-33   Installation of strongbacks. (*Source: Dayton Superior Corporation*)

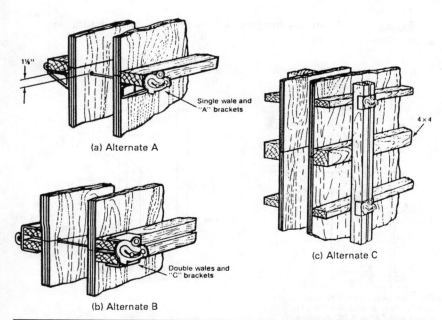

1⅛"

Single wale and
"A" brackets

(a) Alternate A

4 × 4

(c) Alternate C

Double wales and
"C" brackets

(b) Alternate B

FIGURE 9-34 Alternates for attachment of upper and lower plywood panels. (*Source: Dayton Superior Corporation*)

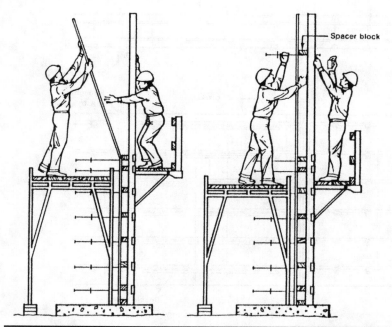

Spacer block

FIGURE 9-35 Installation of second lift of plywood. (*Source: Dayton Superior Corporation*)

Figure 9-36 illustrates forming details for walls with haunches and of three-way walls.

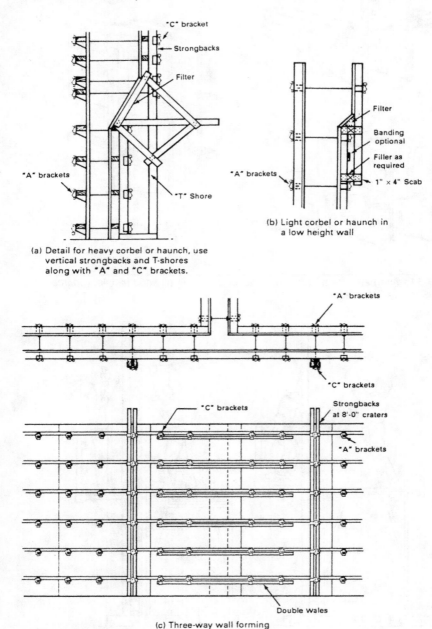

(a) Detail for heavy corbel or haunch, use vertical strongbacks and T-shores along with "A" and "C" brackets.

(b) Light corbel or haunch in a low height wall

(c) Three-way wall forming

FIGURE 9-36   Details for forming walls with haunches and three-way walls. (*Source: Dayton Superior Corporation*)

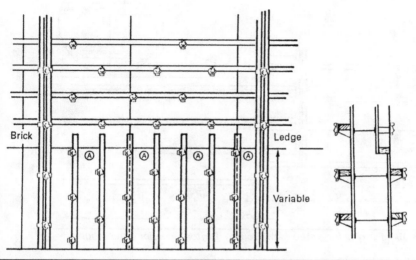

FIGURE 9-37 Forming assembly to provide a brick ledge. (*Source: Dayton Superior Corporation*)

Forms for a Concrete Wall Requiring a Ledge for Brick

Brick ledges may be formed with 2×4 lumber placed in either a vertical or a horizontal position, as shown in Figure 9-37. By adding shims of the required thickness to a 2×4, ledges of varying thickness may be formed. Intermediate vertical 2×4s only need to project slightly above the 2×4s forming the brick ledge.

Forms for a Stepped Concrete Wall

Figure 9-38 illustrates a method of forming a wall with a step in the elevation. Using the Jahn brackets and tie extender for the attachment of strongbacks allows 2×4 wales to run free if tie holes do not line up at stepdowns. Where tie alignment is fairly close, the Jahn brackets may be used, as shown in the insert.

Modular Panel Systems

Figure 9-39 illustrates a section of a modular panel that can be attached to an adjacent panel to produce a continuous wall-forming system. Panel sections are also located on the opposite side of a concrete wall. When several of these sections are joined together and form ties are installed, they produce a quickly assembled and economical system of forms for walls, permitting repetitive reuses. After the concrete has attained sufficient rigidity, the ends of the ties

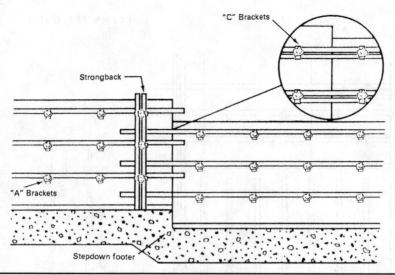

Figure 9-38 Forming for a step wall. (*Source: Dayton Superior Corporation*)

"C" Brackets

Strongback

"A" Brackets

Stepdown footer

Figure 9-39
Section of modular
panel. (*Source:
Ellis Construction
Specialties, Ltd.*)

are released, and the forms are removed intact by sections and moved to another location along the wall. The panel sections are light weight and have a handle, which allows them to be handled easily by a single worker.

Modular panels are available from several manufacturers and may be rented or purchased. The panels are usually 24-in. wide, in lengths of 3, 4, 5, 6, and 8 ft. Fillers are used to adjust the horizontal length of the wall forms to precise dimensions. Fillers are available in 1, 1½, and 2-in. increments, with lengths of 3, 4, 5, 6, and 8 ft. Job-built fillers may be fabricated with ¾-in. plywood attached to standard filler angles.

Adjacent sections are attached by two identical wedge bolts that lock the panels together to form a continuous modular panel system. One horizontal wedge bolt is inserted horizontally through the loop of the form tie, then the second wedge bolt is inserted vertically to secure adjacent panels. Figure 9-40 illustrates the details of the device that is used to lock the form ties into position for this system.

FIGURE 9-40
Attachment of adjacent panels. (*Source: Ellis Construction Specialties, Ltd.*)

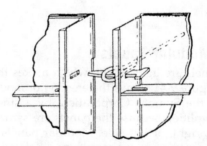

(a) Horizontal wedge-bolt is first installed through the loop or flat tie.

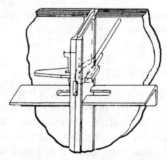

(b) Vertical wedge bolt is then installed to secure two adjacent panels together

FIGURE 9-41
Hand setting
modular panels.
(*Source: Symons
Corporation*)

Hand Setting Modular Panels

Modular panels are lightweight, which makes them easy to handle and erect. Figure 9-41 shows the erection of modular panels, manufactured by the Symons Corporation, by a single worker. Setting panels is simplified because the panels are symmetrical; there is no top, bottom, right, or left side. Modular panels are ideal for hand setting applications where many forming details are required. The only hardware required is the interlocking wedge bolts, which simultaneously connect panels and secure ties.

Inside and outside corners are easily formed by all-steel corners fillers, as shown in Figure 9-42. Slots in the corner fillers match those in panels for easy wedge bolt attachment.

Gang-Forming Applications

Because modular panels are lightweight with good strength, they are ideal for gang forming. In conventional gang forming, large sections, or complete details of modular panels, are assembled first and then moved into position for pouring the concrete. Although gang forming basically uses the same components as hand setting, it offers several advantages. Gangs are easily assembled on the ground and then moved into place. Because the unit is also stripped as a gang, the gang can be used over again without rebuilding, which saves time and material.

FIGURE 9-42
Fillers for corners.
(*Source: Symons
Corporation*)

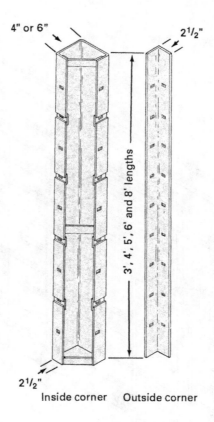

4" or 6"

2¹/₂"

3', 4', 5', 6' and 8' lengths

2¹/₂"

2¹/₂"

Inside corner   Outside corner

Both gang form bolts and wedge bolts are used to connect panels and gang form ties. Gang bolts connect panel siderails and secure the longer ends of gang form ties. The longer end on a gang form tie allows the tie to break back and permits stripping of the gang without disassembly. Like wedge bolts, gang form bolts make separating adjoining gang sections easy.

Figure 9-43 shows a gang form assembly. Panels may be stacked, one above another, to form tall walls. Stacked panels mate tightly to minimize form joints. Lift brackets are used as attachment points for rigging and safe handling of large gangs. For some applications, walers and strongbacks are used only for alignment. However, some applications use high-capacity ties with walers and strongbacks to gather the load of the panels. Therefore, the load is actually placed on the ties through the walers and strongbacks. Because of the strength and rigidity of the system, ties may be placed farther apart than they would normally be in conventional gang forming. Recommendations of manufacturers should be followed for safe assembly and application of modular panels for gang forms.

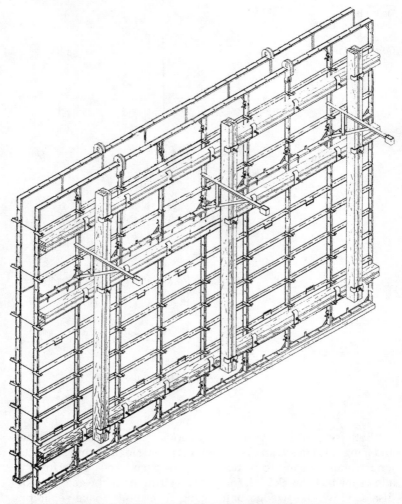

FɪɢURE 9-43 Gang-forming applications. (*Source: Symons Corporation*)

Gang Forms

Figure 9-44 illustrates a section of gang forms that can be used with a matching section located on the opposite side of a concrete wall. A group of these sections joined together with the form ties installed produces a quickly assembled and economical system of forms for walls. After the concrete has attained sufficient rigidity, the ends of the ties are released, and the forms are removed intact by sections and moved by a crane to another location along the wall.

The sections are fabricated with the all-steel forming systems by the Economy Forms Corporation (EFCO) under the trade names of

FIGURE 9-44 Wall-forming system. (*Source: EFCO*)

E-BEAM and SUPER STUD. These construction support products are used in conjunction with contractor-supplied plywood to produce a lightweight modular wall-forming system. This system weighs approximately 12 lb per sq ft, including pipe braces and scaffold brackets. E-BEAMs bolt directly to the SUPER STUD vertical stiffbacks through predrilled holes on 2-in. centers that line up with the SUPER STUD hole pattern. Thus, the forms may be assembled with little hardware or clamping devices in minimal time.

Plywood may be attached to the E-BEAM with a standard air nailing gun, or case-hardened nails can be driven by hand through the plywood and into the E-BEAM. The withdrawal strength is virtually the same as for nails driven with an actuated gun. Self-drilling and tapping screws driven by a drywall screw gun also provide a satisfactory fastening device.

A large variety of accessories is available for bracing and adjustments of the SUPER STUD system. Two cross-sectional dimensions, 6 by 6 in. and 9 by 9 in., are available in four standard modular lengths, which vary from 2 to 12 ft. The SUPER STUD system with accessories offers the versatility of a giant erector set (Figure 9-45).

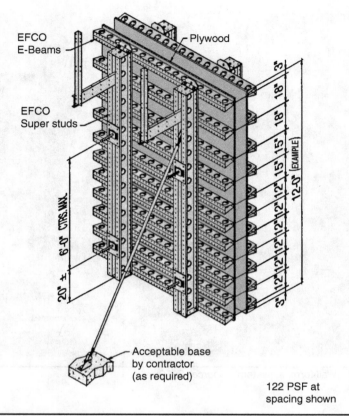

EFCO
E-Beams

Plywood

EFCO
Super studs

Acceptable base
by contractor
(as required)

122 PSF at
spacing shown

Figure 9-45   Detail assembly for 12 ft wall. (*Source: EFCO*)

Forms for Curved Walls

Figure 9-46 illustrates a section of gang forms assembled and used to construct circular concrete walls using the forming system manufactured under the trade name of REDI-RADIUS by EFCO. The system is designed to form fixed- or variable-radius walls. The radius form system will adjust to any radius 10 ft or greater in diameter, providing a smooth and continuous curved concrete finish.

The system can be handled with a crane in large gang sections and requires minimum bracing and alignment. The forming system consists of large interchangeable form panels that can be easily field assembled to meet any radius, concave or convex, from the 10-ft minimum radius to a straight wall. The panels have a built-in tie bearing, which combined with a special tapered tie allows quick erection and stripping of the forms.

The standard panel module is 6 by 12 ft, which permits a form area of 72 sq ft in one panel. Thus, the all-steel face sheet of a single

FIGURE 9-46 Gang forms for constructing a curved wall. (*Source: EFCO*)

panel provides a large area with unblemished concrete surface. Panel sizes range from 2 to 5 ft in height and from 4 to 12 ft in length.

The curvature is obtained by placing the panel on a wood template of the desired radius. The inside or outside radius is then obtained by tightening the bolts on the tension straps (Figure 9-47).

FIGURE 9-47 Tightening bolts on tension strap to obtain the desired radius for a curved wall. (*Source: EFCO*)

Jump Form System

In conventional gang forming, a crane serves two purposes: lifting and supporting. Only a small fraction of that time is spent in productive lifting. Most of the 15 to 20 minutes that a crane spends cycling, a conventional gang form goes into idly supporting the form. Jump-Form, the trade name of Patent Construction Systems, is a forming system that dramatically reduces the crane time. It is used for gang forming successive vertical lifts including shearwalls, core walls, piers, battered walls, lift stations, and control towers.

The JumpForm system eliminates the supporting function, reducing crane time to as little as 5 minutes on a typical lift. This is accomplished by two unique features, gangs that tilt back and an automatic release/attach device. JumpForm's gangs retract, allowing workers to conveniently mount the next lift's landing brackets while the Jump-Form remains securely supported on the wall. This provides a clear advantage over conventional gang forms, which must be fully removed from the wall by the crane, before mounting the landing brackets.

JumpForm's unique release/attach device works in conjunction with the landing brackets to eliminate the need for any additional wall attachments. All ties can be released, the gang stripped, and other preparations made before attaching a crane. Once the crane is hooked up and begins lifting, the mechanism automatically releases the JumpForm from the wall, then automatically engages at the next lift to fully support the system. The crane is released immediately and preparations begin for the next pour.

An entire lift of JumpForms can be flown and secured in rapid succession. Safety is improved because the release/attach mechanism operates remotely; therefore, workers do not need to be on the system to affect either the final release or the reattachment.

Workers can quickly bring gangs to a near true vertical position by adjusting the braces to a reference mark noted from the initial lift, as illustrated in Figure 9-48. Transit crews guide only the final adjustment of the pipe braces, which saves time compared to the wedges and stand-off jacks used in conventional gang forming. The pipe braces minimize bonding resistance by effectively peeling the gangs away from the wall for easy stripping. As the gangs retract, inserts for the next lift becomes accessible from the JumpForm system's upper level work platform. Workers can now mount the landing brackets before releasing JumpForm from the wall, an especially convenient feature when constructing shearwalls.

Workers can mount the next lift's landing brackets on both sides of the wall while the JumpForm system is still in place. This minimizes crane operations because the forms stay on the wall throughout the formwork phase. Because JumpForm does not need the support of intersecting floors, it is well suited to the construction of tall bridge piers and similar floorless structures.

Figure 9-48 Workers adjusting pipe braces for JumpForm system. (*Source: Patent Construction Systems*)

All gangs, platforms, and other key components stay intact throughout the forming process. Gangs, platforms supports, and vertical members feature lightweight aluminum joists and walers. Joists and walers are available in standard lengths and use fast connecting hardware. There is minimal cutting and nailing compared to all-wood systems. JumpForm accepts gangs from 8 to 16-ft high and from 8 to 44 ft wide.

The JumpForm system enhances worker safety. All work is performed from guarded platforms, including a 5-ft-wide operating platform. No one needs to be on the forms during crane handling. A locking feature that is built into the release/attach mechanism automatically engages when the gang is in the vertical position.

This enables the gang to resist uplift wind forces and remain fully engaged in its mounted position.

A typical JumpForm cycle begins by removing all form ties and anchor positioning bolts. Next, the gang is stripped by turning adjusting screws on the pipe braces. With the inserts now accessible from the upper level platform, the landing brackets are attached for the next lift. At the same time, the wind anchors and finish concrete patching is performed while working from the lower platform. The crane is then hooked up to the gang and workers leave the JumpForm. Next, the JumpForm is lifted and flown to the next level where it automatically attaches to the landing brackets. Workers then return to the JumpForm and immediately release the crane. Wind anchors are attached at either the tie location or the landing bracket inserts. Then, after cleaning and oiling the gang from the upper platform of the opposite gang, the gang is plumbed using adjustable pipe braces. Crews can now set all steel and boxouts, working from the upper level of the opposing form. The JumpForm cycle is completed by setting the ties and embedments.

Self-Lifting Wall-Forming System

A craneless forming system for constructing core walls on high-rise buildings of any size or complexity is available from Patent Construction System under the trade name, Self-Climber. This wall-forming system consists of a single, highly mechanized system to support the gang forms for the entire core. By activating hydraulic jacks from a central control panel, the Self-Climber raises one floor at a time. In a short interval, often less than 30 minutes, all forms reach the next level and preparations begin for the next pour. This is an advantage over conventional gang forming, where crane-handling increases the crew size and adds a day or more to the cycling operations. Generally, a system like this is cost effective only for buildings 15 to 20 stories or taller.

The superstructure of the wall-forming system consists of lateral and transverse beams that support all of the Self-Climber's forms and platforms. The hydraulic jacks push the superstructure to the next level, raising all forming and platform areas. Full-height gang forms feature large plywood sheets for a smooth finish. For architectural finish, form liners or rustication strips can be used.

A level of scaffold platforms hangs just below the forms to create comfortable, efficient working spaces; while the platform atop the superstructure provides a convenient work and storage area. All platforms travel with the forms. Nothing is dismantled between raises, an important safety consideration.

All of the jacks are readily accessible, mounted behind the forms. Each jack is self-contained and has its own pump and reservoir system to allow for individual control during a raise. There is not a large number of hydraulic lines or individual electric cords to get in the way of workers.

One side of each pair of forms is suspended from a trolley system, so workers can retract and easily reset the forms. Plumbing and leveling of the forms is accomplished by adjusting screw legs on the form support brackets. There is ample space for workers to operate and no masts or similar obstructions to hinder concrete placement. Crew members have a heightened sense of security, so productivity does not suffer as the core's height increases.

The top platform of the Self-Climber can be designed to store tools and building materials, such as reinforcing steel and conduit. The top platform also offers protection from the rain. In the winter the entire system can be enclosed and equipped with heaters, so forming can continue during adverse weather.

Insulating Concrete Forms

Insulating concrete forms (ICF) are units of rigid plastic foam that hold concrete in place until it is cured. The units are assembled, stacked, and then filled with reinforced concrete. The units are lightweight and easy to erect. Bracing and alignment systems are provided by most manufacturers of ICFs. This forming system can be used for beams, columns, or walls. Unlike conventional formwork, the insulating concrete–forming units are left in place, which results in high energy efficiency.

Wall can be constructed up to 10 feet high using ICFs. Walls constructed with ICFs can decrease the required capacity of heat-and-air equipment by 50%, compared to traditional wood-framed walls. The units provide backing for interior and exterior finishes. Gypsum wall board may be attached directly to the inside of the wall form. The exterior of the wall may be finished with tongue-and-groove siding, cedar shingles, or brick veneer.

The insulating foam is made from expanded polystyrene or extruded polystyrene material. The three basic form types are hollow foam block, foam planks held together by plastic ties, and panels with integral foam and plastic ties.

Block units are the smallest, typically 8 in. thick, 1 ft high, and 4 ft long. They are molded with special edges that interconnect the blocks. The connections may be tongue-and-groove, interlocking teeth, or raised squares. Plank units range in sizes from 1 by 4 ft to 1 by 8 ft and are connected by nonfoam material. The long faces of foam are shipped as separate pieces, which resemble wooden planks. The planks are equipped with crosspieces as part of the wall-setting sequence. Panel units are the largest, with sizes up to 4 by 12 ft. Their foam edges are flat and panels are interconnected by nonfoam fasteners. The panels are connected with integral foam or plastic ties.

There are many manufacturers of ICFs. Each manufacturer provides detailed information about selection, assembly, and erection of their products. The following illustrate the basic steps for constructing a wall using ICFs.

1. Set temporarily braces along first course to align and secure ICFs.

2. Place insulating concrete form blocks on concrete footing.

3. Complete one course around the entire perimeter of the structure.

4. Set horizontal and vertical reinforcing steel as required.

5. Stager subsequent courses so vertical joints do not align between courses.

6. Ensure vertical and horizontal cavities are aligned.

7. Cut openings as required, or cut after the entire wall is built.

8. Install bucks/opening blockouts for doors and windows.

9. Brace forms along walls and openings to ensure plumb and square alignment.

10. Adequate bracing is required along walls and at window and door openings.

11. Foam seal joints to secure blocks until concrete is poured.

12. Pour concrete in 2- to 4-ft lifts following manufacturer's instructions.

13. Typically high-flow concrete is pumped to ensure flow into interior spaces.

14. Manufacturer's instruction must be followed to prevent a blowout

References

1. "National Design Specification for Wood Construction," ANSI/AF&PA NDS-2005, American Forest & Paper Association, Washington, DC, 2005.
2. "Design Values for Wood Construction," Supplement to the National Design Specification, National Forest Products Association, Washington, DC, 2005.
3. "Plywood Design Specification," APA—The Engineered Wood Association, Tacoma, WA, 1997.
4. "Concrete Forming," APA—The Engineered Wood Association, Tacoma, WA, 2004.
5. "Plywood Design Specification," APA—The Engineered Wood Association, Tacoma, WA, 2004

CHAPTER 10

Forms for Columns

General Information

Concrete columns are usually one of five shapes: square, rectangular, L-shaped, octagonal, or round. Forms for the first four shapes are generally made of Plyform sheathing backed with either 2×4 or 2×6 vertical wood battens. Column clamps surround the column forms to resist the concrete pressure acting on the sheathing. Forms for round columns are usually patented forms fabricated of fiber tubes, plastic, or steel. However, all shapes of columns can be made of fiberglass forms.

An analysis of the cost of providing forms, including materials, labor for erecting and removing, and number of reuses, should be made prior to selecting the materials to be used.

Pressure on Column Forms

Determining the lateral pressure of the freshly placed concrete against the column forms is the first step in the design of column forms. Because forms for columns are usually filled rapidly, frequently in less than 60 minutes, the pressure on the sheathing will be high, especially for tall columns.

As presented in Chapter 3, the American Concrete Institute recommends that formwork be designed for its full hydrostatic lateral pressure as given by the following equation:

$$P_m = wh \qquad (10\text{-}1)$$

where P_m is the lateral pressure in pounds per square foot, w is the unit weight of newly placed concrete in pounds per cubic foot, and h is the depth of the plastic concrete in feet. For concrete that is placed rapidly, such as in columns, h should be taken as the full height of the form. There are no maximum and minimum values given for the pressure calculated from Eq. (10-1).

283

For the limited placement condition of concrete with a slump 7 in. or less and normal internal vibration to a depth of 4 ft or less, formwork for columns may be designed for the following lateral pressure:

$$P_m = C_w C_c [150 + 9,000R/T] \qquad (10\text{-}2)$$

where P_m = calculated lateral pressure, lb per sq ft
C_w = unit weight coefficient
C_c = chemistry coefficient
R = rate of fill of concrete in form, ft per hr
T = temperature of concrete in form, degrees Fahrenheit
Minimum value of P_m is $600C_w$, but in no case greater than wh
Applies to concrete with a slump of 7 in or less
Applies to normal internal vibration to a depth of 4 ft or less

Values for the unit weight coefficient C_w in Eq. (10-2) are shown in Table 3-1 and the values for the chemistry coefficient C_c are shown in Table 3-2. The minimum pressure in Eq. (10-2) is $600C_w$ lb per sq ft, but in no case greater than wh.

This equation should be used with discretion. For example, the pressure should not exceed wh, where w is the unit weight of concrete and h is the depth in feet below the upper surface of freshly placed concrete. Thus, the maximum pressure at the bottom of a form 6 ft tall with 150 lb per cu ft concrete will be (150 lb per cu ft)(6 ft) = 900 lb per sq ft, regardless of the rate of filling the form. However, using Eq. (10-2) for a rate of placement of 6 ft per hour at a temperature of 90°F, the calculated value is 750 lb per sq ft. For a rate of placement of 6 ft per hour and a temperature of 60°F, using Eq. (10-2), the calculated value is 1,050 lb per sq ft. As stated previously, the American Concrete Institute recommends Eq. (10-1) for the design of concrete formwork.

Designing Forms for Square or Rectangular Columns

Figure 10-1 illustrates a form for a representative square column, using sheathing and patented column clamps. If the thickness of the sheathing is selected, the design consists of determining the maximum safe spacing of the column clamps, considering the pressure from the concrete and the strength of the sheathing. The strength of the sheathing must be adequate to resist bending and shear stresses, and it must be sufficiently rigid so that the deflection will be within an acceptable amount, usually less than $l/360$ or $\frac{1}{16}$ in.

As illustrated in Figure 10-1(b), it is assumed that the magnitude of the pressure on a given area of sheathing will vary directly

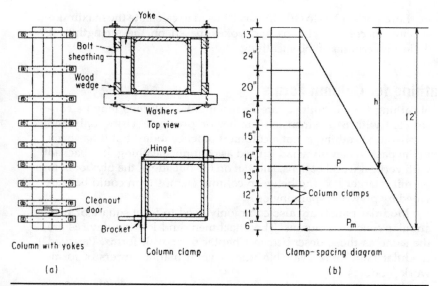

FIGURE 10-1   Forms for square column.

with the depth of the area below the surface of the concrete. For concrete weighing 150 lb per cu ft, the maximum pressure is given by Eq. (10-1).

Chapter 5 presented the equations for determining the allowable span length of form members based on bending, shear, and deflection. The terms in the equation include the physical properties, the allowable stresses, and the uniform load w, measured in lb per ft. For clarity, the equations from Chapter 5 are repeated here.

For lumber, the allowable span lengths with multiple supports are as follows:

For bending, $l_b = [120F_b S/w]^{1/2}$ (10-3)

For shear, $l_v = 192F_v bd/15w + 2d$ (10-4)

For deflection, $l_\Delta = [1{,}743EI/360w]^{1/3}$ for $\Delta = l/360$ (10-5)

For deflection, $l_\Delta = [1{,}743EI/16w]^{1/4}$ for $\Delta = \frac{1}{16}$ in. (10-6)

For Plyform, the allowable span lengths over multiple supports are as follows:

For bending, $l_b = [120F_b S_e/w_b]^{1/2}$ (10-7)

For shear, $l_s = 20F_s(Ib/Q)/w_s$ (10-8)

For deflection, $l_d = [1{,}743EI/360w_d]^{1/3}$ for $\Delta = l/360$ (10-9)

For deflection, $l_d = [1{,}743EI/16w_d]^{1/4}$ for $\Delta = \frac{1}{16}$ in. (10-10)

Equations (10-3) to (10-10) may be used to calculate the maximum spacing of column clamps and yokes, based on the strength and deflection criteria of the sheathing.

Sheathing for Column Forms

Sheathing for job-built square or rectangular columns may be constructed with S4S dimension lumber or plywood. Using only S4S lumber as sheathing is not a common practice today, but it has been in the past. When plywood is used as column sheathing, it is backed with vertical wood battens, laid flat on the outside of the plywood, to permit a larger spacing between column clamps than could be provided by plywood spanning only.

Modular panels are also commonly used to form square or rectangular concrete columns. The attachment and locking devices are the same as those described in Chapter 6 for wall forms. Patented modular panels are available from several manufacturers of formwork products.

For round columns, the common types of forms are fiber tubes, fiberglass, or steel forms. Fiber tubes are lightweight and easily handled, but may only be used once. Fiberglass column forms may be reused, but may occasionally require some repair or maintenance. Steel column forms are heavier than fiber tubes or fiberglass forms, but may be reused many times without requiring repairs. Each of these column forming systems is discussed in the following sections.

Maximum Spacing of Column Clamps Using S4S Lumber Placed Vertical as Sheathing

Consider a vertical strip of sheathing 12 in. wide. The safe spacing of column clamps is l in. For concrete whose unit weight is 150 lb per cu ft, the pressure of the freshly placed concrete against the sheathing at depth h will be:

$$\text{From Eq. (10-1), } P_m = wh$$

where P_m is the pressure in lb per sq ft, and h is measured in feet. For a 12-in. (1-ft) wide strip of sheathing, the uniform load will be:

$$w = (150h \text{ lb per sq ft})(1 \text{ ft})$$
$$= 150h$$

where w is the pressure, lb per lin ft, on a strip of sheathing 12 in. wide.

The sheathing for the column forms must have adequate strength to resist this pressure, based on the support spacing of the column clamps. Thus, the spacing of the column clamps is determined by the allowable span length of the sheathing.

Example 10-1

Consider only S4S lumber as sheathing for a square concrete column form. The column will be 6 ft high and the concrete unit weight is 150 lb per cu ft. The pressure on the bottom of the form will be:

From Eq. (10-1), P_m = (150 lb per cu ft)(6 ft)
$= 900$ lb per sq ft

The uniform load on a 12-in. (1-ft) strip of sheathing will be:

$$w = (900 \text{ lb per sq ft})(1 \text{ ft})$$
$$= 900 \text{ lb per lin ft}$$

Consider using 2-in.-thick No. 2 grade Douglas Fir-Larch lumber laid flat for sheathing. Assume the lumber will be dry and a short-duration loading, less than 7 days. The safe spacing of the column clamps is l in. and can be calculated as follows. Using the actual thickness of 1.5 in. for the S4S dimension lumber, the physical properties for a 12-in.-wide strip of the lumber laid flat can be calculated:

From Eq. (5-9), $S = bd^2/6$
$= (12 \text{ in.})(1.5 \text{ in.})^2/6$
$= 4.5 \text{ in.}^3$
From Eq. (5-8), $I = bd^3/12$
$= (12 \text{ in.})(1.5 \text{ in.})^3/12$
$= 3.375 \text{ in.}^4$

The NDS permits an increase in tabulated design value when the member is laid flat (refer to Table 4-7). However, the flat-use adjustment factor will not be used in this design. Adjusting the tabulated design value from Table 4-3 by 25% for short-duration loading, less than 7 days, the indicated allowable stresses for a dry condition are

For bending, F_b = 1.25(900) = 1,125 lb per sq in.
For shear, F_v = 1.25(180) = 225 lb per sq in.
Modulus of elasticity, E = 1.0(1,600,000) = 1,600,000 lb per sq in.

The maximum allowable spacing of column clamps can be calculated from Eqs. (10-3) to (10-6) as follows:

For bending, $l = [120F_b S/w]^{1/2}$
$= [120(1,125)(4.5)/900]^{1/2}$
$= 25.9 \text{ in.}$

For shear, $l = 192F_vbd/15w + 2d$

$\qquad = 192(225)(12)(1.5)/15(900) + 2(1.5)$

$\qquad = 60.6$ in.

For deflection, $l = [1,743EI/360w]^{1/3}$ $\qquad$ for $\Delta = l/360$

$\qquad = [(1,743)(1,600,000)(3.375)/360(900)]^{1/3}$

$\qquad = 30.7$ in.

For deflection, $l = [1,743EI/16\,{}^{3}\!/_{16}w]^{1/4}$ $\qquad$ for $\Delta = {}^{1}\!/_{16}$ in.

$\qquad = [1,743(1,600,000)(0.3.375)/16(900)]^{1/4}$

$\qquad = 28.4$ in.

For this 2-in. S4S sheathing, the maximum permissible span length is 25.9 in., governed by bending. Therefore, the spacing of the column clamps should not exceed this amount.

For this design a 25-in. spacing of clamps can be used. The column clamps or yokes must be strong enough to resist the loads transmitted to them by the sheathing. The design of yokes is discussed later in this chapter.

Table 10-1 shows the maximum span lengths for the S4S lumber, laid flat in the vertical direction, to carry the entire pressure in the vertical direction between column clamps. These span lengths are based on using clamps or yokes that are strong enough to resist the loads transmitted to them by the sheathing.

Support spacings shown in Table 10-1 are calculated using Eqs. (10-3) to (10-6) for the concrete pressures and allowable stresses shown at the bottom of the table. The spacings are the maximum as governed by bending, shear, or deflection.

Plywood Sheathing with Vertical Wood Battens for Column Forms

A common practice for fabricating forms is to place vertical battens of 2×4 dimension lumber on the outside of the plywood sheathing, as illustrated in Figure 10-2. The 2×4s are placed flat against the plywood and act as studs, similar to the design of wall forms, as discussed in Chapter 9. For some job conditions, 2×6 battens are used in lieu of 2×4 battens.

The concrete pressure acts against the plywood panel, which must transfer the pressure in a horizontal direction to the adjacent vertical wood battens. The vertical wood battens must then transfer the loads, transferred to them from the plywood sheathing, in a vertical direction between column clamps.

The plywood must have sufficient strength and rigidity to transfer the concrete pressure between the wood battens. Equations (10-7) to (10-10) can be used to determine the maximum distance between the wood battens, which is equivalent to the maximum span length of plywood.

Depth below Top of Form, ft	Maximum Pressure, 150h, lb per sq ft	Maximum Span Length for S4S Lumber Laid Flat Vertical	
		Nominal Thickness	
		1-in.	2-in.
4	600	15	31
5	750	14	28
6	900	12	25
7	1050	11	23
8	1200	11	22
9	1350	10	21
10	1500	10	20
11	1650	9	19
12	1800	9	19
13	1950	8	17
14	2100	8	17
15	2250	8	16
16	2400	7	15
17	2550	7	15
18	2700	7	15
19	2850	7	14
20	3000	7	14
21	3150	6	13
22	3300	6	13
23	3450	6	13
24	3600	6	12
25	3750	6	12

Notes:
1. For S4S lumber, values based on F_b = 1,125 psi, F_v = 225 psi, and E = 1,600,000 lb per sq in., with no splits.
2. Values shown for S4S lumber laid flat between column clamps, without plywood.
3. Concrete pressure based on concrete of 150 lb per cu ft unit weight. Spacing is governed by bending, shear, or deflection. Maximum deflection of $l/360$, but not more than $1/16$ in.
4. Spacing based on continuous members with four or more supports, and on Eqs. (10-3) to (10-6).

TABLE 10-1 Maximum Span Lengths of S4S Lumber Laid Flat for Column Sheathing, in.

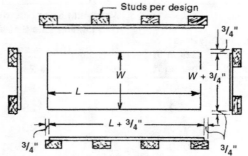

Note: Dimensions based on use of ³/₄" Plyform

(a) Plan view of rectangular column form

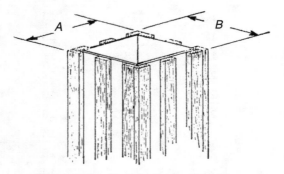

(b) Isometric view of square column form

FIGURE 10-2 Column sheathing of plywood with vertical wood battens.

Tables for Determining the Maximum Span Length of Plyform Sheathing

Table 10-2 gives the maximum span lengths using Eqs. (10-7) to (10-10) for Class I Plyform with short-duration load, less than 7 days. The values are shown for single span, two spans, and multiple spans applications. The maximum span lengths for the Plyform sheathing shown in Table 10-2 applies to conditions where the panels are installed in the vertical direction; therefore, the panels are placed in the weak direction for transferring the concrete pressure horizontally between vertical wood battens. The span length is the maximum permissible horizontal distance between the vertical battens that are used to support the panels.

Table 10-3 gives the maximum span lengths using Eqs. (10-7) to (10-10) for Class I Plyform with short-duration load, less than 7 days. The values are shown for single span, two spans, and multiple spans applications. The maximum span lengths for the Plyform sheathing

Depth below Top of Form, ft	Maximum Concrete Pressure ($P_m = wh$), lb per sq ft	Maximum Spacing, in.								
		Single Span (2 supports)			Two Spans (3 supports)			Multiple Spans (4 or more supports)		
		Thickness, in.			Thickness, in.			Thickness, in.		
		⅝	¾	1	⅝	¾	1	⅝	¾	1
4	600	6	7	11	7	9	14	7	9	13
5	750	5	7	10	7	9	12	6	8	12
6	900	5	6	9	6	8	11	6	8	11
7	1050	5	6	9	5	7	10	6	7	11
8	1200	4	6	8	5	6	10	5	6	10
9	1350	4	6	8	5	6	8	5	6	9
10	1500	4	5	8	4	5	8	4	5	8
11	1650	4	5	8	4	5	7	4	5	7
12	1800	4	5	7	4	5	7	4	5	7
13	1950	4	5	7	4	4	7	4	4	7
14	2100	4	5	7	—	4	6	4	4	6
15	2250	4	5	7	—	4	6	—	4	6
16	2400	4	4	7	—	4	6	—	4	6
17	2550	—	4	6	—	4	5	—	4	5
18	2700	—	4	6	—	4	5	—	4	5
19	2850	—	4	6	—	4	5	—	4	5
20	3000	—	4	6	—	—	5	—	—	5

Notes:
1. For Plyform, values based on $F_b = 1,930$ psi, $F_s = 72$ psi, and $E = 1,650,000$ psi, with face grain installed in the weak direction, physical properties for stress applied perpendicular to face grain.
2. Concrete pressure based on concrete with 150 lb per cu ft unit weight. Spacings are governed by bending, shear, or deflection. Maximum deflection limited to $l/360$, but not more than $1/16$ in.

TABLE 10-2   Maximum Spans of Plyform with Face Grain Parallel to Supports

shown in Table 10-3 applies to conditions where the panels are installed in the horizontal direction; therefore, the panels are placed in the strong direction for transferring the concrete pressure horizontally between vertical wood battens. The span length is the maximum permissible horizontal distance between the vertical battens that are used to support the panels.

Depth below Top of Form, ft	Maximum Concrete Pressure ($P_m = wh$), lb per sq ft	Maximum Spacing, in.								
		Single Span (2 supports)			Two Spans (3 supports)			Multiple Spans (4 or more supports)		
		Thickness, in.			Thickness, in.			Thickness, in.		
		⅝	¾	1	⅝	¾	1	⅝	¾	1
4	600	9	11	14	10	11	14	11	13	16
5	750	9	10	13	9	10	12	10	11	14
6	900	8	9	12	8	9	12	9	10	13
7	1050	8	8	11	7	8	11	8	9	12
8	1200	7	8	10	7	8	10	8	9	11
9	1350	7	7	9	7	7	9	7	8	10
10	1500	6	7	9	6	7	9	7	8	10
11	1650	6	7	9	6	7	9	6	7	9
12	1800	6	7	8	6	7	8	6	7	9
13	1950	6	6	8	5	6	8	6	7	8
14	2100	5	6	7	5	6	7	5	6	7
15	2250	5	6	7	5	6	7	5	6	7
16	2400	5	6	7	5	6	7	5	6	7
17	2550	5	5	7	5	5	6	5	5	6
18	2700	5	5	7	4	5	6	4	5	6
19	2850	5	5	7	4	5	6	4	5	6
20	3000	5	5	7	4	5	6	4	5	6

Notes:
1. For Plyform, values based on $F_b = 1{,}930$ psi, $F_s = 72$ psi, and $E = 1{,}650{,}000$ psi, with face grain installed in the strong direction, physical properties for stress applied parallel to face grain.
2. Concrete pressure based on concrete with 150 lb per cu ft unit weight. Spacings are governed by bending, shear, or deflection. Maximum deflection limited to $l/360$, but not more than $1/16$ in.

TABLE 10-3   Maximum Spans of Plyform with Face Grain Across Supports

Maximum Spacing of Column Clamps Using Plyform with Vertical Wood Battens

The maximum spacing of column clamps will be limited by the strength of the sheathing that must transfer the concrete pressure between clamps. As discussed in the preceding section, the sheathing for square or rectangular columns usually consists of plywood with vertical wood battens to provide increased spacing of column clamps. The strength and deflection equations presented at the beginning of this chapter may be used to determine the maximum permissible spacing of column clamps, which is equivalent to the allowable span length of the composite plywood with the wood battens.

Example 10-2

Forms for a 20-in.-square column consist of ¾-in.-thick Plyform, backed with vertical 2 × 4 No. 2 grade Southern Pine wood battens laid flat, similar to the forming system shown in Figure 10-2. A plan view of one side of the 20-in. square column form is shown in Figure 10-3. Concrete, having a unit weight of 150 lb per cu ft, will be placed in the 10-ft-high column form. Class I Plyform, which is ¾-in. thick, will be placed in the horizontal direction (face grain across supports) as sheathing.

The maximum pressure of the concrete on the forms will occur at the bottom of the column form. This maximum pressure can be calculated as follows:

From Eq. (10-1), $P_m = wh$

$$= (150 \text{ lb per cu ft})(10 \text{ ft})$$
$$= 1{,}500 \text{ lb per sq ft}$$

This pressure acts against the Plyform, which must transfer it in a horizontal direction to the adjacent vertical wood battens. For a 20-in. column formed with ¾-in. Plyform backed with 2 × 4 vertical wood battens laid flat, the clear span distance between the 2 × 4s will be 5.875 in., as shown in Figure 10-3. Thus, the Plyform must have adequate strength and rigidity to transfer the 1,500 lb per sq ft concrete pressure over the 5.875-in. span length.

For this example, the Plyform will be placed with the face grain in the horizontal direction. A review of Table 10-3 indicates that ¾-in. Class I Plyform can sustain a pressure of 1,500 lb per sq ft with support spacing up to 7 in. when the panels are placed in the horizontal

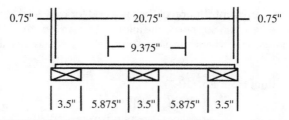

Figure 10-3 Plan view of one side of 20-in.-square column form constructed of ¾-in. Plyform with 2 × 4 vertical battens.

(strong direction) across wood battens. Therefore, the Plyform sheathing selected for this application will be satisfactory. The length of the Plyform for one side of the column form will be 20.75 in., as shown in Figure 10-3.

From Tables 4-2, 4-4, and 4-7 the allowable stresses for the No. 2 grade Southern Pine 2 × 4 vertical wood battens laid flat and short-duration loading can be calculated as follows:

Allowable bending stress, $F_b = C_D \times C_{fu} \times$ (reference design
bending value)
$$= (1.25) \times (1.1) \times (1,500 \text{ lb per sq in.})$$
$$= 2,062.5 \text{ lb per sq in.}$$

Allowable shear stress, $F_v = C_D \times$ (reference design shear value)
$$= (1.25) \times (175 \text{ lb per sq in.})$$
$$= 218.7 \text{ lb per sq in.}$$

Modulus of elasticity, $E = 1,600,000$ lb per sq in.

The three vertical wood battens must transfer the concrete pressure between column clamps. The center batten is the most heavily loaded and must sustain the pressure to the midpoint of the adjacent battens, a distance of 9.375 in. The uniform load between column clamps on the center batten may be calculated as follows:

$$w = (1,500 \text{ lb per sq ft}) (9.375/12 \text{ ft})$$
$$= 1,172 \text{ lb per lin ft}$$

The physical properties of section modulus and moment of inertia of the 2 × 4 wood battens laid flat can be obtained from Table 4-1 as follows:

From Table 4-1:
Section modulus, $S = 1.313$ in.^{3}
Moment of inertia, $I = 0.984$ in.^{4}

Neglecting any contribution of the plywood for transferring the concrete pressure between supports, the allowable span length of the 2 × 4 wood battens based on bending, shear, and deflection may be calculated as follows:

For bending, from Eq. (10-3),
$$l_b = [120F_b S/w]^{1/2}$$
$$= [120(2,062.5)(1.313)/1,172]^{1/2}$$
$$= 16.6 \text{ in.}$$

For shear, from Eq. (10-4),

$$l_v = 192F_vbd/15w + 2d$$
$$= 192(218.7)(3.5)(1.5)/15(1,172) + 2(1.5)$$
$$= 15.5 \text{ in.}$$

For deflection not to exceed $l/360$, from Eq. (10-5),

$$l_\Delta = [1,743EI/360w]^{1/3}$$
$$= [1,743(1,600,000)(0.984)/360(1,172)]^{1/3}$$
$$= 18.6 \text{ in.}$$

For deflection not to exceed $\frac{1}{16}$ in., from Eq. (10-6),

$$l_\Delta = [1,743EI/16w]^{1/4}$$
$$= [1,743(1,600,000)(0.984)/16(1,172)]^{1/4}$$
$$= 19.6 \text{ in.}$$

Below is a summary of the allowable span lengths of the 2×4 vertical wood battens:

For bending, the maximum span length = 16.6 in.
For shear, the maximum span length = 15.5 in.
For deflection, the maximum span length = 18.6 in.

For this example, shear governs the maximum span length of the wood battens. The spacing of the column clamps at the bottom of the form must not exceed 15.5 in. For constructability, the spacing would likely be placed at 12 in. Figure 10-4 shows a plan view of the assembled forms.

The pressure on the forms is greatest at the bottom, linearly decreasing to zero at the top of the form, see Eq. (10-1). Therefore, the spacing of the column clamps may be increased in the upper portion of the forms. It may be desirable to place the Plyform panel in the horizontal direction at the bottom of the form, and then place the panels in the vertical direction in the upper portion of the form where the concrete pressure is not as great. The procedure illustrated in this example may be used to calculate the allowable spacing of column clamps at any height of the column forms, based on the strength and deflection of the sheathing.

The column clamps must be strong enough to resist the loads transmitted to them by the sheathing. There are numerous manufacturers of patented column clamps for forming concrete columns. These manufacturers generally specify the safe spacing for their particular clamps. The following sections provide information on patented clamps.

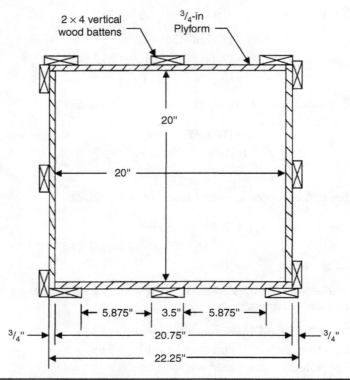

Figure 10-4 Assembled forms for 20-in.-square column with ¾-in. Plyform backed with 2 × 4 vertical wood battens.

Column Clamps for Column Forms

Column clamps that provide support for the sheathing of column forms may be constructed of wood yokes or steel clamps. Wood yokes are not as commonly used today as they have been in the past.

Several types of attachments and locking devices are available from numerous manufacturers, which may be used to secure the composite plywood and stud forming system. These formwork accessories include patented steel clamps, which are attached with wedge bolts, and cam-locks, which attach horizontal walers on the outside of the vertical studs.

Design of Wood Yokes for Columns

Wood yokes may be made from 2 × 4, 3 × 4, or 4 × 4, or larger pieces of lumber, assembled around a column, as illustrated in Figure 10-5. Two steel bolts are installed through holes to hold the yoke members

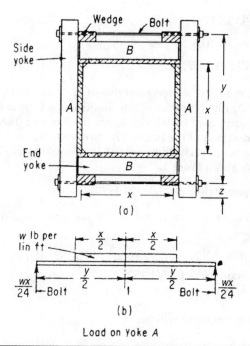

(a)

w lb per
lin ft

$\frac{x}{2}$ $\frac{x}{2}$

$\frac{wx}{24}$ Bolt $\frac{y}{2}$ $\frac{y}{2}$ Bolt $\frac{wx}{24}$

(b)

Load on Yoke A

FIGURE 10-5   Column forms with wood yokes.

in position. The members are designed as side yokes (A) and end yokes (B). The members must be strong enough to resist the forces transmitted to them from the concrete through the sheathing.

End yoke B is held against the sheathing by two hardwood wedges, which act as end supports for a simple beam subjected to a uniform load. Consider the two wedges to be x in. apart, producing a span equal to x. Let the spacing of the yokes be equal to l in. Let P equal the pressure on the sheathing in pounds per square foot. The uniform load on the yoke will then be:

$$w = Pl/12 \text{ lb per lin ft}$$

The bending moment at the center of the yoke will be:

$$M = wx^2/96 \text{ in.-lb}$$
$$= (Pl/12)(x)^2/96$$
$$= Plx^2/1,152 \qquad\qquad (a)$$

The resisting moment will be:

$$M = F_b S \qquad\qquad (b)$$

Equating (a) and (b) gives:

$$F_bS = Plx^2/1{,}152$$
$$S = Plx^2/1{,}152F_b \qquad\qquad (10\text{-}11)$$

where S is the required section modulus of end yoke B, in.[3].

Side yoke A, illustrated in Figure 10-5(b), is a simple beam, subjected to a uniform load equal to w lb per lin ft over a distance equal to x in the midsection of the beam. The two bolts, spaced y in. apart, are the end supports, each subjected to a load equal to $wx/24$ lb. Summing moments about Point 1,

$$M = [wx/24][y/2] - [wx/24][x/4]$$

Substituting $Pl/12$ for w in the above equation, the applied moment is:

$$
\begin{aligned}
M &= [Plx/12(24)][y/2] - [Plx/(12)(24)][(x/4)] \\
&= Plxy/576 - Plx^2/1{,}152 \\
&= Plx(2y - x)/1{,}152 \qquad\qquad (c)
\end{aligned}
$$

The resisting moment will be:

$$M = F_bS \qquad\qquad (d)$$

Equating (c) and (d) gives:

$$F_bS = Plx(2y - x)/1{,}152$$

Solving for S, we get:

$$S = Plx(2y - x)/1{,}152F_b \qquad\qquad (10\text{-}12)$$

where S is the required section modulus for side yoke A, in.[3]

Example 10-3

Determine the minimum size wood yoke member B required for the given conditions:

$$P = 900 \text{ lb per sq ft.}$$
$$l = 12 \text{ in.}$$
$$x = 16 \text{ in.}$$

Using S4S lumber with an allowable bending stress F_b of 1,093 lb per sq in., the required section modulus can be calculated as follows:

From Eq. (10-11),
$$
\begin{aligned}
S &= Plx^2/1{,}152\,F_b \\
&= (900)(12)(16)^2/1{,}152(1{,}093) \\
&= 2.19 \text{ in.}^3
\end{aligned}
$$

Table 4-1 indicates that a 2 × 4 S4S member with the 4-in. face perpendicular to the sheathing will have a section modulus of 3.06 in.^{3}, which is greater than the required 2.19 in.^{3} Therefore, a 2 × 4 S4S member will be adequate. Also, a 3 × 4 S4S member, with the 3-in. face perpendicular to the sheathing, will be adequate.

Example 10-4

Determine the minimum size wood yoke member A required for the S4S lumber with an allowable bending stress F_b of 1,093 lb per sq in. and an allowable shear stress F_v of 237 lb per sq in. for the following conditions:

$$P = 900 \text{ lb per sq ft}$$
$$l = 12 \text{ in.}$$
$$x = 16 \text{ in.}$$
$$y = 28 \text{ in.}$$

The required section modulus can be calculated as follows:

From Eq. (10-12), $S = Plx[2y - x]/1{,}152F_b$
$$= 900(12)(16)[2(28) - 16]/1{,}152(1{,}093)$$
$$= 5.48 \text{ in.}^3$$

Table 4-1 indicates that a 3 × 4 S4S member, with a section modulus of 7.15 in.^{3}, will be required. Check the member for the unit stress in shear. The shear force is:

$$V = [Plx]/[(2)(12)(12)]$$
$$= 900(12)(16)/288$$
$$= 600 \text{ lb}$$

The applied shear stress is:

$$f_v = 3V/2bd$$
$$= 3(600)/2(12.25)$$
$$= 73 \text{ lb per sq in.}$$

Because the allowable shear stress F_v of 237 lb per sq in. is greater than the applied shear stress f_v, the S4S member is adequate in shear.

Each bolt is subjected to a tensile stress equal to 600 lb. A ⅜-in.-diameter bolt is strong enough, but a ½-in.-diameter bolt should be used to provide sufficient area in bearing between the bolt and the wood wedges.

Determine the minimum length of z in Figure 10-5 to provide adequate area in shear between the bolt and the end of the yoke. The area in shear is z times twice the width of member A,

$$A = (z)(2)(3.5)$$
$$= 7z$$

The required area, based on an allowable shear stress of 237 lb per sq in. is

$$A = (600\ lb)/(237\ lb\ per\ sq\ in.)$$
$$= 2.53\ sq\ in.$$

Equating the two areas A and solving for z gives:

$$7z = 2.53$$
$$z = 2.53/7$$
$$= 0.36\ in.$$

However, the length of z should be at least 3 in. to eliminate the possibility of splitting a member.

Steel Column Clamps with Wedges

Figure 10-6 illustrates a steel column clamp for forming square or rectangular columns. One set of clamps consists of two hinged units, which permits fast assembly with positive locking attachments by steel wedges. This clamp is available in a 36-in. and a 48-in. overall side bar length and can be adjusted to any fraction of an inch. The 36-in. size clamp with flat 2 × 4s can form up to a 24-in.-square column. The 48-in. size with the same lumber will form a 36-in.-square column. Both clamps will form rectangular columns, ranging from 8 to 24 in. for the smaller size clamps and from 12 to 36 in. for the larger size clamps.

Steel wedges are used to attach and secure the column clamps. The clamps automatically square the column as the wedge is tightened. To prevent twisting of the column forms, the clamps should be alternated 90° as shown in Figure 10-6.

Column forms are often made with ¾-in.-thick Plyform and either 2 × 4s or 2 × 6s placed flat against the outside of the Plyform. The spacing of column clamps is governed by deflection of both the lumber and the clamps. Table 10-4 provides clamp spacings for this steel column clamp. Column clamp deflection is limited to $l/270$ and a concrete temperature of 70°F. The spacings

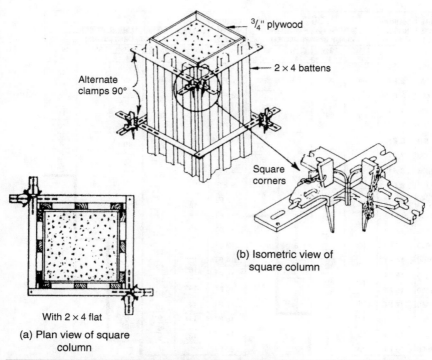

³/₄" plywood

2 × 4 battens

Alternate
clamps 90°

Square
corners

(b) Isometric view of
square column

With 2 × 4 flat

(a) Plan view of square
column

FIGURE 10-6 Column form with steel clamps and steel wedges. (*Source: Dayton Superior Corporation*)

shown are for the column clamps only. Additional checks of the sheathing must be performed to determine the limiting clamp spacing based on the strength and deflection and of the sheathing (see Example 10-2).

Example 10-5

Use Table 10-4 to determine the spacing and number of column clamps for a 20-in. square column 10 ft high. The sheathing will be ¾-in.-thick Plyform with 2 × 4 vertical wood battens.

Using the full hydrostatic pressure of the concrete on the forms, the maximum lateral pressure will be:

From Eq. (3-1), $P = wh$

$$= (150 \text{ lb per cu ft})(10 \text{ ft})$$

$$= 1,500 \text{ lb per sq ft}$$

TABLE 10-4 Column Clamp Spacing Chart

CONCRETE PRESSURE LBS/SQ.FT.	COLUMN HEIGHT FT.	36"* OVERALL LENGTH OF SIDEBAR — CLAMP OPENING (INCHES)				COLUMN HEIGHT FT.	48"* OVERALL LENGTH OF SIDEBAR — CLAMP OPENING (INCHES)				
		14½"	19½"	24½"	28½"		20½"	25½"	30½"	35½"	40½"
0	0	4"	4"	4"	4"	0	4"	4"	4"	4"	4"
150	1 FT.					1 FT.					
300	2 FT.	33"	33"	33"	26"	2 FT.	33"	33"	33"	28"	22"
450	3 FT.				20"	3 FT.				21"	18"
600	4 FT.	29"	29"	26"		4 FT.	29"	29"	27"		11"
750	5 FT.				14"	5 FT.				15"	10"
900	6 FT.			17"	11"	6 FT.		23"	19"	12"	8"
1050	7 FT.	23"	23"		10"	7 FT.	26"			10"	8"
1200	8 FT.			14"	8"	8 FT.			15"	10"	6"
1350	9 FT.	20"	19"	12"	8"	9 FT.	20"	20"	13"	8"	6"
1500	10 FT.	18"	17"	10"	8"/7"/6"/6"	10 FT.	18"	17"	12"	8"/7"	6"
1650	11 FT.		14"	9"	6"	11 FT.		14"	10"	7"	6"
1800	12 FT.	18"		8"		12 FT.	18"		10"	6"	6"
1950	13 FT.	16"	12"	8"/7"		13 FT.	16"	14"	9"		
2100	14 FT.	15"	12"	7"		14 FT.		12"	8"		
2250	15 FT.	14"	10"	7"		15 FT.	16"	12"	8"/7"		
2400	16 FT.		10"	6"		16 FT.	14"	10"	7"		
2550	17 FT.	14"	10"	6"/6"		17 FT.	14"	10"	6"		
2700	18 FT.	14"	9"			18 FT.	14"	9"	6"		
2850	19 FT.	13"	8"/8"			19 FT.	12"	8"	6"		
3000	20 FT.	9"	8"			20 FT.	6"	7"	6"		

Note:

The column clamp opening will be the width of the column plus twice the thickness of the 2 × 4 stud and the ¾-in. plywood sheathing: 20 + 2(1.5 + 0.75) = 24.5 in. Therefore, from Table 10-4 select the 36-in. column clamp with a 24.5-in. opening. As shown in Table 10-4, the 36-in. column clamp with a 1,500 lb per sq ft pressure and 24.5-in. column opening requires seven column clamps. The lower 10-in. spacing must be adjusted because the bottom clamp requires 6-in. clearance above the slab. The total number of required clamps is 7, spaced at 6, 8, 12, 14, 17, 26, and 33 in.

The ¾-in. Plyform sheathing backed with 2 × 4 vertical wood battens must have adequate strength and rigidity to transfer the concrete pressure between the column clamps. Example 10-2 shows the calculations to ensure adequacy of the sheathing and vertical wood battens.

Concrete Column Forms with Patented Rotating Locking Device

Figure 10-7 illustrates a column forming system with ¾-in. plywood sheathing, vertical 2 × 6 studs, and steel clamps that are secured by a patented device with a handle that rotates to open and close the column clamp. The patented cam-over column clamp has a claw that engages the pin at the closing corner and quickly and securely locks the column form together.

The form can be job built and gang formed with no loose pieces. The column clamps are fabricated from steel angles with ½-in.-diameter holes spaced at 1 in. on centers along the clamp for adjustment of the column clamp to various sizes. The studs are placed flat against the plywood at a maximum spacing of 12 in. on centers. For special load conditions, 4 × 4 studs may also be used in lieu of the 2 × 4 studs.

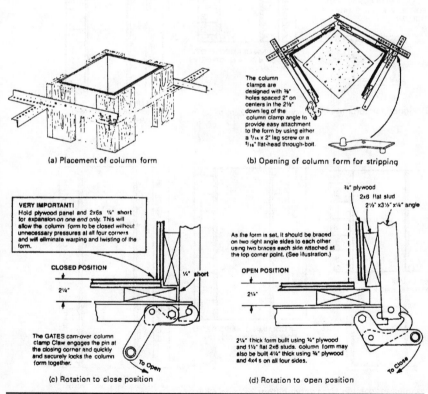

(a) Placement of column form

The column clamps are designed with ¾" holes spaced 2" on centers in the 2½" down leg of the column clamp angle to provide easy attachment to the form by using either a ³/₁₆ x 2" lag screw or a ⁵/₁₆" flat-head through-bolt.

(b) Opening of column form for stripping

VERY IMPORTANT!
Hold plywood panel and 2x6s ¼" short for expansion on one end only. This will allow the column form to be closed without unnecessary pressures at all four corners and will eliminate warping and twisting of the form.

CLOSED POSITION

2¼"

¼" short

The GATES cam-over column clamp Claw engages the pin at the closing corner and quickly and securely locks the column form together.

To Open

(c) Rotation to close position

As the form is set, it should be braced on two right angle sides to each other using two braces each side attached at the top corner point. (See illustration.)

¾" plywood
2x6 flat stud
2½" x3½" x¼" angle

OPEN POSITION

2¼"

2¼" thick form built using ¾" plywood and 1½" flat 2x6 studs. column form may also be built 4¼" thick using ¾" plywood and 4x4 s on all four sides.

To Close

(d) Rotation to open position

FIGURE 10-7 Steel column clamps with rotating locking device. (*Source: Gates & Sons, Inc.*)

The corner diagonal opposite from the locking corner acts as the hinge point for easy opening and resetting of the column form. A special device, consisting of a plate with pins, is available that may be installed in the opposite corners of the column clamp for squaring corners and to help stabilize the column form while setting and stripping.

Examples of the spacing of column clamps for this system of column forms are shown in Figure 10-8 for column heights of up to 16 ft. For greater heights or special applications, the manufacturer should be

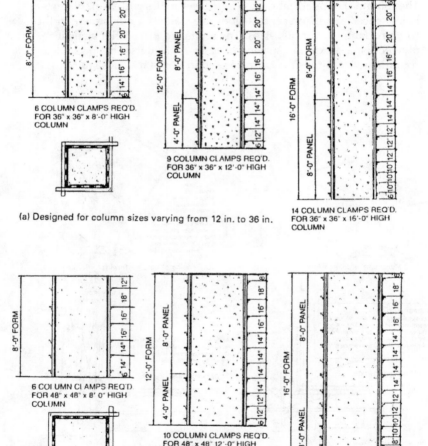

FIGURE 10-8   Illustration of typical column clamp spacings using the Gates & Sons, Inc. column-forming system.

consulted. Typical clamp spacings shown in Figure 10-8 are for column forms constructed of ¾- in. plywood and 2 × 6 studs placed flat, or 4 × 4 studs. Spacing may vary due to job conditions, temperature changes, vibration, rate of placement of concrete, and type of lumber used for construction of the form. The location of the studs should not exceed 12 in. on centers or a 6.5-in. span between members in all cases.

Column Forms Using Jahn Brackets and Cornerlocks

Figure 10-9 illustrates a column forming system that secures the plywood and vertical studs with horizontal walers. Jahn brackets and cornerlocks, described in Chapter 9 for wall forms, are used to fasten horizontal walers.

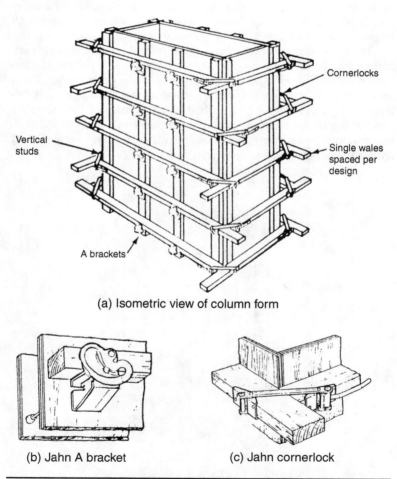

Cornerlocks

Vertical studs

Single wales spaced per design

A brackets

(a) Isometric view of column form

(b) Jahn A bracket

(c) Jahn cornerlock

FIGURE 10-9   Column form with Jahn brackets and cornerlocks. (*Source: Dayton Superior Corporation*)

The bracket has an eccentric action that securely holds the form-work. The cornerlock is used at outside corners to secure the ends of the walers. It slips into place with its handle perpendicular to the wale. Only two nails are needed for attachment while barbed plates grip the side of the 2×4s for positive nonslip action. The locking handle has a cam action, drawing the wales together at true right angles. No special tools are needed for either installation or stripping.

Modular Panel Column Forms

Modular panel forms provide several ways to form columns of various shapes and sizes. Outside corners and panels or fillers can be combined to form square or rectangular columns. Generally modular panels provide a fast and more accurate column form than job-built forms. Figure 10-10 shows a column form erected with modular panels.

The form consists of four panels that are fastened together at each corner. Wedge bolts are used to secure the corners of the forms. Column hinges are used to allow the panels to swing open for quick and easy stripping of the forms after the concrete has been placed. A column lifting bracket may be attached to the corners of the forms for positioning of the column form. Form liners can be combined with the column forms for special architectural finishes. Figure 10-11 shows the details of the connections.

Figure 10-10   Modular panels for column forms. (*Source: Symons Corporation*)

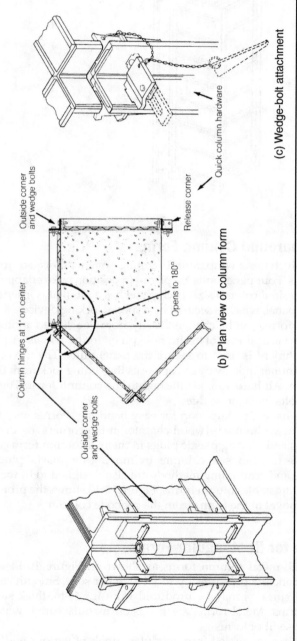

(c) Wedge-bolt attachment

Quick column hardware

Outside corner
and wedge bolts

Release corner

Opens to 180°

Column hinges at 1' on center

Outside corner
and wedge bolts

(b) Plan view of column form

FIGURE 10-11 Connections for modular panel column forms. (*Source: Symons Corporation*)

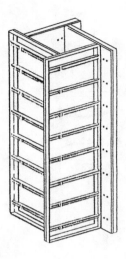

FIGURE 10-12
Adjustable
wraparound
column forms.
(*Source: Gates &
Sons, Inc.*)

Adjustable Wraparound Column Forms

Figure 10-12 illustrates a wraparound forming system fabricated from modular panels. Four panel units are placed in position by overlapping to obtain the desired column size. Each panel has a pin that is inserted to its adjacent panel, which is secured by a special locking device.

The panel forms can be changed easily from one size to another by changing a vertical row of ³⁄₁₆-in. hole plugs in the panels. A pin and lock attachment is used to secure the panel units. The tap of a carpenter's hammer quickly opens or closes the sliding lock from the connecting pin. All hardware is attached to the column form. There are no loose bolts, nuts, or wedges.

The forms have a pickup loop for easy handling. Forms may be ordered with or without hardwood chamfer at the corners. By drilling extra holes and placing plastic plugs in them, a column form can be used for 18- to 24-in. size columns by moving the plastic plugs. Bracing plates and scaffolding methods are also designed to fit these wraparound adjustable column forms. Figure 10-13 shows the proper method of removal of the forms from the concrete column.

All-Metal Forms for Rectangular Forms

Lightweight all-metal column forms as shown in Figure 10-14 are available for concrete pressures up to 1,200 lb per sq ft. Smooth concrete finished surfaces may be produced from the ⅛-in.-thick steel face of the forms. Metal forms are precision manufactured, which provides precise fit of forms.

The forms are assembled from modular panels of various widths and lengths. Modules in heights of 1, 2, 4, and 8 ft are available in

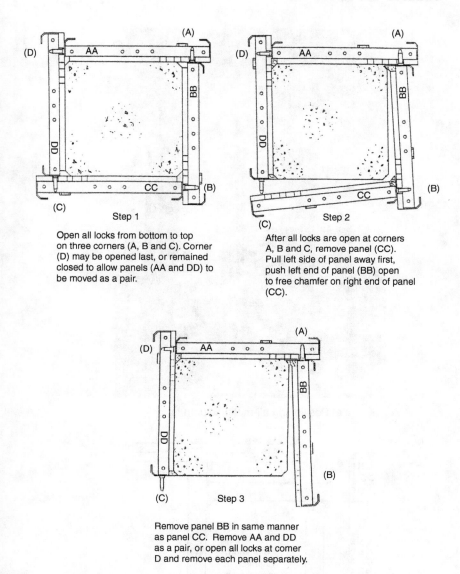

Step 1

Open all locks from bottom to top on three corners (A, B and C). Corner (D) may be opened last, or remained closed to allow panels (AA and DD) to be moved as a pair.

Step 2

After all locks are open at corners A, B and C, remove panel (CC). Pull left side of panel away first, push left end of panel (BB) open to free chamfer on right end of panel (CC).

Step 3

Remove panel BB in same manner as panel CC. Remove AA and DD as a pair, or open all locks at corner D and remove each panel separately.

Figure 10-13 Removal of adjustable wraparound column forms. (*Source: Gates & Sons, Inc.*)

column sizes of 10 to 30 in., in increments of 2 in. The modules may be stacked to the desired column height and attached together to secure adjacent modules. Only three bolts are required per 8-ft-high form. No outside angle corners are required. Figure 10-15 shows combinations of the modular system to accommodate various column heights.

(a) Positioning all-metal column form

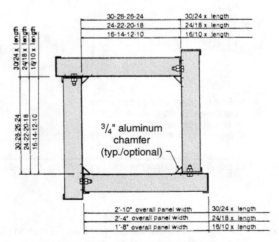

(b) Typical plan view pouring position

FIGURE 10-14   All-metal modular concrete column forms. (*Source: EFCO*)

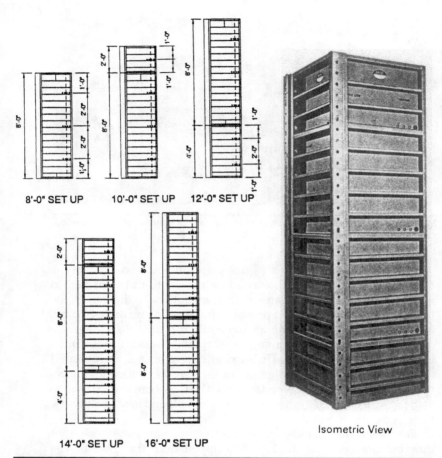

8'-0" SET UP   10'-0" SET UP   12'-0" SET UP

14'-0" SET UP   16'-0" SET UP

Isometric View

Figure 10-15 Setup arrangements for all-metal modular concrete columns. (*Source: EFCO*)

Fiber Tubes for Round Columns

Forms that are used frequently for round concrete columns are available under various trade names. An example of these types of forms is the Sonotube shown in Figure 10-16, manufactured by the Sonoco Products Company. This form is made from many layers of high-quality fiber, spirally wound and laminated with a special adhesive.

Fiber form tubes are manufactured for column sizes from 6 to 48 in., in increments of 2 in. The tube should be braced every 8 ft. For a rate of concrete placement of less than 15 ft per hr and concrete pressures up to 3,000 lb per sq ft, the tubes will not cause the form to buckle, swell, or lose shape. The concrete can be vibrated as required, but care should be taken to prevent the vibrator from damaging the form. Concrete should be placed with a tremie.

FIGURE 10-16   Placement of concrete in round fiber tube columns. (*Source: Sonoco Products Company*)

The forms are fabricated as one-piece units, so no assembly is required. They can be cut or sawed on the job to fit beams and to allow for utility outlets. Because of their exceptional light weight, the forms are easily handled and provide a fast forming method. Placing, bracing, pouring, and finishing require minimal time.

Fiber tube forms have a variety of special uses. Sections can be used in forming pilasters, and half-round, quarter-round, and obround columns. In forming such columns, the form section is used in connection with the plywood or wooden form. The forms can also be used to stub piers for elevated ramps, outdoor signs, pole and fence-pole bases, flagstones, and round steps.

Various inside coatings of fiber tubes are available, which offer different finishes after the tube is stripped from the column. The coated tube produces a visible spiral seam on the stripped column. A seamless tube with a specially finished inner ply is available which minimizes, but does not completely eliminate, the spiral seam appearance on the finished concrete column. A high-quality tube that is fitted with a plastic liner is available, which imparts a smoother architectural finish to round columns.

The forms should be stripped as soon as possible after concrete has set. The recommended time is 24 to 48 hours and should not exceed 5 days. The forms can be stripped by setting a saw blade to the thickness of the form and making two vertical cuts, then removing the form. An alternate method is to slit the form 12 in. from the top with a sharp knife, then use a broad-bladed tool and pry the form off the column.

Steel Forms for Round Columns

Steel forms, consisting of sheet steel attached to prefabricated steel shapes, such as angles, are frequently used for round columns where the number of reuses justifies the initial cost or where special conditions require their use. Steel round forms develop an exceptionally smooth, hard surface that is free of voids and with a minimum number of seams.

Vertical and horizontal seams allow easy opening and closing of the column forms with each pour. Angles are permanently attached to the outside of the forms along each vertical and horizontal seam. The angles from two mating seams are attached by speed-bolts.

Full-circle forms for column sizes up to 48 in. are usually assembled by joining two half-circles, using steel bolts to join the component parts (Figure 10-17). Quarter-round sections are available for larger-diameter columns. Also, sections may be stacked on top of each other and bolted together to provide forms of any desired height.

Standard column diameters from 12 to 72 in. and heights in lengths of 1, 2, 4, and 8 ft are commonly available. Diameters from 12 to 36 in. are available in 2-in. increments. Diameters from 36 to 72 in.

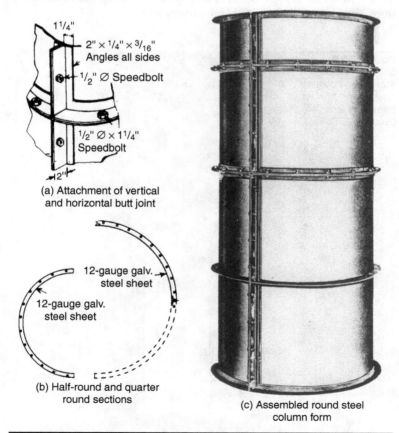

(a) Attachment of vertical and horizontal butt joint

(b) Half-round and quarter round sections

(c) Assembled round steel column form

Figure 10-17 Round steel column form. (*Source: Deslauriers, Inc.*)

are available in 6-in. increments. In some locations, steel forms may be rented or purchased.

One-Piece Steel Round Column Forms

Round steel column forms are available as single one-piece units for forming. Because the forms are fabricated as one piece, the finished surface of the column does not show seam marks.

These types of forms are slipped over the top of the reinforcing steel before placement of concrete, as illustrated in Figure 10-18. Round column forms are available in 12- to 96-in. diameters. Steel ribs keep the forms round and facilitate bracing and aligning the forms. There is no loose hardware, which simplifies the reassembly for the next pour.

Column capitals are handled easily, as illustrated in Figure 10-19. Steel capitals are available for use with forms whose diameters vary from 12 to 42 in. in increments of 2 in., and whose top diameters vary from 3 ft 6 in. to 6 ft 0 in. in increments of 6 in.

FIGURE 10-18
Slipping one-piece steel round column form over reinforcement.
(*Source: EFCO*)

Figure 10-19 Steel column forms with capitals. (*Source: EFCO*)

Plastic Round Column Forms Assembled in Sections

Column forms consisting of 24-in.-diameter sections 1 ft long are available, as illustrated in Figure 10-20. These forms are made of durable plastic with a constant radius. They are manufactured in half-round sections that may be bolted together in units to form column heights up to 10 ft, designed with a 2,000 lb per sq ft form pressure.

Figure 10-20
Plastic column
forms assembled
in sections.
(*Source:
Deslauriers, Inc.*)

Because they are lightweight, they are easily handled by one person. The interlocking joints, both horizontal and vertical, minimize grout, leakage, and discoloration.

Spring-Open Round Fiberglass Forms

Figures 10-21 and 10-22 illustrate a one-piece cylindrical column form molded of fiberglass-reinforced polyester for a consistently smooth concrete finish. It is fabricated as one piece with one vertical joint that springs open for easy removal of the form. The vertical joint is secured by wedge bolts. A round bracing collar is attached to the form for alignment and to set an anchor template at the bottom of the form.

The interior of the column form is sanded and finished to produce a smooth surface. This form is reusable and is available for rent in diameters of 12, 16, 18, 24, 30, and 36 in. and in lengths ranging from 4 to 16 ft in even 2-ft increments. The longer forms and other diameters up to 48 in. are available for purchase only.

Columns up to 16 ft high can be poured with standard 150 lb per cu ft density concrete. The thickness of the fiberglass material increases with the diameter of the column form.

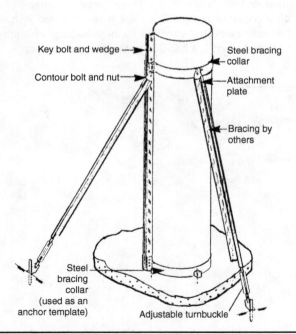

Key bolt and wedge

Steel bracing collar

Contour bolt and nut

Attachment plate

Bracing by others

Steel bracing collar (used as an anchor template)

Adjustable turnbuckle

Figure 10-21   Erection of spring-open round fiberglass form. (*Source: Symons Corporation*)

FIGURE 10-22 Opening form after pour. (*Source: Symons Corporation*)

Fiberglass column capital forms are also available, manufactured to the same specifications as the round column forms. The capitals are molded in two half-sections with vertical flanges designed for quick alignment.

One-Piece Round Fiberglass Column Forms

Figure 10-23 illustrates a one-piece round column form manufactured of fiberglass-reinforced plastic. Diameters are available from 12 to 28 in., and lengths up to 20 ft, for form pressures of 2,250 lb per sq ft. These forms will not dent, sag, rot, or weather, and require little maintenance.

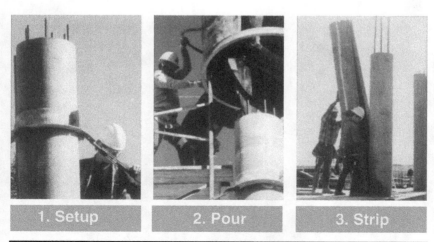

FIGURE 10-23 One-piece fiberglass column forms. (*Source: Molded Fiber Glass Concrete Forms Company*)

Because they are lightweight, they are easy to handle and simple to place and strip. These forms contain only one vertical seam and are supplied with bracing collars and "fast" bolts that are designed for repeated use. The units may be nested together, thus requiring less storage space.

References

1. Dayton Superior Corporation, Dayton, OH.
2. Deslauriers, Inc., LaGrange Park, IL.
3. EFCO, Des Moines, IA.
4. Ellis Construction Specialties, Ltd,, Oklahoma City, OK.
5. Gates & Sons, Inc., Denver, CO.
6. Molded Fiber Glass Concrete Forms Company, Independence, KS.
7. Sonoco Products Company, Hartsville, SC.
8. Symons Corporation, Elk Grove Village, IL.

CHAPTER 11

Forms for Beams and Floor Slabs

Concrete Floor Slabs

There are many types of concrete floor slabs, including, but not limited to, the following:

1. Concrete slabs supported by concrete beams
2. Concrete slabs of uniform thickness with no beams, designated as flat slabs
3. Fiberglass dome forms for two-way concrete joist systems
4. Metal-pan and concrete-joist-type slabs
5. Cellular-steel floor systems
6. Corrugated-steel forms and reinforcement floor systems
7. Concrete slabs on steel floor lath

The forms used to support each of these slabs or floor systems will depend on the type of slab, as described in the following sections.

In previous chapters of this book, calculations are shown to determine the minimum size and spacing of members for formwork. For example, 2-in.-thick lumber was used for studs and wales for wall forms. Formwork for floor slabs is somewhat different than formwork for walls or columns. Slab forms are elevated; therefore, they require some type of vertical support. Also, a full crew of laborers will be working on the formwork; therefore, 2 × 4 the need for safety is important. Thus, for constructability and safety it is desirable to use 4 × 4 joists on 4 × 6 stringers for slab forms because 4-in.-thick members are easier for workers to walk on, and place, than 2-in.-thick members. Also, the 4-in.-thick member will not roll-over or fall sideways during installation compared to 2-in. thick members.

Safety of Slab-Forming Systems

Forms for concrete beams and slabs should provide sufficient strength and rigidity at the lowest practical cost, considering materials, labor, and any construction equipment used in making, erecting, and removing them. Consideration must be given to both the static dead load and any impact loads that may be applied to the forming system. The forming system must provide adequate resistance to lateral forces that may be imposed, in addition to the vertical loads from concrete, workers, tools, and equipment.

In many instances, the failure of formwork is a result of improper or inadequate shoring for slabs. The shores that support slab-forming systems must have sufficient load capacities, and they must be securely fastened at the bottom and top ends to prevent movement or displacement while they are in use. It is especially important to attach both ends of the shores to the slab form because it is possible for the slab form to lift off the top of a shore due to unbalanced loading during placement of concrete. Also, a shore system may shift due to inadequate support at the bottom of a shore. Two-way horizontal and diagonal braces should be installed to brace the shores adequately for slab formwork.

Loads on Concrete Slabs

Prior to designing the forms to support a concrete slab, it is necessary to know the magnitude of the loads that the forms must support. The loads that will be applied to the slab forms include the dead weight of the reinforcing steel and the freshly placed concrete, the dead load of the formwork materials, and the live load of workers, tools, and equipment. Chapter 3 presents loads on concrete formwork.

The effect of impact loads should also be included in designing forms, as that caused by motor-driven concrete buggies, concrete falling from a bucket, or pumping of concrete. The effect of concrete falling from a bucket is discussed in Chapter 7.

The unit weight of concrete for most structures is 145 to 150 lb per cu ft. It is common to refer to normal concrete as having a unit weight of 150 lb per cu ft. The load from the concrete on a sq ft of the floor decking will be:

$$p = w_c \, (h)$$

where p = load, lb per sq ft
   w_c = unit weight of concrete, lb per cu ft
   h = thickness of concrete, ft

For example, for a 6-in.-thick slab of 150 lb per cu ft concrete, the vertical dead load of the concrete on the floor decking will be:

$$p = w_c(h)$$
$$= 150 \text{ lb per cu ft } (6/12 \text{ ft})$$
$$= 75 \text{ lb per sq ft}$$

The live load will include workers and buggies used to place the concrete, plus materials that may be stored on the slab. It is common practice to assume a live load varying from 50 to 75 lb per sq ft of floor or more, depending on the anticipated conditions. The live loads on formwork are discussed in Chapter 3.

The following is an illustration of the loads of formwork on a 6-in. slab with a 5.0 lb per sq ft dead load of formwork materials and a live load of 50 lb per sq ft:

Dead load of freshly placed concrete	=	75 lb per sq ft
Dead load of formwork materials	=	5 lb per sq ft
Live load of workers and tools	=	50 lb per sq ft
Total load	=	130 lb per sq ft

Definition of Terms

Sometimes there are variations in the terminology used in the construction industry. Figure 11-1 illustrates the components of a job-built slab-forming system for a beam and slab concrete structure. The following terms are used in this book:

1. *Decking* is the solid plywood panels that form the floor of the formwork against which the fresh concrete is placed. Sheets of Plyform, the plywood manufactured especially for concrete formwork, are commonly used. The decking may also be patented fiberglass domes or steel panel forms. It provides resistance to the vertical pressure of the freshly placed concrete.

2. *Joists* are the members under the decking that provide support for the floor decking. Joists are usually single members

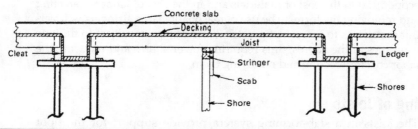

FIGURE 11-1 Forms for concrete beams and slab with intermediate stringers.

of wood material, usually 4 in. thick with a depth that depends on the loads applied to the form, the species and grade of lumber, and span length of the joist. Joists may also be metal beams or trusses.

3. *Stringers* are members under the joists that provide support for the joists. Stringers are usually single members of wood material, usually 4 in. thick with a depth that depends on the loads applied from the joists, the species and grade of lumber, and span length of the stringer. Stringers may also be metal beams or trusses.

4. *Shores* are members that support the joists and stringers and beam bottoms for beam-slab forming systems. Shores may be single wood posts, single steel joists, or shoring frames.

Design of Forms for Concrete Slabs

The steps in the design of forms to support concrete slabs include the following:

1. Determine the total unit load on the floor decking, including the effect of impact, if any.

2. Select the kind and net thickness of the floor decking.

3. Determine the safe spacing of floor joists, based on the strength and permissible deflection of the decking.

4. Select the floor joists, considering the load, kind, size, and length of the joist.

5. Select the kind, size, and length of the stringers to support the joist.

6. Select the kind, size, and safe spacing of the shores, considering the load, the strength of the stringer, and the safe capacity of the shores.

Usually the most economical forms result when the joists are spaced for the maximum safe span of the decking. Similarly, reasonably large joists, which permit long spans, thus requiring fewer stringers, will be economical in the cost of materials and in the cost of labor for erecting and removing the forms. The use of reasonably large stringers will permit the shores to be spaced greater distances apart, subject to the safe capacities of the shores, thus requiring fewer shores and reducing the labor cost for erecting and removing them.

Spacing of Joists

The joists in a slab-forming system provide support for the floor decking. Therefore the maximum spacing of joists is determined by the allowable span length of the decking that rests on the joists. In the

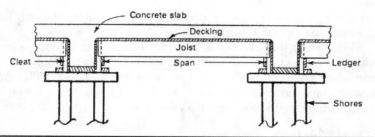

FIGURE 11-2 Forms for concrete beams and slab with single-span joists.

design of slab forms, the thickness and grade of the floor decking are often selected based on the availability of materials. Plyform is commonly used as decking for slab forms.

Figure 11-2 illustrates a system of wood forms for a beam and slab type concrete floor with single-span joists; that is, there are no intermediate stringers of support between the ends of the joists. This system is satisfactory for joists with relatively short spans. However, for longer spans the required joist sizes may be too large for single spans. For this condition it may be necessary to place intermediate stringers under the joists, perpendicular to the joists, to provide support to the joists. This is further discussed in later sections of this chapter.

The slab decking must have adequate strength and rigidity to resist bending stresses, shear stresses, and deflection between joists. The equations for determining the allowable span length of plywood decking for uniformly distributed pressures were presented in Chapter 5 and are reproduced in Table 11-1 for clarity. The allowable stresses for

		Bending	Deflection
Bending l_b, in.	Shear l_s, in.	l_d, in. for $\Delta = l/360$	l_d, in. for $\Delta = {}^1/_{16}$ in.
$[120F_bS_e/w_b]^{1/2}$ Eq. (5-41)	$20F_s(Ib/Q)/w_s$ Eq. (5-48)	$[1{,}743EI/360w_d]^{1/3}$ Eq. (5-57a)	$[1{,}743EI/16w_d]^{1/4}$ Eq. (5-57b)

Notes:
1. See Tables 4-9 and 4-11 for physical properties of plywood and Plyform, respectively.
2. See Tables 4-10 and 4-12 for allowable stresses of plywood and Plyform, respectively.
3. Units for stresses and physical properties are pounds and inches.
4. Units for w are lb per linear ft for a 12-in.-wide strip of plywood.

TABLE 11-1 Equations for Calculating Allowable Span Lengths for Floor Decking with Three or More Spans, Based on Bending, Shear, and Deflection Due to Uniformly Distributed Pressure, in.

bending F_b and shear F_s are in pounds per square inch. The unit for allowable span length is inches.

Because the slab decking is normally placed over multiple joists (more than four) the equations for allowable span lengths of three or more spans will apply. The physical properties of plywood and Plyform sheathing are shown in Tables 4-9 and 4-11 in Chapter 4. Allowable stresses are shown in Table 4-9 for plywood and Table 4-11 for Plyform.

Example 11-1

Determine the maximum spacing of joists for a concrete floor slab that will be formed with ¾-in.-thick Plyform decking. The total design load on the slab will be 150 lb per sq ft. The deflection criterion limits the permissible deflection to $l/360$ but not to exceed ¹⁄₁₆ in.

The face grain of the Plyform will be placed across the joists; therefore, the physical properties for stress applied parallel to face grain will apply. From Tables 4-9 and 4-11, the physical properties and allowable stresses for ¾-in.-thick Class I Plyform can be obtained.

The physical properties of ¾-in. Plyform are

Approximate weight = 2.2 lb per sq ft

Moment of inertia, $I = 0.199$ in.^{4}

Effective section modulus, $S_e = 0.455$ in.^{3}

Rolling shear constant, $Ib/Q = 7.187$ in.^{2}

The allowable stresses of Class I Plyform are

Bending, $F_b = 1,930$ lb per sq in.

Shear, $F_s = 72$ lb per sq in.

Modulus of elasticity, $E = 1,650,000$ lb per sq in.

Because the Plyform will be placed over multiple supports, the equations for three or more spans will apply. The spacing of the joists is governed by the maximum span length based on bending, shear, and deflection of the Plyform decking as follows.

For bending stress in the Plyform decking, the maximum span length from Table 11-1,

$$l_b = [120 F_b S_e / w_b]^{1/2}$$
$$= [120(1,930)(0.455)/150]^{1/2}$$
$$= 26.5 \text{ in.}$$

For rolling shear stress in Plyform decking, the maximum span length from Table 11-1,

$$l_s = 20 F_s (Ib/Q)/w_s$$
$$= 20(72)(7.186)/150$$
$$= 68.9 \text{ in.}$$

For deflection not to exceed $l/360$, the maximum span length from Table 11-1,

$$l_d = [1,743EI/360w_d]^{1/3}$$
$$= [1,743(1,650,000)(0.199)/360(150)]^{1/3}$$
$$= 21.9 \text{ in.}$$

For deflection not to exceed $\frac{1}{16}$ in., the maximum span length from Table 11-1,

$$l_d = [1,743EI/16w_d]^{1/4}$$
$$= [1,743(1,650,000)(0.199)/16(150)]^{1/4}$$
$$= 22.1 \text{ in.}$$

Summary for the ¾-in.-Thick Plyform Decking

For bending, the maximum span length of the Plyform = 26.5 in.
For shear, the maximum span length of the Plyform = 68.9 in.
For deflection, the maximum span length of the Plyform = 21.9 in.

In this example, deflection controls the maximum span length of the Plyform decking. Therefore, the joists that support the Plyform must be placed at a spacing that is no greater than 21.9 in. For constructability, the spacing would likely be placed at 18 in. on centers.

Use of Tables to Determine Maximum Spacing of Joists

Tables have been developed by APA—The Engineered Wood Association and by companies in the construction industry that supply formwork products, which provide the allowable span length of plywood for a given concrete pressure. When the design of formwork is based on design tables of manufacturers or associations in the construction industry, it is important to follow their recommendations.

Examples of these types of tables are presented in Chapter 4 for Plyform, which is manufactured by the plywood industry specifically for use in formwork for concrete. Tables 11-2 through 11-5 are reproduced here from Chapter 4 for determining the support spacing of Plyform for various panel thicknesses and uniform pressures. Thus, the spacing of joists for formwork decking can be determined based on the support spacing of the Plyform in the tables.

When computing the allowable pressure of concrete on plywood, the center-to-center distances between supports should be used to determine the pressure based on the allowable bending stress in the fibers.

When computing the allowable pressure of concrete on plywood as limited by the permissible deflection of the plywood, it is proper to

Plywood Thickness, in.	Support Spacing, in.						
	4	**8**	**12**	**16**	**20**	**24**	**32**
$^{15}/_{32}$	2,715	885	355	150	—	—	—
	(2,715)	(885)	(395)	(200)	115	—	—
$^{1}/_{2}$	2,945	970	405	175	100	—	—
	(2,945)	(970)	(430)	(230)	(135)	—	—
$^{19}/_{32}$	3,110	1,195	540	245	145	—	—
	(3,110)	(1,195)	(540)	(305)	(190)	(100)	—
$^{5}/_{8}$	3,270	1,260	575	265	160	—	—
	(3,270)	(1,260)	(575)	(325)	(210)	(100)	—
$^{23}/_{32}$	4,010	1,540	695	345	210	110	—
	(4,010)	(1,540)	(695)	(325)	(270)	(145)	—
$^{3}/_{4}$	4,110	1,580	730	370	225	120	—
	(4,110)	(1,580)	(730)	(410)	(285)	(160)	—
$1^{1}/_{8}$	5,965	2,295	1,370	740	485	275	130
	(5,965)	(2,295)	(1,370)	(770)	(535)	(340)	(170)

Notes:
1. Courtesy APA—The Engineered Wood Association, "Concrete Forming," 2004.
2. Deflection limited to $l/360$th of the span, $l/270$th for values in parentheses.
3. Plywood continuous across two or more spans.

TABLE 11-2 Recommended Maximum Pressure on Plyform Class I—Values in Pounds per Square Foot with Face Grain Across Supports

use the clear span between the supports plus ¼ in. for supports whose nominal thickness is 2 in., and the clear span between the supports plus ⅝ in. for supports whose nominal thickness is 4 in.

The recommended concrete pressures are influenced by the number of continuous spans. For face grain across supports, assume three continuous spans up to 32-in. support spacing and two spans for greater spacing. For face grain parallel to supports, assume three spans up to 16 in. and two spans for 20 and 24 in. These are general rules as recommended by APA—The Engineered Wood Association.

There are many combinations of frame spacings and plywood thicknesses that may meet the structural requirements for a particular job. However, it is recommended that only one thickness of plywood be used and then the frame spacing varied for different pressures. Plyform can be manufactured in various thicknesses, but it is good practice to base design on $^{19}/_{32}$, $^{5}/_{8}$, $^{23}/_{32}$, and ¾-inch Class I Plyform because they are the most commonly available thicknesses.

Tables 11-2 through 11-5 give the recommended maximum pressures of concrete on Class I and Structural I Plyform decking. Calculations for

Plywood Thickness, in.	Support Spacing, in.					
	4	8	12	16	20	24
15/32	1,385	390	110	—	—	—
	(1,385)	(390)	(150)	—	—	—
½	1,565	470	145	—	—	—
	(1,565)	(470)	(195)	—	—	—
19/32	1,620	530	165	—	—	—
	(1,620)	(530)	(225)	—	—	—
5/8	1,770	635	210	—	—	—
	(1,770)	(635)	(280)	120	—	—
23/32	2,170	835	375	160	115	—
	(2,170)	(835)	(400)	(215)	(125)	—
¾	2,325	895	460	200	145	—
	(2,325)	(895)	(490)	(270)	(155)	(100)
1⅛	4,815	1,850	1,145	710	400	255
	(4,815)	(1,850)	(1,145)	(725)	(400)	(255)

Notes:
1. Courtesy APA—The Engineered Wood Association, "Concrete Forming," 2004.
2. Deflection limited to $l/360$th of the span, $l/270$th for values in parentheses.
3. Plywood continuous across two or more spans.

TABLE 11-3   Recommended Maximum Pressure on Plyform Class I—Values in Pounds per Square Foot with Face Grain Parallel to Supports

these pressures were based on deflection limitations of $l/360$ and $l/270$ of the span or on the shear or bending strength, whichever provided the most conservative (lowest load) value.

When computing the allowable pressure of concrete on plywood as limited by the allowable unit shearing stress and shearing deflection of the plywood, use the clear span between the supports.

Size and Span Length of Joists

Figure 11-2 illustrated single-span joists for the support of decking for concrete slabs. The selection of the size, length, and spacing of the joists will involve one of the following:

1. Given the total load on the decking, the spacing of the joists and the size and grade of the joists, determine the maximum span for the joists.

2. Given the total load on the decking and the spacing of the joists, determine the minimum-size joists required.

Plywood Thickness, in.	Support Spacing, in.						
	4	**8**	**12**	**16**	**20**	**24**	**32**
15⁄32	3,560	890	360	155	115	—	—
	(3,560)	(890)	(395)	(205)	115	—	—
½	3,925	980	410	175	100	—	—
	(3,925)	(980)	(435)	(235)	(135)	—	—
19⁄32	4,110	1,225	545	245	145	—	—
	(4,110)	(1,225)	(545)	(305)	(190)	(100)	—
5⁄8	4,305	1,310	580	270	160	—	—
	(4,305)	(1,310)	(580)	(330)	(215)	100	—
23⁄32	5,005	1,590	705	350	210	110	—
	(5,005)	(1,590)	(705)	(400)	(275)	(150)	—
¾	5,070	1,680	745	375	230	120	—
	(5,070)	(1,680)	(745)	(420)	(290)	(160)	—
1⅛	7,240	2,785	1,540	835	545	310	145
	(7,240)	(2,785)	(1,540)	(865)	(600)	(385)	(190)

Notes:
1. Courtesy APA—The Engineered Wood Association, "Concrete Forming," 2004.
2. Deflection limited to $l/360$th of the span, $l/270$th for values in parentheses.
3. Plywood continuous across two or more spans.

TABLE 11-4 Recommended Maximum Pressures on Structural I Plyform—Values in Pounds per Square Foot with Face Grain Across Supports

An analysis may be made to determine the economy of using either grade No. 1 or No. 2 lumber for joists, considering the difference in the cost of the lumber and the potential salvage value of each grade. Joist sizes for concrete floor formwork are usually selected as 4-in.-thick members for stability, to allow workers to place joists across stringers without the joists falling sideways.

The economy of member sizes should be investigated prior to making the selection. The cost of joists includes the cost of the lumber plus the labor cost of fabricating and erecting the joists, less the salvage value of the lumber after the forms are stripped. The bending strength of a joist of a given grade and species of lumber is directly related to its section modulus. For example, a 4 × 6 S4S joist is 2.46 times stronger than a 4 × 4 S4S joist, whereas the ratio for quantity of lumber for a given length is 1.57. The labor cost of fabricating and erecting a 4 × 6 joist may be little, if any, more than that for a 4 × 4 joist. For these reasons, it may be more economical to use 4 × 6 joists instead of 4 × 4 joists if the additional strength of the former size can be fully utilized.

Plywood Thickness, in.	Support Spacing, in.					
	4	8	12	16	20	24
15/32	1,970 (1,970)	470 (530)	130 (175)	— —	— —	— —
1/2	2,230 (2,230)	605 (645)	175 (230)	— —	— —	— —
19/32	2,300 (2,300)	640 (720)	195 (260)	— (110)	— —	— —
5/8	2,515 (2,515)	800 (865)	250 (330)	105 (140)	— (100)	— —
23/32	3,095 (3,095)	1,190 (1,190)	440 (545)	190 (255)	135 (170)	— —
3/4	3,315 (3,315)	1,275 (1,275)	545 (675)	240 (315)	170 (210)	— (115)
1 1/8	6,860 (6,860)	2,640 (2,640)	1,635 (1,635)	850 (995)	555 (555)	340 (355)

Notes:
1. Courtesy APA—The Engineered Wood Association, "Concrete Forming," 2004.
2. Deflection limited to $l/360$th of the span, $l/270$th for values in parentheses.
3. Plywood continuous across two or more spans.

TABLE 11-5 Recommended Maximum Pressures on Structural I Plyform—Values in Pounds per Square Foot with Face Grain Parallel to Supports

The joists must have adequate strength and rigidity to resist bending stresses, shear stresses, and deflection between supports. The equations for determining the allowable span lengths of wood members were presented in Chapter 5 and are reproduced in Table 11-6 for clarity. The allowable stresses for bending F_b and shear F_v are in pounds

Span Condition	Bending l_b, in.	Shear l_v, in.	Bending Deflection	
			l_Δ, in. for $\Delta = l/360$	l_Δ, in. for $\Delta = 1/16$ in.
Single spans	$[96F_bS/w]^{1/2}$ Eq. (5-30)	$16F_vbd/w + 2d$ Eq. (5-32)	$[4,608EI/1,800w]^{1/3}$ Eq. (5-33a)	$[4,608EI/80w]^{1/4}$ Eq. (5-33b)

TABLE 11-6 Equations for Calculating Allowable Span Lengths for Single-Span Wood Beams Based on Bending, Shear, and Deflection for Beams with Uniformly Distributed Loads, in.

per square inch. The unit for allowable span length is inches. Because joists are usually simply supported, the equations for single spans will apply.

The examples that follow illustrate a method of determining each of the three values previously listed. The results are based on resistance to bending, shear, and deflection for a simple span that is supported at each end. If a joist extends over more than two supports, it can support a larger load than the one specified. It is assumed that the joists are braced adequately. Table 11-6 provides the equations for determining the allowable span lengths of wood members based on bending, shear, and deflection. There equations can be used for single span joists as illustrated in Figure 11-2.

Example 11-2

Given the total load on the decking, the spacing of the joists, and the size and grade of the joists, determine the maximum span for the joist. For this example, the total load on the joist is 150 lb per sq ft and the joist spacing is 20 in. on centers. The size and grade of the joists are 4×4 S4S No. 1 grade Douglas Fir-Larch.

From Table 4-1 the physical properties of 4×4 lumber are

Section modulus, $S = 7.15$ in.^{3}

Moment of inertia, $I = 12.51$ in.^{4}

The decking will be attached to the top of the joists to stabilize the joists. The lumber selected for the joists is No. 1 grade Douglas Fir-Larch. Assume a dry condition and short load-duration, less than 7 days. Adjusting the reference design values in Table 4-3 for size C_F in Table 4-3a and short load-duration (C_D) in Table 4-4, the allowable stresses will be:

Bending stress, $F_b = C_F(C_D)$ (reference value for bending stress)

$= 1.5(1.25)(1,000)$

$= 1,875$ lb per sq in.

Shear stress, $F_v = C_D$ (reference value for shear stress)

$= 1.25(180)$

$= 225$ lb per sq in.

Modulus of elasticity, $E = 1,700,000$ lb per sq in.

The uniform linear load transmitted from the decking to the joist can be calculated by multiplying the slab load by the joist spacing as follows:

$$w = (150 \text{ lb per sq ft})(20/12 \text{ ft})$$
$$= 250 \text{ lb per lin ft}$$

The allowable span lengths for the joists with single spans can be calculated for bending, shear, and deflection as follows.

For bending in the joist, the maximum span length from Table 11-2,

$$l_b = [96F_bS/w]^{1/2}$$
$$= [96(1,875)(7.15)/250]^{1/2}$$
$$= 71.7 \text{ in.}$$

For shear stress in the joist, the maximum span length from Table 11-2,

$$l_v = 16F_vbd/w + 2d$$
$$= 16(225)(3.5)(3.5)/250 + 2(3.5)$$
$$= 183.4 \text{ in.}$$

For deflection not to exceed $l/360$, the maximum span length from Table 11-2,

$$l_\Delta = [4,608EI/1,800w]^{1/3}$$
$$= [4,608(1,700,000)(12.51)/1,800(250)]^{1/3}$$
$$= 60.1 \text{ in.}$$

For deflection not to exceed $\frac{1}{16}$ in., the maximum span length from Table 11-2,

$$l_\Delta = [4,608EI/80w]^{1/4}$$
$$= [4,608(1,700,000)(12.51)/80(250)]^{1/4}$$
$$= 47.0 \text{ in.}$$

Summary for the 4 × 4 Joist

For bending, the maximum span length of the joist = 71.7 in.

For shear, the maximum span length of the joist = 183.4 in.

For deflection, the maximum span length of the joist = 47.0 in.

For this example, deflection controls the maximum span length of the joist. Therefore, the joist must be supported at a distance of not greater than 47.0 in., or about 4 ft.

Example 11-3

The total load on a decking is 175 lb per sq ft. The decking will be supported by 7-ft-long single-span joists spaced at 18 in. on centers. Consider 6-in.-wide No. 1 grade Southern Pine and determine the minimum section modulus required. The lumber will be used at a jobsite where no adjustment factors are necessary; dry condition, normal temperature load-duration, etc. For this condition, the allowable bending stress for 6-in.-wide No. 1 Southern Pine can be obtained from Table 4-2 as F_b =1,650 lb per sq in.

Rewriting Eq. (5-30) in Table 11-2 to solve for the section modulus in terms of the uniformly distributed load w, the span length l_b, and the allowable stress F_b, the equation for the required section modulus can be obtained as follows. From Eq. (5-30) in Table 11-2:

$$l_b = [96F_b S/w]^{1/2}$$

Rewriting this equation for the section modulus,

$$S = w(l_b)^2/96F_b$$

Substituting the values for load, span, and allowable stress,

$$w = (175 \text{ lb per sq ft})(18/12 \text{ ft}) = 262.5 \text{ lb per lin ft}$$
$$l_b = 7 \text{ ft, or 84 in.}$$
$$F_b = 1,650 \text{ lb per sq in.}$$

The required section modulus is

$$\begin{aligned} S &= w(l_b)^2/96F_b \\ &= 262.5(84)^2/96(1,650) \\ &= 11.7 \text{ in.}^3 \end{aligned}$$

Referring to Table 4-1, it will be noted that this section modulus may be provided by one of the following sizes of lumber:

$$3 \times 6 \text{ S4S for which S} = 12.60 \text{ in.}^3$$
$$4 \times 6 \text{ S4S for which S} = 17.65 \text{ in.}^3$$

Both the 3×6 and 4×6 are adequate for bending. The d/b is ≤ 2 for both members; therefore, no lateral support is required to satisfy beam stability (refer to Table 4-6). This member should also be checked for shear and deflection, as illustrated in Chapter 5.

Use of Tables to Determine the Maximum Spans for Lumber Framing Used to Support Plywood

Tables 11-7 and 11-8 give the maximum spans for lumber framing members, such as studs and joists that are used to support plywood subjected to pressure from concrete. The spans listed in Table 11-7 are based on using grade No. 2 Douglas Fir or grade No. 2 Southern Pine. The spans listed in Table 11-8 are based on using grade No. 2 Hem-Fir.

Equivalent Uniform Load, lb per ft	Douglas Fir #2 or Southern Pine #2 Continuous over 2 or 3 Supports (1 or 2 Spans) Nominal Size Lumber							Douglas Fir #2 or Southern Pine #2 Continuous over 4 or more Supports (3 or more Spans) Nominal Size of Lumber						
	2 × 4	2 × 6	2 × 8	2 × 10	4 × 4	4 × 6	4 × 8	2 × 4	2 × 6	2 × 8	2 × 10	4 × 4	4 × 6	4 × 8
200	48	73	92	113	64	97	120	56	81	103	126	78	114	140
400	35	52	65	80	50	79	101	39	58	73	89	60	88	116
600	29	42	53	65	44	64	85	32	47	60	73	49	72	95
800	25	36	46	56	38	56	72	26	41	52	63	43	62	82
1000	22	33	41	50	34	50	66	22	35	46	56	38	56	73
1200	19	30	38	46	31	45	60	20	31	41	51	35	51	67
1400	18	28	35	43	29	42	55	18	28	37	47	32	47	62
1600	16	25	33	40	27	39	52	17	26	34	44	29	44	58
1800	15	24	31	38	25	37	49	16	24	32	41	27	42	55
2000	14	23	29	36	24	35	46	15	23	30	39	25	39	52
2200	14	22	28	34	23	34	44	14	22	29	37	23	37	48
2400	13	21	27	33	21	32	42	13	21	28	35	22	34	45
2600	13	20	26	31	20	31	41	13	20	27	34	21	33	43
2800	12	19	25	30	19	30	39	12	20	26	33	20	31	41

TABLE 11-7 Maximum Spans, in Inches, for Lumber Framing Douglas-Fir No. 2 or Southern Pine No. 2

Equivalent Uniform Load, lb per ft	Douglas Fir #2 or Southern Pine #2 Continuous over 2 or 3 Supports (1 or 2 Spans) Nominal Size Lumber							Douglas Fir #2 or Southern Pine #2 Continuous over 4 or more Supports (3 or more Spans) Nominal Size of Lumber						
	2 × 4	2 × 6	2 × 8	2 × 10	4 × 4	4 × 6	4 × 8	2 × 4	2 × 6	2 × 8	2 × 10	4 × 4	4 × 6	4 × 8
3000	12	19	24	29	18	29	38	12	19	25	32	19	30	39
3200	12	18	23	28	18	28	37	12	19	24	31	18	29	38
3400	11	18	22	27	17	27	35	12	18	23	30	18	28	36
3600	11	17	22	27	17	26	34	11	18	23	30	17	27	35
3800	11	17	21	26	16	25	33	11	17	22	29	16	26	34
4000	11	16	21	25	16	24	32	11	17	22	28	16	25	33
4200	11	16	20	25	15	24	31	11	17	22	28	16	24	32
4400	10	16	20	24	15	23	31	10	16	22	27	15	24	31
4600	10	15	19	24	14	23	30	10	16	21	26	15	23	31
4800	10	15	19	23	14	22	29	10	16	21	26	14	23	30
5000	10	15	18	23	14	22	29	10	16	21	25	14	22	29

Notes:
1. Courtesy APA—The Engineered Wood Association, "Concrete Forming," 2004.
2. Spans are based on the 2001 NDS allowable stress values: $C_D = 1.25$, $C_r = 1.0$, $C_M = 1.0$.
3. Spans are based on dry, single-member allowable stresses multiplied by a 1.25 duration-of-load factor for 7-day loads.
4. Deflection is limited to $l/360$th of the span with $l/4$ in. maximum.
5. Spans are measured center-to-center on the supports.

TABLE 11-7 Maximum Spans, in Inches, for Lumber Framing Douglas-Fir No. 2 or Southern Pine No. 2 (*Continued*)

Equivalent Uniform Load, lb per ft	Hem-Fir #2 Continuous over 2 or 3 Supports (1 or 2 Spans) Nominal Size Lumber							Hem-Fir #2 Continuous over 4 or more Supports (3 or more Spans) Nominal Size Lumber						
	2 × 4	2 × 6	2 × 8	2 × 10	4 × 4	4 × 6	4 × 8	2 × 4	2 × 6	2 × 8	2 × 10	4 × 4	4 × 6	4 × 8
200	45	70	90	110	59	92	114	54	79	100	122	73	108	133
400	34	50	63	77	47	74	96	38	56	71	87	58	86	112
600	28	41	52	63	41	62	82	29	45	58	71	48	70	92
800	23	35	45	55	37	54	71	23	37	48	61	41	60	80
1000	20	31	40	49	33	48	64	20	32	42	53	37	54	71
1200	18	28	36	45	30	44	58	18	28	37	47	33	49	65
1400	16	25	33	41	28	41	54	16	26	34	43	29	45	60
1600	15	23	31	39	25	38	50	15	24	31	40	26	41	54
1800	14	22	29	37	23	36	48	14	22	30	38	24	38	50
2000	13	21	28	35	22	34	45	14	23	28	36	22	35	46
2200	13	20	26	33	20	32	42	13	21	27	34	21	33	43
2400	12	19	25	32	19	30	40	12	20	26	33	20	31	41
2600	12	19	25	30	18	29	38	12	20	25	32	19	30	39
2800	12	18	24	29	18	28	36	12	19	24	31	18	28	37

TABLE 11-8 Maximum Spans, in Inches, for Lumber Framing Hem-Fir No. 2

Equivalent Uniform Load, lb per ft	Hem-Fir #2 Continuous over 2 or 3 Supports (1 or 2 Spans) Nominal Size Lumber							Hem-Fir #2 Continuous over 4 or more Supports (3 or more Spans) Nominal Size Lumber						
	2×4	2×6	2×8	2×10	4×4	4×6	4×8	2×4	2×6	2×8	2×10	4×4	4×6	4×8
3000	11	18	23	28	17	26	35	11	18	24	30	17	27	36
3200	11	17	22	27	16	25	34	11	18	23	29	17	26	34
3400	11	17	22	27	16	25	32	11	17	22	29	16	25	33
3600	11	17	21	26	15	24	31	11	17	22	28	16	24	32
3800	10	16	21	25	15	23	31	10	17	22	28	15	24	31
4000	10	16	20	24	14	23	30	10	16	21	27	15	23	30
4200	10	15	10	24	14	22	29	10	16	21	27	14	22	30
4400	10	15	19	24	14	22	28	10	16	21	26	14	22	29
4600	10	15	19	23	13	21	28	10	15	20	26	14	21	28
4800	10	14	19	22	13	21	27	10	15	20	25	13	21	28
5000	10	14	18	22	13	20	27	10	15	20	24	13	21	27

Notes:

1. Courtesy American Plywood Association, "Concrete Forming," 2004.
2. Spans are based on the 2001 NDS allowable stress values, $C_D = 1.25$, $C_r = 1.0$, $C_M = 1.0$.
3. Spans are based on dry, single-member allowable stresses multiplied by a 1.25 duration-of-load factor for 7-day loads.
4. Deflection is limited to $l/360$th of the span with $l/4$ in. maximum.
5. Spans are measured center-to-center on the supports.

TABLE 11-8 Maximum Spans, in Inches, for Lumber Framing Hem-Fir No. 2 (*Continued*)

The allowable stresses are based on a load-duration of 7 days maximum, and not more than 19% moisture content. The deflections are limited to $l/360$ with maxima not to exceed ¼ in. The values are for lumber framing directly supporting the plywood. Although Tables 11-7 and 11-8 are for single members, these tables can be adapted for use with multiple members.

Stringers

Figure 11-2 illustrated a system of forms using joists supported at the ends only. Although this system may be satisfactory for joists with relatively short spans, the sizes of joists required for longer spans are so large that it is economical to install one or more rows of stringers at intermediate points under the joists, as illustrated in Figure 11-3. Also, stringers must be used to support joists for flat slabs. The spacings, sizes, and spans of the stringers should be selected to obtain the maximum economy for the form system, considering the cost of materials and labor and the salvage value of the lumber.

As illustrated in Figure 11-3, the strip of decking whose load is supported by a joist is one whose width extends from midway between supports on opposite sides of a stringer. This width is designated by the dimension a. Because the loads acting on a stringer are transmitted from the joists, they are concentrated. For this reason, the calculation for the bending moment in a stringer is somewhat complicated, and the magnitude of the moment will vary with the relative position of the joists with respect to the shores that support the stringers. If the reduced bending moment resulting from the continuous spans of a stringer over several shores is neglected, this should provide a safety factor large enough to offset the effect of considering the loads from the joists to be uniformly distributed along the stringer and will simplify the design of a stringer. This allows the designer to use the equations previously presented for uniformly distributed loads for calculating the bending, shear, and deflection of stringers.

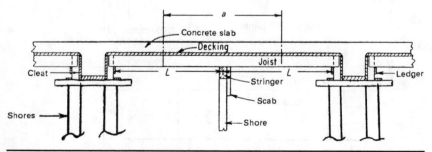

FIGURE 11-3 Forms for concrete beams and slab with intermediate stringer.

A stringer must be wide enough to transmit the loads in bearing from the joists without exceeding the allowable unit stress. Also, it must be large enough in transverse section to resist the bending moment and horizontal shearing stress without exceeding the safe limits. The deflection will usually be small and often will not govern the design.

Ledgers

As illustrated in Figures 11-2 and 11-3, the ends of joists may be supported by ledgers, which are nailed to cleats, usually 2 × 4 members, nailed to the side forms for beams. The ledgers must be able to support the joists.

If the ledger supports this entire load between the joist and the ledger, it must have an adequate bearing capacity. It is common practice to nail the ends of the joists to the cleats attached to the forms for the sides of the beams. If this method is used to construct the forms, the spacing of the cleats on the beam form should be the same as for the joists.

Forms for Flat-Slab Concrete Floors

Figure 11-4 illustrates a method of using prefabricated panels for the decking for flat-slab concrete floors, where the concrete columns are spaced at 12 ft in one direction and at 16 ft in the other direction. The same method may be used, with modified form panels, for other spacings of columns.

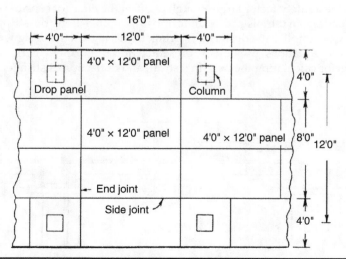

FIGURE 11-4 Form panels for flat-slab concrete floor.

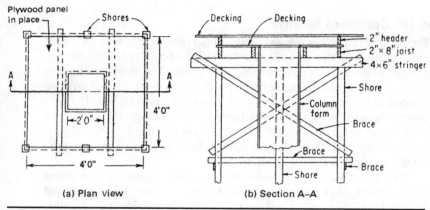

(a) Plan view   (b) Section A–A

FIGURE 11-5   Forms for a drop panel.

Figure 11-5 illustrates a method of constructing forms for a drop panel to provide a greater thickness for the concrete slab adjacent to a column. The decking may be made of plywood, assembled in two equal parts, with a joint along section line *A-A*.

The forms for drop panels are erected and rigidly braced first, then the forms for the main slabs are erected. Using the method illustrated, the forms for the columns are stripped first, followed by the forms for the drop panels, then the forms for the balance of the slab. The forms may be erected to permit some panels to remain in position, supported by shores, as long as such support is required.

Figure 11-6 illustrates a method of constructing forms for the main slab. Two-way diagonal braces, extending from the bottoms to the tops of adjacent shores, should be installed to provide adequate strength and rigidity.

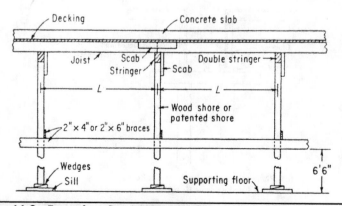

FIGURE 11-6   Forms for a flat-slab concrete floor.

Forms for Concrete Beams

Figure 11-7 illustrates several methods of constructing forms for concrete beams. As shown in Figure 11-7(a), the beam bottom, or soffit, may be made from one or more dressed planks 2 in. thick and having the required width. If more than one plank is used, they should be cleated together on the underside at intervals of about 3 ft. The side forms are made of plywood panels or boards, held together with 2 × 4 cleats. The ledgers attached to the vertical blocking carry the joist load to the shore. The kickers at the bottom of the side form resist the lateral pressure from the concrete.

The beam bottom may also be plywood, backed by dimension lumber laid in the flat position, as illustrated in Figure 11-7(b). Plywood panels are placed on the top of the runners to provide a continuous

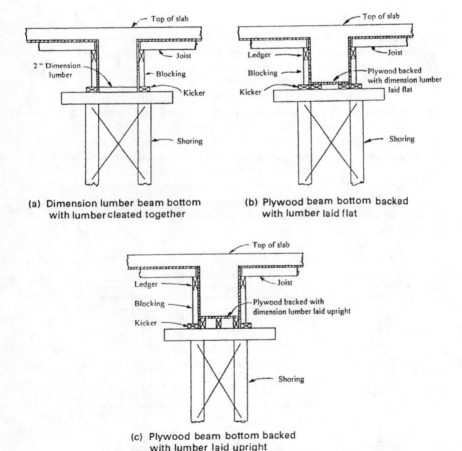

(a) Dimension lumber beam bottom with lumber cleated together

(b) Plywood beam bottom backed with lumber laid flat

(c) Plywood beam bottom backed with lumber laid upright

FIGURE 11-7 Forms for a concrete beam and slab.

smooth surface for the beam bottom. As illustrated in Figure 11-7(c), the beam bottom may also be plywood, backed with dimension lumber placed in the vertical direction to provide more resistance to bending and deflection.

The side forms are made from plywood panels, held together by form ties. Side forms may also be made with boards, held together with 2×4 cleats. The ledgers attached to the vertical 2×4 cleats assist in supporting the ends of the joists.

Spacing of Shores under Beam Bottoms

The spacing of shores under beam bottoms will be limited by the strength of the beam bottom in bending, shear, and deflection as well as by the capacities of the shores. Although the spacing of the shores along the beam forms may be governed by the strength or deflection of the beam bottoms, the shores must sustain the total vertical load from the beam forms plus the dead and live loads from the slab to the midpoint of the adjacent rows of shores on either side of the beam forms.

For the concrete beam form in Figure 11-7(a), the vertical load on a 12-in. strip of beam bottom can be used to determine the allowable span length of beam bottom using the equation presented in Chapter 5. For the concrete beam form illustrated in Figure 11-7(b) and 11-7(c), the plywood transfers the load in a direction perpendicular to the direction of the beam, between the horizontal wood runners. The runners must transfer the load between the shores. A strip of the beam bottom, from the center points between spacings of the runners, may be analyzed for bending, shear, and deflection. This procedure discards any contribution of the plywood for transferring load between shores, which is a conservative approach.

The pressure of concrete against beam forms acts in two directions, one horizontally outward against the beam sides and one vertically downward against the bottom of the beam forms. The horizontal pressure begins at zero at the top of the beam sides and increases linearly to a maximum at the bottom of the side. The vertical pressure is uniformly distributed across the bottom of the beam form.

The maximum horizontal pressure on the beam sides, which acts at the bottom of the beam, can be calculated by multiplying the unit weight of the freshly placed concrete by the height of the beam, the distance from the top of the concrete to the bottom of the beam. Thus, the pressure on the beam sides is similar to the pressure on a column form because the concrete is placed rapidly in beam forms.

Example 11-4

Determine the safe spacing of shores based on bending, shear, and deflection of the beam bottom, as illustrated in Figure 11-7(a). The beam is 16 in. wide and the total depth from the top of the concrete slab to the bottom of the beam is 24 in. Consider a 50 lb per sq ft live load on the formwork. The unit weight of the concrete is 150 lb per cu ft and the permissible deflection is $l/360$, not to exceed $\frac{1}{16}$ in.

The 16-in. wide beam bottom will be fabricated from 2 × 12 S4S lumber. Because the 16-in beam bottom is wider than an 11¼-in. standard 2 × 12, two pieces of lumber will be used. One piece of lumber will be a full size 2 × 12 and the second piece will be 4.75-in. cut from another 2 × 12 to make up the full 16-in. beam bottom. The lumber will be cleated together to form the beam bottom.

The lumber will be No. 1 grade Southern Pine. Assume a dry condition of lumber, moisture content less than 19%, and a short load-duration of less than 7 days. Because the lumber will be laid flat, the flat use adjustment factor, C_{fu}, may be used. From Tables 4-2, 4-4, and 4-7, the allowable stresses will be:

Allowable bending stress, $F_b = C_D(C_{fu})$(reference design bending value)
$$= 1.25(1.2)(1,250 \text{ lb per sq in.})$$
$$= 1,875 \text{ lb per sq in.}$$

Allowable shear stress, $F_v = C_D$ (reference design shear value)
$$= 1.25(175 \text{ lb per sq in.})$$
$$= 218 \text{ lb per sq in.}$$

Modulus of elasticity, $E = 1,700,000$ lb per sq in.

The uniformly distributed load acting vertically against the 16-in. wide beam bottom between shores may be calculated as follows:

Concrete, 150 lb per cu ft (24/12 ft)(16/12 ft) = 400.0 lb per lin ft
Assumed dead load, 4 lb per sq ft(16/12 ft)   =   5.3 lb per lin ft
Live load, 50 lb per sq ft (16/12 ft)   =   66.7 lb per lin ft
Total load, w = 472.0 lb per lin ft

The physical properties of the section modulus and the moment of inertia for a piece of 16-in.-wide and 2-in.-thick dimension lumber laid flat between shores can be calculated as follows:

From Eq. (5-9), the section modulus can be calculated as follows:

Section modulus, $S = bd^2/6$
$$= 16 \text{ in.}(1.5 \text{ in.})^2/6$$
$$= 6.0 \text{ in.}^3$$

From Eq. (5-8), the moment of inertia can be calculated.

Moment of inertia, $I = bh^3/12$
$$= (16 \text{ in.})(1.5)^3/12$$
$$= 4.5 \text{ in.}^4$$

The allowable span length of the beam bottom based on bending, shear, and deflection may be calculated as follows:

For bending, the maximum span length
from Eq. (5-34), $l_b = [120F_bS/w]^{1/2}$
$$= [120(1,875)(6.0)/472]^{1/2}$$
$$= 53.4 \text{ in.}$$

For shear, the maximum span length
from Eq. (5-36), $l_v = 192F_vbd/15w + 2d$
$$= 192(218)(16)(1.5)/15(472 + 2(1.5)$$
$$= 144.8 \text{ in.}$$

For deflection not to exceed $l/360$, the maximum span length
from Eq. (5-37a), $l_\Delta = [1,743EI/360w]^{1/3}$
$$= [1,743(1,700,000)(4.5)/360(472)]^{1/3}$$
$$= 42.8 \text{ in.}$$

Deflection not to exceed $\frac{1}{16}$ in., the maximum span length
from Eq. (5-37b), $l_\Delta = [1,743EI/16w]^{1/4}$
$$= [1,743(1,700,000)(4.5)/16(472)]^{1/4}$$
$$= 36.4 \text{ in.}$$

Following is a summary of the allowable span lengths of the 2-in.-thick beam bottom:

For bending, the maximum span length = 53.4 in.
For shear, the maximum span length = 144.8 in.
For deflection, the maximum span length = 36.4 in.

For this example, deflection governs the maximum span length of the beam bottom. Therefore, the shores must be spaced at a distance not to exceed 36.4 in. to prevent excessive deflection of the beam bottoms. For constructability, the spacing would likely be at 36 in. on centers. The total load on the shore from the concrete beam will be (472 lb per lin ft)(36.0/12 ft) or 1,416 lb, plus the dead and live loads from the concrete slab on either side of the centerline of the concrete beam to the midpoint of the row of shores perpendicular to the direction of the concrete beam.

Example 11-5

Determine the safe spacing of shores based on bending, shear, and deflection of the beam bottom as illustrated in Figure 11-7(b). The beam is 16 in. wide and the total depth from the top of the concrete slab to the bottom of the beam is 24 in. Consider a 50 lb per sq ft live load on the formwork. The unit weight of the concrete is 150 lb per cu ft and the permissible deflection is $l/360$, not to exceed $\frac{1}{16}$ in.

Class I Plyform, which is ¾ in. thick, will be used for the beam bottom, backed by two 2 × 4 runners of dimension lumber laid in the flat position. The runners will be No. 2 grade Spruce-Pine-Fir laid flat in a dry condition with short load-duration of less than 7 days. The applicable adjustment factors will be size C_F, load-duration C_D, and flat-use C_{fu}. From Tables 4-3, 4-3a, 4-4, and 4-7, the allowable stresses can be calculated as follows:

Allowable bending stress, $F_b = C_F(C_D)(C_{fu})$ (reference bending stress)

$$= 1.5(1.25)(1.1)(875 \text{ lb per sq in.})$$

$$= 1,804 \text{ lb per sq in.}$$

Allowable shear stress, $F_v = C_D$ (reference shear stress)

$$= 1.25(135 \text{ lb per sq in.})$$

$$= 168 \text{ lb per sq in.}$$

Modulus of elasticity, $E = 1,400,000$ lb per sq in.

The uniformly distributed pressure acting vertically against a 1-sq-ft area of the Plyform beam bottom may be calculated as follows:

Concrete, 150 lb per cu ft (24/12 ft) = 300 lb per sq ft
Assumed dead load of form material =   4 lb per sq ft
Live load, 50 lb per sq ft   = _50 lb per sq ft_
Total load, w = 354 lb per sq ft

The Plyform must transfer this 354 lb per sq ft pressure between the flat runners, perpendicular to the direction of the concrete beam. The unsupported length of Plyform will be the distance between the runners, perpendicular to the direction of the beam. This distance is calculated as follows:

Total width across the concrete beam bottom = 16 in.
Subtracting the width of 2 runners, 2(3.5 in.) = _7 in._
Unsupported length of plywood =   9 in.

An examination of Tables 11-2 and 11-3 for Class I Plyform indicates that the ¾-in. Plyform can safely support the 354 lb per sq ft load over an unsupported length that is greater than 9 in. Therefore, the ¾-in. thickness is satisfactory. The load transferred from the runners must now be transferred between the shores.

Discarding the vertical support provided by the side members of the beam forms, the 2 × 4 runners that are laid flat must transfer the total load of the beam bottom between shores. The total uniformly distributed linear load of the beam bottom between supports may be calculated as follows:

> Dead load of concrete,
> $$150 \text{ lb per cu ft } (24/12)(16/12) = 400.0 \text{ lb per ft}$$
> Assumed dead load of formwork,
> $$4 \text{ lb per sq ft } (16/12) = 5.3 \text{ lb per ft}$$
> Live load on beam bottom,
> $$50 \text{ lb per sq ft } (16/12) = \underline{66.7 \text{ lb per ft}}$$
> $$\text{Total load, } w = 472.0 \text{ lb per ft}$$

The two 2×4 runners laid flat must safely transfer this 472 lb per ft load between shores. Because there are two runners, each runner must sustain one-half of this load, or 236 lb per ft. The physical properties of section modulus and moment of inertia of the runners can be obtained from Chapter 4 as follows:

From Table 4-1:

Section modulus, $S = 1.313 \text{ in.}^3$
Moment of inertia, $I = 0.984 \text{ in.}^4$

The allowable span length of the 2×4 runners between shores based on bending, shear, and deflection may be calculated as follows:

> For bending, the maximum span length
> from Eq. (5-34), $l_b = [120 F_b S / w]^{1/2}$
> $$= [120(1,804)(1.313)/236]^{1/2}$$
> $$= 34.7 \text{ in.}$$

> For shear, the maximum span length
> from Eq. (5-36), $l_v = 192 F_v bd / 15w + 2d$
> $$= 192(168)(3.5)(1.5)/15(236) + 2(1.5)$$
> $$= 50.8 \text{ in.}$$

> For deflection not to exceed $l/360$, the maximum span length
> from Eq. (5-37a), $l_\Delta = [1,743 EI / 360w]^{1/3}$
> $$= [1,743(1,400,000)(0.984)/360(236)]^{1/3}$$
> $$= 30.4 \text{ in.}$$

> For deflection not to exceed $\frac{1}{16}$ in., the maximum span length
> from Eq. (5-37b), $l_\Delta = [1,743 EI / 16w]^{1/4}$
> $$= [1,743(1,400,000)(0.984)/16(236)]^{1/4}$$
> $$= 28.2 \text{ in.}$$

Below is a summary of the allowable span lengths of the 2-in.-thick beam bottom:

For bending, the maximum span length = 34.7 in.

For shear, the maximum span length = 50.8 in.

For deflection, the maximum span length = 28.2 in.

For this example, deflection governs the maximum span length of the beam bottom. Therefore, the shores must be placed at a distance not to exceed 28.2 in. to prevent excessive deflection of the beam bottoms. For constructability, the shores will be placed at 24 in. on centers.

Example 11-6

Determine the safe spacing of shores based on bending, shear, and deflection of the beam bottom as illustrated in Figure 11-7(c). The beam is 16 in. wide and the total depth from the top of the concrete slab to the bottom of the beam is 24 in. Consider a 50 lb per sq ft live load on the formwork. The unit weight of the concrete is 150 lb per cu ft and the permissible deflection is $l/360$, not to exceed $\frac{1}{16}$ in.

Class I Plyform, which is ¾ in. thick, will be used for the beam bottom, backed by three 2×4 runners of dimension lumber placed in the upright position. The runners will be No. 2 grade Hem-Fir in a dry condition with a short load-duration of less than 7 days. The applicable adjustment factors will be size C_F and load-duration C_D. From Tables 4-3, 4-3a, and 4-4 the allowable stresses can be calculated as follows:

$$\text{Allowable bending stress, } F_b = C_F(C_D) \text{ (reference bending stress)}$$
$$= 1.5(1.25)(850 \text{ lb per sq in.})$$
$$= 1,593 \text{ lb per sq in.}$$
$$\text{Allowable shear stress, } F_v = C_D \text{ (reference shear stress)}$$
$$= 1.25(150 \text{ lb per sq in.})$$
$$= 187 \text{ lb per sq in.}$$
$$\text{Modulus of elasticity, } E = 1,300,000 \text{ lb per sq in.}$$

The uniformly distributed pressure acting vertically against a 1 sq ft area of the Plyform beam bottom may be calculated as follows:

Concrete, 150 lb per cu ft (24/12 ft) = 300 lb per sq ft
Assumed dead load of formwork   =   4 lb per sq ft
Live load, 50 lb per sq ft   =   50 lb per sq ft
Total load, w = 354 lb per sq ft

The Plyform must transfer this 354 lb per sq ft pressure between the runners, perpendicular to the direction of the concrete beam.

The unsupported length of Plyform will be the distance between the plywood and the runners. This distance is calculated as follows:

Total width across the concrete beam bottom = 16.0 in.
Subtracting the width of 3 runners, 3(1.5 in.) = 4.5 in.
Total clearance of Plyform between runners = 11.5 in.
Thus, the net unsupported length of Plyform is 11.5/2 = 5.75 in.

An examination of Tables 11-2 and 11-3 for Class I Plyform indicates the ¾-in. Plyform can safely support the 354 lb per sq ft load over an unsupported length that is greater than 5.75 in. Therefore, the ¾-in. 4 × 4¹⁄₁₆ thickness is satisfactory. The load transferred from the runners must now be transferred between the shores.

Discarding the vertical support provided by the side members of the beam sides, the three 2 × 4 runners that are upright must transfer the dead load and the live load of the beam bottom between shores. However, the center runner is the most heavily loaded. The center runner must sustain the pressure from its center to the midpoint between the adjacent runners on each side as follows:

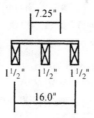

Total width of Plyform beam bottom = 16.0 in.
Width of three runners, 3(1.5 in.)   = 4.5 in.
Net distance between runners   = 11.5 in.

The width of the effective pressure carried by the center runner is one-half of this distance plus the width of the center runner, or 11.5/2 + 1.5 = 7.25 in. The center runner must sustain the pressure over this portion of the beam bottom. The uniform load on the center runner in the direction of the beam may be calculated as follows:

Concrete dead load,
150 lb per cu ft (24/12)(7.25/12) = 181.3 lb per ft
Assumed dead load of forms,
4 lb per sq ft (7.25/12) =   2.4 lb per ft
Live load on beam bottom,
50 lb per sq ft (7.25/12) =   30.2 lb per ft
Total load = 213.9 lb per ft
For calculations, use w = 214 lb per lin ft

The physical properties of section modulus and moment of inertia of the 2×4 runners can be obtained from Table 4-1 in Chapter 4 as follows:

Section modulus of two members, $S = 3.06$ in.^{3}

Moment of inertia of two member, $I = 5.36$ in.^{4}

The allowable span length of the 2×4 runners between shores based on bending, shear, and deflection may be calculated as follows:

For bending, the maximum span length
from Eq. (5-34), $l_b = [120F_bS/w]^{1/2}$
$$= [120(1{,}593)(3.06)/214]^{1/2}$$
$$= 52.2 \text{ in.}$$

For shear stress, the maximum span length
from Eq. (5-37), $l_v = 192F_vbd/15w + 2d$
$$= 192(187)(1.5)(3.5)/15(214) + 2(3.5)$$
$$= 65.7 \text{ in.}$$

For deflection not to exceed $l/360$, the maximum span length
from Eq. (5-37a), $l_\Delta = [1{,}743EI/360w]^{1/3}$
$$= [1{,}743(1{,}300{,}000)(5.36)/360(214)]^{1/3}$$
$$= 54.0 \text{ in.}$$

For deflection not to exceed $\frac{1}{16}$ in., the maximum span length
from Eq. (5-37b), $l_\Delta = [1{,}743EI/16w]^{1/4}$
$$= [1{,}743(1{,}300{,}000)(5.36)/16(214)]^{1/4}$$
$$= 43.3 \text{ in.}$$

Below is a summary of the allowable span lengths of the 2-in.-thick beam bottom:

For bending, the maximum span length = 52.2 in.

For shear, the maximum span length = 65.7 in.

For deflection, the maximum span length = 43.3 in.

For this example, deflection governs the maximum span length of the beam bottom. Therefore, the shores must be placed at a distance not to exceed 43.3 in. to prevent excessive deflection of the beam bottoms. For constructability the spacing would likely be at 40 in. on centers. The 2×4 runners have a $d/b \leq 2$; therefore, beam stability is satisfied (refer to Table 4-6).

Forms for Exterior Beams

Figure 11-8 illustrates a method of constructing forms for an exterior beam. A brace should be installed from the exterior end of each shore head to the top of a stud. The 2×4 wale along the tops of the studs ensures better alignment of the forms between braces.

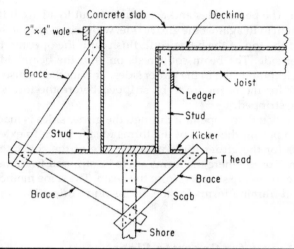

FIGURE 11-8 Forms for a spandrel beam.

Form Details for Beams Framing into Girders

Figure 11-9 illustrates two methods of framing the ends of beam sides and soffits into the forms for girders. If Plan A is used, the opening through the side of a girder form is made equal to the dimensions of

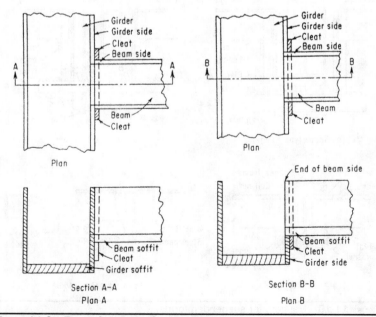

FIGURE 11-9 Forms for beams framing into girders.

the beam. The beam sides and the soffit are cut to a length that will permit them to fit against the girder side, with the insides of the beam sides and the top of the beam soffit flush with the opening through the girder side. The beam soffit rests on, and the beam sides bear against, cleats attached to the girder sides. This method is used where the forms for the beam are to be stripped before the forms for the girder are stripped.

Using Plan B, the opening through the girder sides is made large enough to permit the ends of the forms for the beam to extend into the forms for the girder flush with the face of the concrete girder. This plan is used where the forms are to be stripped from the girder before the forms are stripped from the beam. The same methods may be used in framing forms for the beams and girders into forms for columns.

Suspended Forms for Concrete Slabs

Figure 11-10 illustrates four types of steel hangers used to suspend forms from steel beams.

The *snap-type hanger* illustrated in Figure 11-10(a) is a single rod bent to pass over the upper flange of a steel beam and long enough to permit the lower ends to extend below the beam soffit far enough to

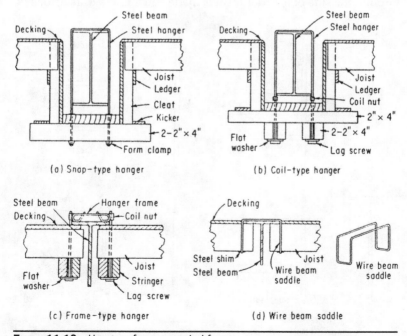

FIGURE 11-10   Hangers for suspended forms.

support the stringers. The stringers are supported by slotted wedge-type form clamps.

The *coil-type hanger* illustrated in Figure 11-10(b) consists of two bent rods with two coil nuts welded to the ends. The lag screws, which engage these nuts, support stringers, as illustrated. A flat steel washer should be used with each screw.

The *frame-type hanger*, illustrated in Figure 11-10(c), is a type used to support forms for concrete slabs where it is not necessary to encase the steel beams in concrete. Lag screws, inserted from below, engage the coil nuts, which are welded to the hanger frames. A flat steel washer should be used with each screw.

The *wire beam saddle hanger*, illustrated in Figure 11-10(d), is a type used to support joists where the total loads from concrete are relatively small. Because there is no means of adjusting the lengths of these hangers, they should be ordered for the exact sizes required. A steel shim should be placed under the bottom of the joist, as illustrated, to increase the area in bearing between the joist and the hanger.

Designing Forms for Concrete Slabs

There are several approaches to the design for slab forms. One approach is to select the material to be used for formwork, then use the equations in Tables 5-3 and 5-4 in Chapter 5 to determine the spacing of the materials; including decking, joists, stringers, and shores. Another approach is to use tables that have been developed by a company that furnishes formwork systems and accessories. Still another approach is to use patented forming systems that have been developed by companies that rent or sell formwork for concrete slabs. The latter approach is discussed in Chapter 12. In the latter two approaches, the materials and recommendations of the company that supplies the formwork and its accessories must be followed.

Forms for concrete slabs should provide sufficient strength and rigidity at the lowest practical cost, considering materials, labor, and any construction equipment used in making, erecting, and removing them. The designer should know the magnitudes of the forces that act on the component parts and the strengths of these parts in resisting the forces. Chapter 3 presented pressures and loads that act on formwork.

The strength of lumber varies with the species and grade used for formwork. The allowable unit stresses in bending, horizontal shear, compression perpendicular to the grain, and compression parallel to the grain are presented in Chapter 4. Each design should be based on information that applies to the allowable unit stresses for the species and grade of lumber used for the formwork.

As presented in Chapter 5, many calculations are required to determine the adequate strength of formwork members. It requires considerable time to analyze each beam element of formwork to determine the applied and allowable stresses for a given load condition, and to compare the calculated deflection with the permissible deflection. For each beam element, a check must be made for bending, shear, and deflection.

For simplicity in the design of formwork, it is common practice to rewrite the equations for stresses and deflection in order to calculate the permissible span length of a member in terms of member size, allowable stress, and loads on the member. Table 5-3 summarizes the equations for calculating the allowable span lengths of wood members and Table 5-4 provides a summary of the equations for calculating the allowable span lengths of plywood members. These two tables enable the designer to design formwork in an organized manner. Usually the steps in designing concrete slab forms follow the sequence listed below:

1. Determine the total load on the floor decking, including the effect of impact, if any. See Chapter 3 for static loads on formwork and Chapter 7 for impact loads.

2. Select the kind, grade, and thickness of the material to be used for the decking, see Tables 4-9 and 4-10 for plywood and Tables 4-11 and 4-12 for Plyform, the plywood manufactured especially for formwork. Then determine the maximum safe spacing of the joists, considering the allowable unit fiber stress in bending, the horizontal shear stress, and the permissible deflection in the decking. Table 5-4 lists the equations for calculating the allowable span lengths of plywood based on bending and shear stresses, and the deflection criteria.

3. Select the grade and size of joists to be used. More than one size may be considered. See Table 4-1 for the physical properties of lumber and Tables 4-2 and 4-3 for reference design values of wood members, noting that adjustment factors are given in Tables 4-4 through 4-8. Determine the maximum safe spacing of stringers, based on the allowable unit stresses in bending and shear and the permissible deflection whichever limits the span length for the joists. Table 5-3 provides the equations for calculating the allowable span lengths of lumber based on stresses and deflection.

4. Select the grade and size of lumber to be used for stringers, see Tables 4-2 and 4-5. Determine the maximum safe span lengths of stringers, based on allowable unit stresses in bending and shear and the permissible deflection, whichever limits the span length for the stringers. Table 5-3 lists the equations to calculate the allowable span lengths. A check must also be made for the bearing stress between the stringers and joists.

5. Select the size and spacing of shores that will safely withstand the forces transmitted to them by the stringers. An alternate method is to select the shoring system, and then determine the maximum safe

spacing along the stringers or joists, checking to be certain that this spacing does not exceed that determined in steps 3 and 4.

Design of Formwork for Flat-Slab Concrete Floor with Joists and Stringers

Design formwork for a flat-slab concrete floor as illustrated in Figure 11-11. The design includes the decking, joists, and stringers. Requirements for shoring of formwork are discussed in the following section. The following design criteria will apply:

1. Thickness of concrete slab, 6 in.
2. Concrete unit weight, 150 lb per cu ft
3. Deflection limited to $l/360$, but not greater than $\frac{1}{16}$ in.
4. All form lumber will be S4S No. 2 grade Southern Pine

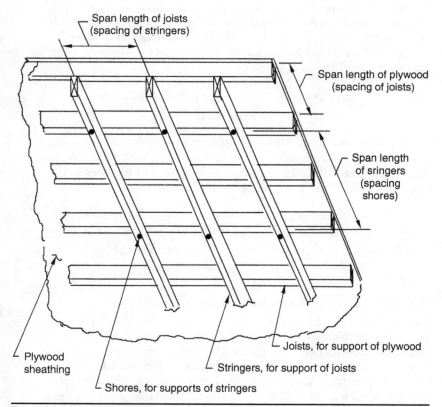

Span length of joists (spacing of stringers)

Span length of plywood (spacing of joists)

Span length of sringers (spacing shores)

Joists, for support of plywood

Plywood sheathing

Stringers, for support of joists

Shores, for supports of stringers

FIGURE 11-11 Allowable span lengths and components of a slab form.

5. Decking for the slab form will be ¾-in. Plyform Class I

6. Assume wet condition and load-duration less than 7 days

The design will be performed in the following sequence: determine the dead loads and live loads on the slab forms; determine the spacing of joists based on the strength and deflection of Plyform decking; determine the spacing of stringers based on the strength and deflection of joists; check the bearing strength between joists and stringers; determine the spacing of shores based on the strength and deflection of stringers; and check the bearing strength between stringers and shores.

Loads on Slab Forms

The loads that will be applied to the slab forms include the dead weight of the freshly placed concrete and reinforcing steel, the dead load of the formwork materials, and the live load of workers, tools, and equipment. If the live load will produce impact, as that caused by motor-driven concrete buggies or by concrete falling from a bucket, the effect of the impact should be considered in designing the forms. For this design it is assumed that there will be no impact. The effect of the impact of concrete falling from a bucket is discussed in Chapter 7.

The load from the concrete slab on a square foot of plywood decking can be calculated as follows:

Uniform load per sq ft = 150 lb per cu ft (6/12 ft)

= 75 lb per sq ft

The dead load due to the weight of formwork material is assumed as 5 lb per sq ft. The recommended minimum live load according to ACI Committee 347 is 50 lb per sq ft. Following is a summary of the loads on the slab form that will be used for the design:

Dead load of freshly placed concrete   = 75 lb per sq ft

Assume dead load of formwork materials =   5 lb per sq ft

Live load of workers and tools   = <u>50 lb per sq ft</u>

Design load = 130 lb per sq ft

Plywood Decking to Resist Vertical Load

Because the type and thickness of plywood have already been chosen, it is necessary to determine the allowable span length of the plywood, that is, the distance between joists, such that the bending stress, shear stress, and deflection of the plywood will be within allowable limits. Table 4-11 provides the section properties for the ¾-in.-thick Plyform selected for the design. The Plyform will be placed with the face grain across the supporting joists. Therefore, the properties for stress applied parallel to face grain will apply. The physical properties for a 1-ft-wide strip of the ¾-in.-thick plywood installed with the face grain across supports from Table 4-11 are as follows:

Approximate weight = 2.2 lb per sq ft
Moment of inertia, $I = 0.199$ in.^{4}
Effective section modulus, $S_e = 0.455$ in.^{3}
Rolling shear constant, $Ib/Q = 7.187$ in.^{2}

Table 4-12 gives the allowable stresses for the Plyform Class I as follows:

Allowable bending stress, $F_b = 1,930$ lb per sq in.
Allowable rolling shear stress, $F_s = 72$ lb per sq in.
Modulus of elasticity, $E = 1,650,000$ lb per sq in.

Bending Stress in Plyform Decking
Because the plywood will be installed over multiple supports, three or more, Eq. (5-41) can be used to determine the allowable span length of the plywood based on bending.

From Eq. (5-41), $l_b = [120F_b S_e/w_b]^{1/2}$
$$= [120(1,930)(0.455)/(130)]^{1/2}$$
$$= 28.5 \text{ in.}$$

Rolling Shear Stress in Plyform Decking
The allowable span length of the plywood based on rolling shear stress can be calculated.

From Eq. (5-48), $l_s = 20F_s(Ib/Q)/w_s$
$$= 20(72)(7.187)/(130)$$
$$= 79.6 \text{ in.}$$

Bending Deflection in Plyform Decking
For a maximum permissible deflection of $l/360$, but not greater than $1/16$ in., and plywood with three or more spans, Eqs. (5-57a) and (5-57b) can be used to calculate the allowable span lengths for bending deflection.

From Eq. (5-57a), $l_d = [1,743EI/360w_d]^{1/3}$ for $\Delta = l/360$
$$= [1,743(1,650,000)(0.199)/360(130)]^{1/3}$$
$$= 23.0 \text{ in.}$$

From Eq. (5-57b), $l_d = [1,743EI/16w_d]^{1/4}$ for $\Delta = 1/16$-in.
$$= [1,743(1,650,000)(0.199)/16(130)]^{1/4}$$
$$= 22.9 \text{ in.}$$

Summary for the ⅞-in.-thick Plyform Decking

For bending, the maximum span length of the Plyform = 28.5 in.

For shear, the maximum span length of the Plyform = 79.6 in.

For deflection, the maximum span length of the Plyform = 22.9 in.

For this design, deflection controls the maximum span length of the Plyform decking. The joists that support the plywood must be placed at a spacing that is no greater than 22.9 in. For uniformity, choose a joist spacing of 20.0 in. Another alternative would be to reduce the spacing to 16 in. because an extra joist is needed at the Plyform joint. If the deflection criteria were $l/270$, or not to exceed ⅛ in., the calculated allowable span length is 25.4 in., which would allow a larger spacing of joists at 24 in. on centers. Also, APA—The Engineered Wood Association allows a clear span for the span length, plus ⅝ in. for 4-in. framing lumber.

Joists for Support of Plyform

The joists support the Plyform and the stringers must support the joists. Therefore, the maximum allowable spacing of stringers will be governed by the bending, shear, and deflection in the joists.

Consider using 4×4 S4S joists whose actual size is 3½ in. by 3½ in. From Table 4-1, the physical properties can be obtained for 4×4 S4S lumber:

Area, $A = 12.25$ in.^{2}

Section modulus, $S = 7.15$ in.^{3}

Moment of inertia, $I = 12.51$ in.^{4}

The lumber selected for the forms is No. 2 grade Southern Pine (refer to Table 4-2). The reference design values for Southern Pine are already size adjusted. The allowable stresses can be obtained by adjusting the tabulated values by the adjustment factors for short-duration loading and a wet condition. Table 4-4 provides the adjustments for short-duration loads and Table 4-5 for application under a wet condition. Following are the allowable stresses that will be used for the design:

$$\text{Allowable bending stress, } F_b = C_D \times C_M \times \text{(reference bending stress)}$$
$$= (1.25)(0.85)(1,500)$$
$$= 1,593 \text{ lb per sq in.}$$
$$\text{Allowable shear stress, } F_v = C_D \times C_M \times \text{(reference shear stress)}$$
$$= (1.25)(0.97)(175)$$
$$= 212 \text{ lb per sq in.}$$
$$\text{Modulus of elasticity, } E = C_M \times \text{(reference modulus of elasticity)}$$
$$= (0.9)(1,600,000)$$
$$= 1,440,000 \text{ lb per sq in.}$$

Bending in 4 × 4 S4S Joists

Each joist will support a horizontal strip of decking 20-in.-wide, which will produce a uniform load of 130 lb per sq ft × 20/12 ft = 217 lb per lin ft along the entire length of the joist. The allowable span length of the joists, that is, the distance between stringers, based on bending can be determined as follows:

From Eq. (5-34), $l_b = [120F_b S/w]^{1/2}$
$$= [120(1,593)(7.15)/217]^{1/2}$$
$$= 79.3 \text{ in.}$$

Shear in 4 × 4 S4S Joists

The allowable span length based on shear in the joists can be calculated.

From Eq. (5-36), $l_v = 192F_v bd/15w + 2d$
$$= 192(212)(3.5)(3.5)/15(217) + 2(3.5)$$
$$= 160.1 \text{ in.}$$

Deflection in 4 × 4 S4S Joists

The permissible deflection is $l/360$, but not greater than $\frac{1}{16}$ in. Using Eqs. (5-37a) and (5-37b), the allowable span lengths due to deflection can be calculated as follows.

From Eq. (5-37a), $l_\Delta = [1,743EI/360w]^{1/3}$ for $\Delta = l/360$
$$= [1,743(1,440,000)(12.51)/360(217)]^{1/3}$$
$$= 73.7 \text{ in.}$$

From Eq. (5-37b), $l_\Delta = [1,743EI/16w]^{1/4}$ for $\Delta = \frac{1}{16}$ in.
$$= [1,743(1,440,000)(12.51)/16(217)]^{1/4}$$
$$= 54.8 \text{ in.}$$

Summary for 4 × 4 S4S Joists

For bending, the maximum span length of joists = 79.3 in.

For shear, the maximum span length of joists = 160.1 in.

For deflection, the maximum span length of joists = 54.8 in.

For this design, deflection governs the maximum span length of the joists. Because the stringers must support the joists, the maximum spacing of the wales must be no greater than 54.8 in. For constructability a spacing of 48.0 in. center to center will be used for stringers.

Stringers for Support of Joists

For this design consider using 4-in.-thick stringers. Thus the actual size of a stringer will be 3½ in. Determine the unit stress in bearing between joists and stringers.

$$\text{Contact area in bearing, } A = 3.5 \text{ in.} \times 3.5 \text{ in.}$$
$$= 12.25 \text{ sq in.}$$

The total load in bearing between a joist and a stringer will be the unit pressure of the concrete acting on an area based on the joist and stringer spacing. For this design, the area will be 20 in. wide by 48 in. long. Therefore, the pressure acting between the joist and the stringer will be:

$$P = 130 \text{ lb per sq ft } [20/12 \times 48/12]$$
$$= 867 \text{ lb}$$

The calculated applied unit stress in bearing, perpendicular to grain, between a joist and a stringer will be

$$f_{c\perp} = 867 \text{ lb}/12.25 \text{ sq in.}$$
$$= 70.8 \text{ lb per sq in.}$$

The allowable compression stress perpendicular to grain for No.2 grade Southern Pine can be obtained by multiplying the reference value for shear perpendicular to grain in Table 4-2 by the adjustment factor for a wet condition ($C_M = 0.67$) in Table 4-5. There is no load-duration adjustment factor in the compression perpendicular to grain.

Allowable compression stress perpendicular to grain is

$$F_{c\perp} = C_M \times \text{(reference value of compression perpendicular to grain)}$$
$$= (0.67)(565 \text{ lb per sq in.})$$
$$= 378.5 \text{ lb per sq in.}$$

The allowable compressive unit stress of 378.5 lb per sq in. is greater than the applied unit stress of 70.8 lb per sq in. Therefore, the unit stress is within the allowable value for the 4×4 No. 2 grade Southern Pine joists.

Size of Stringer Based on Selected 48-in. Spacing

Although the loads transmitted from the joists to the stringers are concentrated, it is generally sufficiently accurate to treat them as uniformly distributed loads, having the same total value as the concentrated loads, when designing forms for concrete slabs. With a 48-in.

spacing and a 130 lb per sq ft pressure, the uniform load on a stringer will be 130 lb per sq ft × 48/12 ft = 520 lb per lin ft. Consider using 4 × 6 S4S stringers.

From Table 4-1, the physical properties for 4 × 6 S4S lumber are

$A = 19.25$ in.^{2}
$S = 17.65$ in.^{3}
$I = 48.53$ in.^{4}

The allowable stresses for the 4 × 6 No. 2 grade Southern Pine can be obtained by adjusting the reference design values in Table 4-2 by the adjustment factors for short-duration loading and a wet condition. Table 4-4 provides the adjustments for short-duration loads, and Table 4-5 gives adjustment factors for wet condition. Following are the allowable stresses that will be used for the design:

Allowable bending stress, $F_b = C_D \times C_M \times$ (reference bending stress)
$= (1.25)(0.85)(1,250)$
$= 1,328$ lb per sq in.

Allowable shear stress, $F_v = C_D \times C_M \times$ (reference shear stress)
$= (1.25)(0.97)(175)$
$= 212$ lb per sq in.

Modulus of elasticity, $E = C_M \times$ (reference modulus of elasticity)
$= (0.9)(1,600,000)$
$= 1,440,000$ lb per sq in.

Bending in 4 × 6 S4S Stringers

Based on bending, the span length of the stringers must not exceed the value calculated.

From Eq. (5-34), $l_b = [120F_bS/w]^{1/2}$
$= [120(1,328)(17.65)/520]^{1/2}$
$= 73.5$ in.

Shear in 4 × 6 S4S Stringers

Based on shear, the allowable span length of the stringers must not exceed the value calculated.

From Eq. (5-36), $l_v = 192F_vbd/15w + 2d$
$= 192(212)(3.5)(5.5)/15(520) + 2(5.5)$
$= 111.4$ in.

Deflection in 4 × 6 S4S Stringers

The permissible deflection is $l/360$, but not greater than $\frac{1}{16}$-in.

Using Eqs. (5-37a) and (5-37b), the allowable span lengths due to deflection can be calculated as follows:

$$
\begin{aligned}
\text{From Eq. (5-37a), } l_\Delta &= [1{,}743EI/360w]^{1/3} \text{ for } \Delta = l/360 \\
&= [1{,}743(1{,}440{,}000)(48.53)/360(520)]^{1/3} \\
&= 86.6 \text{ in.} \\
\text{From Eq. (5-37b), } l_\Delta &= [1{,}743EI/16w]^{1/4} \text{ for } \Delta = \frac{1}{16} \text{ in.} \\
&= [1{,}743(1{,}440{,}000)(48.53)/16(520)]^{1/4} \\
&= 61.9 \text{ in.}
\end{aligned}
$$

Summary for the 4 × 6 S4S Stringers

For bending, the maximum span length of stringers = 73.5 in.

For shear, the maximum span length of stringers = 111.4 in.

For deflection, the maximum span length of stringers = 61.9 in.

For this design, deflection governs the maximum span length of the stringers. Because the shores must support the stringers, the maximum spacing of shores must be no greater than 61.9 in. For simplicity in layout and constructability, a spacing of 60.0 in. center to center will be used.

Shores for Support of Stringers

Based on the strength of the stringers in bending, shear, and deflection, the stringers must be supported every 60 in. by a shore. A shore must resist the concrete pressure that acts over an area consisting of the spacing of the stringers and the span length of the stringers, 48 in. by 60 in. The vertical load on a shore can be calculated as follows:

$$
\begin{aligned}
\text{Area} &= 48 \text{ in.} \times 60 \text{ in.} \\
&= 2{,}880 \text{ sq in.} \\
&= 20.0 \text{ sq ft}
\end{aligned}
$$

$$
\begin{aligned}
\text{Vertical load on a shore} &= 130 \text{ lb per sq ft (20.0 sq ft)} \\
&= 2{,}600 \text{ lb}
\end{aligned}
$$

Thus, a shore is required that will withstand a vertical load of 2,600 lb. There are numerous patented shores available that can sustain this load. For this design, S4S wood shores will be installed with a 10-ft unbraced length. Using a wood shore with an allowable compression stress parallel to grain of 1,650 lb per sq in. and a modulus of elasticity for column stability of 580,000 lb per sq in., Table 5-5 indicates an allowable load of 4,688 lb for a 4 × 4 S4S wood shore.

Bearing between Stringer and Shore

For this design, 4 × 6 S4S stringers will bear on 4 × 4 S4S wood shores. Determine the unit stress in bearing between stringers and shores.

$$\text{Contact area, } A = 3.5 \text{ in.} \times 3.5 \text{ in.}$$
$$= 12.25 \text{ sq in.}$$

The calculated unit stress in bearing, perpendicular to grain, between a stringer and a shore will be:

$$f_{c\perp} = 2,600 \text{ lb}/12.25 \text{ sq in.}$$
$$= 212.2 \text{ lb per sq in.}$$

From Table 4-2, the tabulated design value for unit compression stress, perpendicular to grain, for No. 2 grade Southern Pine is 565 lb per sq in. The allowable design stress can be calculated by multiplying the reference design value by the wet condition adjustment factor ($C_M = 0.67$) (refer to Table 4-5). There is no load-duration adjustment factor in compression stress perpendicular to grain. The allowable bearing stress will be:

$$F_{c\perp} = (0.67)(565 \text{ lb per sq in.})$$
$$= 378.5 \text{ lb per sq in.}$$

This allowable unit stress of 378.5 lb per sq in. is greater than the applied unit stress of 212.2 lb per sq in. for this grade and species of wood. The formwork must also be adequately braced to resist lateral forces on the slab-forming system as presented in Chapter 5.

Design Summary of Forms for Concrete Slab

The allowable stresses and deflection criteria used in the design are based on the following values.

For ¾-in.-thick Plyform Class I:
    Allowable bending stress = 1,930 lb per sq in.
    Allowable rolling shear stress = 72 lb per sq in.
    Modulus of elasticity = 1,650,000 lb per sq in.

For 4 × 4 S4S joists, No. 2 grade Southern Pine:
    Allowable bending stress = 1,593 lb per sq in.
    Allowable shear stress = 212 lb per sq in.
    Modulus of elasticity = 1,440,000 lb per sq in.
    Permissible deflection less than $l/360$, but not greater than ¹⁄₁₆ in.

For 4 × 6 S4S stringers, No. 2 grade Southern Pine:

Allowable bending stress = 1,328 lb per sq in.

Allowable shear stress = 212 lb per sq in.

Modulus of elasticity = 1,440,000 lb per sq in.

Permissible deflection less than $l/360$, but not greater than $\frac{1}{16}$ in.

The forms for the concrete slab should be built to the following conditions:

Item	Nominal Size and Spacing
Decking	¾-in.-thick Plyform Class I
Joists	4 × 4 S4S No. 2 grade Southern Pine at 20 in. on centers
Stringers	4 × 6 S4S No. 2 grade Southern Pine at 48 in. on centers
Shores	60 in. maximum spacing along stringers

Figure 11-12 summarizes the design. Slab forms must be braced adequately to resist any eccentric loads or lateral forces due to wind. The lateral force applied to slab forms is discussed in Chapter 5.

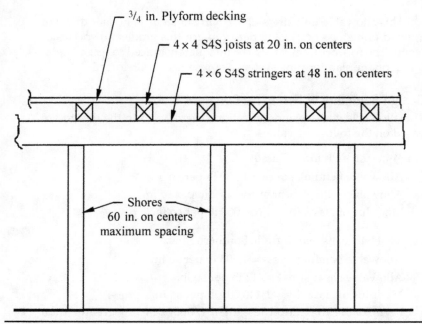

FIGURE 11-12   Summary of design for a slab form.

Solid Slab Thickness, in.	Dead Load, lb per sq ft	P, lb per lin ft, Applied Along Edge of Slab in Either Direction				
		Width of Slab in Direction of Force, ft				
		20	40	60	80	100
4	65	100	100	100	104	130
6	90	100	100	108	144	180
8	115	100	100	138	184	230
10	140	100	112	168	224	280
12	165	100	132	198	264	330
14	190	100	152	228	304	380
16	215	100	172	258	344	430
20	265	106	212	318	424	530

Notes:
1. Reprinted from "Formwork for Concrete," ACI Special Publication No. 4, the American Concrete Institute.
2. Slab thickness given for concrete weighing 150 lb per cu ft; allowance of 15 lb per sq ft for weight of forms. For concrete of a different weight or for joist slabs and beams and slab combinations, estimate dead load per sq ft and work from dead load column, interpolating as needed on a straight-line basis. Do not interpolate in ranges that begin with 100 lb minimum load.
3. Special conditions may require heavier bracing.

TABLE 11-9   Minimum Lateral Force for Design of Slab Form Bracing, lb per lin ft

Minimum Lateral Force for Design of Slab Form–Bracing Systems

The minimum lateral loads required for the design of slab form bracing are presented in Table 11-9. The horizontal load shown in Table 5-8 is in pounds per linear foot, applied along the edge of the slab in either direction as illustrated in Figure 11-13. Where unbalanced loads from the placement of concrete or from starting or stopping of equipment may occur, a structural analysis of the bracing system should be made.

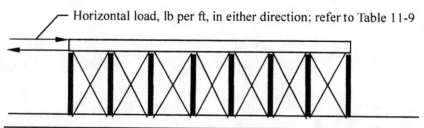

Horizontal load, lb per ft, in either direction; refer to Table 11-9

FIGURE 11-13   4 × 4 Minimum lateral force for bracing slab-forming systems.

References

1. "National Design Specification for Wood Construction," ANSI/AF&PA NDS-2005, American Forest & Paper Association, Washington, DC, 2005.
2. "Design Values for Wood Construction," Supplement to the National Design Specification, American Forest & Paper Association, Washington, DC, 2005.
3. "Concrete Forming," APA—The Engineered Wood Association, Tacoma, WA, 2005.
4. "Plywood Design Specification," APA—The Engineered Wood Association, Tacoma, WA, 1997.
5. "Formwork for Concrete," Special Publication No. 4, American Concrete Institute, Detroit, MI.
6. "Guide to Formwork for Concrete," ACI Committee 347-04, American Concrete Institute, Detroit, MI, 2004.

CHAPTER **12**

Patented Forms for Concrete Floor Systems

Introduction

Many kinds of patented forms for concrete floor systems are available, and the designer usually specifies the particular system that is to be used. Several, but not all, of these systems are illustrated and described in this chapter.

Ceco Flangeforms

As illustrated in Figure 12-1, these forms are available in standard widths of 20 and 30 in. and in lengths of 1, 2, and 3 ft for forming regularly spaced joists. Special filler widths of 10 and 15 in. are available in 1- and 2-ft lengths for filling nonstandard spaces. The tapered Endforms are used at the ends of joists where extra joist width is required. They are furnished only in 20- and 30-in. standard widths and in standard 3-ft lengths. Endcaps are used as closures at the ends of all rows of Flangeforms at beams, bridging joists, and special headers. The available depths are given in Figure 12-1. All dimensions are measured from the outside surfaces of the forms.

As illustrated in Figure 12-2, the forms are assembled on joist soffits. They are also supported by resisting the flanges on joist soffits, such as 2 × 8 lumber laid flat on wood stringers, which in turn are supported by shores. Flange-type forms have several advantages when compared with other types of forms, including the following:

1. The depth of the joist is uniformly fixed by the depth of the form.

2. Forms are installed with all operations such as placing, spacing, and nailing performed above the supporting members.

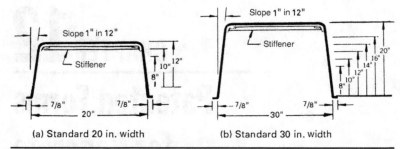

Figure 12-1   Typical cross sections of Flangeforms; all dimensions are outside to outside. (*Source: Ceco Concrete Construction*)

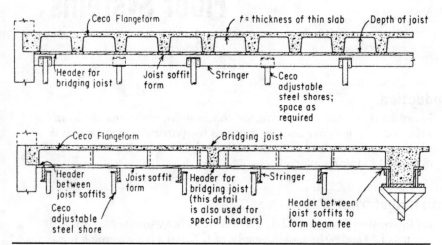

Figure 12-2   Installation of concrete floors using Flangeforms; typical arrangement of centering. (*Source: Ceco Concrete Construction*)

3. The widths of the joists may be varied without changing the sizes of the joist soffits, thus reducing the quantity of form lumber required.

4. By modifying the widths of the joists, the forms may be used for floors with odd dimensions.

Adjustable Steel Forms

As illustrated in Figure 12-3, flangeless types of steel forms are available for use in construction pan-joist types of floor systems. The forms are available in a 20- and 30-in. standard, and a 30- and 40-in. deepset. Figure 12-3 only shows the standard and the 40-in. deepset forms. The dimensions given in the figure are representative of the sizes

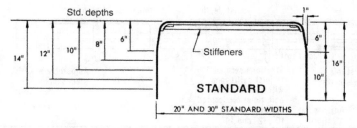

(a) Standard widths of 20 in. and 30 in.

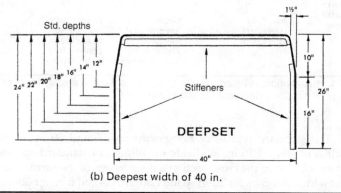

(b) Deepest width of 40 in.

FIGURE 12-3 Typical cross sections of adjustable steel forms; all dimensions are outside to outside. (*Source: Ceco Concrete Construction*)

available for use in constructing this type of floor system. Several lengths are available to permit the forms to be used with slabs having variable dimensions.

As shown in Figure 12-4, the forms are supported at the desired height by nails driven through the vertical surfaces of the forms into the joist soffits. The use of these forms permits the construction of concrete slabs with considerable variety in the dimensions of the joists.

One advantage of using this type of form is that the forms or pans may be stripped as soon as the concrete over the forms has gained the necessary strength, without disturbing the joist soffits and supporting shores. This feature may permit an earlier reuse of this type of pan than is possible with the flange-type pans.

Ceco Longforms

These forms are used with unsupported lengths of up to 12 ft between beam tees or bridging joists. Because the Longforms must be cut to suit the span length of the structure, there are no standard lengths.

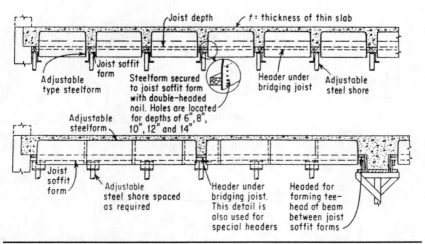

Figure 12-4   Installation using adjustable steel forms; typical arrangement of centering. (*Source: Ceco Concrete Construction*)

The standard widths of Longforms are 20 and 30 in., with filler widths of 10 and 15 in. available for filling nonstandard spaces only. Figure 12-5 gives the cross-sectional dimensions of the forms. Standard joist widths are 5 in. for 20-in.-wide Longforms and 6 in. for 30-in.-wide Longforms. Wider joists can be provided by using special soffit fillers in increments of ½ in. with either 20- or 30-in.-wide Longforms.

Repetition within a given project is important when using standard Longform sizes and lengths. In most cases, it is more economical to vary beam widths, beam tees, or bridging joist widths in order to maintain typical Longform lengths from floor to floor. As illustrated in Figure 12-6, the forms are installed with the ends resting on wood carriers attached to beam forms or to the soffits of bridging joists. The forms are installed with the soffit flanges held together with C clamps. Soffit fillers may be used, as shown in Figure 12-6, to produce wider concrete joists.

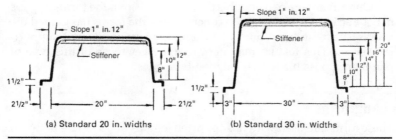

Figure 12-5   Typical cross sections of Ceco Longforms; all dimensions are outside to outside. (*Source: Ceco Concrete Construction*)

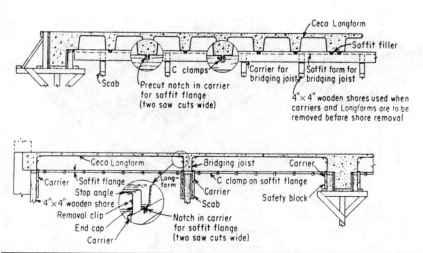

FIGURE 12-6   Installation using Ceco Longforms; typical arrangement of centering. (*Source: Ceco Concrete Construction*)

The carriers that support the ends of the forms may be removed prior to stripping the supporting beam forms, which will permit the removal of the Longforms without disturbing the forms for the beams. Thus, the Longforms may be removed for early reuse without disturbing the shores.

Ceco Steeldomes

These units are used in forming two-way dome slab construction, sometimes described as waffle-type construction. The domes are made of one-piece sheet steel with a flange on each of the four sides. When installed, the flanges of adjacent domes should abut each other, with nails holding them in place. This dome system is available in 2- and 3-ft standard square modules. Figure 12-7 shows the dimensions of standard sizes.

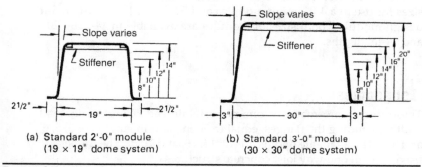

(a) Standard 2'-0" module
(19 × 19" dome system)

(b) Standard 3'-0" module
(30 × 30" dome system)

FIGURE 12-7   Ceco Steeldomes. (*Source: Ceco Concrete Construction*)

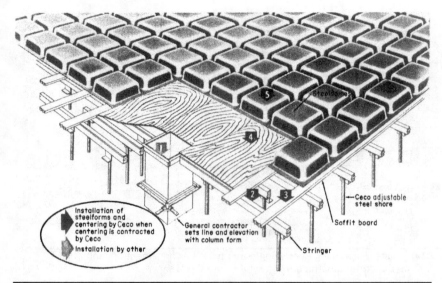

FIGURE 12-8   Installation using Ceco Steeldomes. (*Source: Ceco Concrete Construction*)

Figure 12-8 illustrates a method of supporting the domes—a plywood decking supported in a conventional manner providing a drop-panel adjustment to a concrete column. The sizes and spans of soffit boards and stringers should be determined by using the method discussed in Chapter 11. The domes are removed by introducing compressed air through a hole in each dome into the space above the dome.

Ceco Fiberglassdomes

These domes are available for forming two-way concrete joist floor systems. The two standard size widths available are 41- by 41-in. sizes for use in a 4 ft design modules and 52- by 52-in. for use in 5-ft design modules. Both sizes of modules are available in 14-, 16-, 20-, and 24-in. depths.

Ceco Longdomes

These domes are made of one-piece reinforced fiberglass in total width, including the flanges, equal to 36 in., and in lengths of 3, 5, 6, and 7 ft. The available depths are 12, 14, 16, and 20 in. The domes are used to form one-way joist concrete structures.

Plastic Forms

Plastic forms, which have been used to construct concrete slabs of the one-way and two-way joist types, have several advantages, including the following:

1. They are light, strong, and resistant to impact.
2. They are resistant to denting and warping.
3. They will not corrode or rust.
4. They produce smooth concrete surfaces.

The products of one manufacturer, Molded Fiber Glass Concrete Forms Company, are glass-reinforced plastics molded under heat and pressure in steel molds. This company manufactures several sizes of forms for use in constructing two-way joist or waffle-type slabs. Table 12-1 gives representative dimensions and other information for these forms. The specified dimensions apply to outside surfaces of the forms. The size of voids indicates the width of the void left in the concrete slab, and the overall size is the width of the flange.

The forms are installed, usually on solid decking, with the flanges down and abutting the flanges of adjacent forms. One large-headed nail driven into the decking at the juncture of four flange corners should hold the forms in position. After the concrete has gained the desired strength, the decking is stripped and the pans are removed by injecting a blast of compressed air through each form into the space at the back of the form. If the forms are handled with care, they can be used 25 times or more. If a surface is damaged, it can be restored with a suitable plastic treatment. Figure 12-9 shows these forms being installed on solid decking, and Figure 12-10 illustrates how a form is removed by using compressed air.

Item	19-in. System	24-in. System	30-in. System	41-in. System	52-in. System
Size of voids, in.	19 × 19	24 × 24	30 × 30	41 × 41	52 × 52
Joist center, in.	24 × 24	30 × 30	36 × 36	48 × 48	60 × 60
Joist width, in.	5	6	6	7	8
Range of depth of voids, in.	8, 10, 12, 14, 16	8, 10, 12, 14, 16, 20	8, 10, 12, 14, 16, 18, 20	12, 14, 16, 18, 20, 24	10, 12, 14, 16, 18, 20, 24
Approx. weight of 12- in.-deep void, lb	13	19	21	65	103

Table 12-1 Representative Dimensions and Other Properties of Molded Fiber Glass Forms (MFG Concrete Forms Company)

FIGURE 12-9 Installing plastic molds. (*Source: MFG Concrete Forms Co.*)

FIGURE 12-10 Removing a plastic mold with compressed air. (*Source: MFG Concrete Forms Co.*)

This company also manufactures extra long forms for faster concrete forming and custom forms for special shapes. The custom forms can be formed into almost any shape. Because the forms are light in weight, they are particularly suited for application with flying forms where entire floor sections may be moved from floor to floor in a complete grid format.

Corrugated-Steel Forms

Corrugated-steel sheets are frequently used as forms for concrete floor and roof slabs supported by steel joists, steel beams, and precast concrete joists. The sheets are available in several gauges, widths, and lengths to fit different requirements.

The sheets are laid directly on the supporting joists or beams with the corrugations perpendicular to the supports. Side laps should be one-half of a corrugation, and end laps should be at least 3 in., made over supporting joists or beams. The sheets are fastened to the supports with special clips or spot welds. The edges of adjacent sheets are fastened together with steel clips.

The depth of a concrete slab is measured from the top of the surface to the centroid of the corrugated sheet. The span is the distance from center to center between the supporting members. The maximum safe span will depend on the total load per unit of area, including the weight of the section, the concrete and live loads, and the moment of inertia and section modulus for the section under consideration.

Cellular-Steel Floor Systems

Cellular-steel panels are frequently used for the forms and structural units of floor systems. This system offers a number of advantages, including light weight, the elimination of shores and forms for concrete, rapid construction, and abundant raceways for utility services. Thus, the forming system combines the structural slab and electrical functions within the normal depth of the floor slab.

Each cellular unit has three cells which may be used for telephone, power, and computer cables. Cellular units are available in 24- and 36-in. widths, as illustrated in Figure 12-11. The panels are manufactured from rolled-steel shapes and plates that are electrically welded together in various types. Openings in the cells provide routing for wire cables through the floor slab into the room area.

The panels are prefabricated by the manufacturer to fit any given floor condition and shipped to the project ready for installation without additional fabrication. If storage is necessary, the cellular units may be stacked on wood blocking clear off the ground and tilted longitudinally to ensure against the entrapment of water. All deck bundles have labels or handwritten information clearly visible to identify the material and erection location.

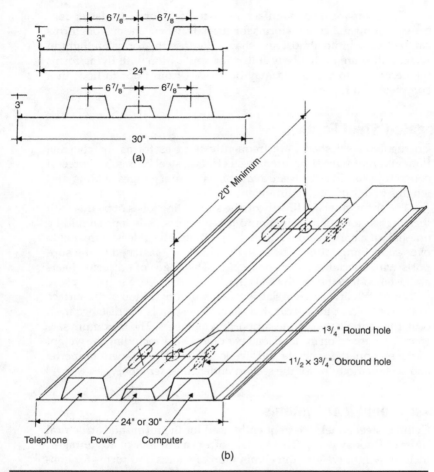

FIGURE 12-11 Cellular units for floor system: (*a*) cellular units in 24- and 30-in. widths; (*b*) isometric view of cellular unit. (*Source: H. H. Robertson Company*)

After alignment, the cellular units are attached to the decking before the placement of concrete for the floor slab. Figure 12-12 shows the installation of a floor cellular system.

Selecting the Proper Panel Unit for Cellular-Steel Floor Systems

Several manufacturers produce units having properties similar to those presented in the preceding section. Because a variety of panel units of cellular-steel panels is available, it is necessary to select a section that has adequate strength and stiffness for a given condition. With the span between supporting steel beams and the total load

FIGURE 12-12 Installation of cellular units for an elevated floor system. (*Source: H. H. Robertson Company*)

on the panels known, and the permissible deflection specified, it is necessary to select a panel unit having the required moment of inertia or section modulus, considering the resistance to bending moment and deflection. The panel unit should be selected for a simple span or a continuous beam, whichever is applicable.

Horizontal Shoring

This is a system of supporting decking, or wood joists and decking for concrete slabs, using telescoping horizontal beams whose lengths can be adjusted over a wide range. The ends of these shores may be supported on permanent steel beams, concrete beams, walls, or temporary stringers and vertical shores. They are available for clear spans in excess of 25 ft, which require no intermediate supports.

Figure 12-13 illustrates a shore consisting of two parts, a lattice-trussed triangular section and a plate section that telescopes into the lattice section to produce the designed length. Horizontal prongs that project from the ends of the shores rest on the supporting members. The shores are made with built-in wedges or other devices that permit them to be set with a camber, if one is desired.

The shores are installed at the desired spacing, the end prongs resting on the supporting members as illustrated in Figure 12-14. The wood joists and decking are installed, and the concrete slab is placed. After the concrete has gained the desired strength, the camber is eliminated and the ends of the shores are pried far enough away from the supporting members to permit them to be lowered. Figure 12-15 illustrates decking installed directly on top of horizontal shoring.

Figure 12-13 Horizontal shores. (*Source: Patent Construction Systems*)

Figure 12-14 Installation of horizontal shores. (*Source: Patent Construction Systems*)

Figure 12-15   Plywood decking installed directly over horizontal shoring. (*Source: Patent Construction Systems*)

Figure 12-16 illustrates a method of supporting horizontal shores on steel-frame construction. Figure 12-16(a) is a Spanall bearing on a beam side form. Ledgers are mounted from the header, which is supported by a coil-tie hanger. The plywood is placed on top of the joists. Figure 12-16(b) shows a Spanall bearing on side forms of fireproofing supported by 4 × 4 headers hung with coil-tie hangers, with plywood placed directly on top of the Spanall.

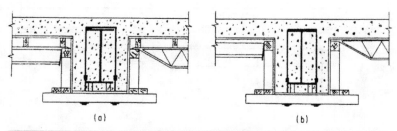

(a)   (b)

Figure 12-16   Methods of supporting horizontal shores on steel frame construction. (*Source: Patent Construction Systems*)

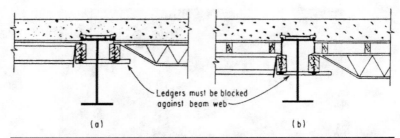

Ledgers must be blocked
against beam web

(a)   (b)

FIGURE 12-17   Additional methods of supporting horizontal shores on steel frame construction. (*Source: Patent Construction Systems*)

Figure 12-17 illustrates another method of supporting horizontal shores on steel frame construction. Figure 12-17(a) is a Spanall bearing on a 4 × 6 ledger that is hung from a steel hanger over the structural steel beam. The plywood is placed directly on top of the horizontal shores. Figure 12-17(b) is a Spanall bearing on a 4 × 6 ledger hung from a steel hanger over a structural steel beam. The plywood is placed on joists that rest on the horizontal shores.

Figure 12-18 illustrates methods of using horizontal shores with forms for a flat slab with drophead construction. Figure 12-18(a) is a typical cross section showing the Spanall and joists supporting the dropheads on the lower ledger while the upper ledger supports the Spanall. The plywood is placed directly on the horizontal shoring.

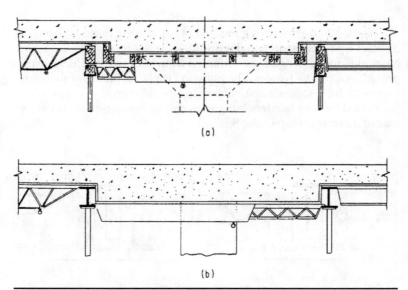

(a)

(b)

FIGURE 12-18   Methods of using horizontal shores with forms for flat-slab with drophead construction. (*Source: Patent Construction Systems*)

Figure 12-18(b) is a typical cross section showing a Spanall supporting a drophead on the bottom flange of a steel beam while the top flange supports the Spanall carrying the slab.

References

1. Ceco Concrete Construction Company, Kansas City, MO.
2. Molded Fiber Glass Concrete Forms (MFG), Independence, KS.
3. H. H. Robertson Company, Ambridge, PA.
4. Patent Construction Systems, Paramus, NJ.

CHAPTER 13

Forms for Thin-Shell Roof Slabs

Introduction

Thin-shell slabs of cylindrical or barrel shapes, with the axes perpendicular to the spans, are frequently used as roofs for structures. For slabs of this kind, the form designer has at least two problems:

1. He or she must determine the elevations of sufficient points along an arc on top of the decking to ensure that the decking will be constructed to the required shape and elevation.

2. He or she must design the members that support the decking, commonly referred to as centering.

The plans furnished by the architect or engineer specify the span, the rise at the center, and the thickness of the concrete shell. This information is required by the designer to determine the geometry of the formwork.

Geometry of a Circle

Because a vertical section through a cylindrical shell is an arc of a circle, the elevation of points along the surface under the shell can be obtained by using the equations that define the geometry of a circle. Figure 13-1 illustrates an arc of a circle, which corresponds to the underside of a thin-shell roof. Although several methods can be used to obtain the desired information, the method developed in this chapter uses the basic algebraic and trigonometric equations to determine the geometry of circular slabs.

Let R = radius of circle
L = span of roof
H = height of rise at center of span
l = one-half the span of the roof

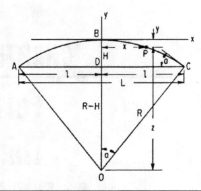

Figure 13-1   Geometry of a circle.

The plans will give the values of H and L, and possibly R. If the value of R is not given, it will be necessary to calculate it. Considering the triangle ODC,

$$\sin a = l/R \tag{a}$$
$$\cos a = (R\text{-}H)/R \tag{b}$$
$$\sin^2 a + \cos^2 a = 1 \tag{c}$$

Substituting the values of $\sin a$ and $\cos a$ in Eq. (c) gives:

$$l^2/R^2 + (R\text{-}H)^2/R^2 = 1$$
$$l^2 + (R\text{-}H)^2 = R^2$$
$$l^2 + R^2 - 2RH + H^2 = R^2$$
$$l^2 - 2RH + H^2 = 0$$
$$2RH = l^2 + H^2$$
$$R = (l^2 + H^2)/2H \tag{13-1}$$

If l is replaced with $L/2$, Eq. (13-1) gives:

$$R = L^2/8H + H/2 \tag{13-2}$$

Example 13-1

Determine the radius of a curve for which the span is 60 ft and the rise is 10 ft. Applying Eq. (13-2) gives:

$$R = L^2/8H + H/2$$
$$= (60 \text{ ft})^2/8(10 \text{ ft}) + (10 \text{ ft})/2$$
$$= 45 \text{ ft} + 5 \text{ ft}$$
$$= 50 \text{ ft}$$

It is necessary to calculate the slope of the tangent to the circle at point C in order to determine whether forms are needed on top of the concrete slab for a portion of the shell.

$$\sin a = l/R$$
$$= (30 \text{ ft})/(50 \text{ ft})$$
$$= 0.60$$

Using the arcsin function on a handheld calculator, the angle a can be determined as follows:

$$\arcsin a = 0.60$$
$$a = 36°\,52'$$

Locating Points on a Circle

Let the curve ABC in Figure 13-1 represent the underside of a shell roof and thus the top of the decking. Assume that the elevations of points A, B, and C are given. It is necessary to determine the elevations of points along the curve ABC, which may be done by determining the distances down from the x axis. Let P be a point on the curve a distance x from the y axis and a distance y below the x axis. The equation for a circle whose center is at the origin of the x axis and the y axis is

$$x^2 + y^2 = R^2 \qquad\qquad\qquad (d)$$

where x and y are the distances from the y axis and the x axis, respectively, to any given point on the circle. Where the circle passes through the origin of the x axis and the y axis, with the center on the y axis and a distance R below the origin, the equation for the circle will be:

$$x^2 + z^2 = R^2 \text{ where } z = R - y \qquad\qquad (e)$$

Substituting this value for z in Eq. (e) gives:

$$x^2 + (R - y)^2 = R^2$$
$$(R - y)^2 = R^2 - x^2$$
$$R = y = +/- \sqrt{R^2 - x^2}$$
$$y = R - \sqrt{R^2 - x^2} \qquad\qquad (13\text{-}3)$$

This equation can be used to determine the distance down from point B of Figure 13-1 to any point P on the curve. Assume that it is desired to determine the distances down from a horizontal line through B to each of 10 points, equally spaced horizontally along the curve BC. Let the horizontal distance from the y axis to a point P be x and let the distance down from the x axis be y.

For $x = 0$, $y = 0$

For $x = l/10$, $y = R - \sqrt{R^2 - (l^2/100)}$

For $x = 2l/10$, $y = R - \sqrt{R^2 - (l^2/25)}$

The sign of x can be plus or minus without changing the value of y. Thus, the curve is symmetrical about the y axis. Table 13-1 gives the equations for determining y for different values of x.

Value of x	Value of y
0	0
$\dfrac{l}{10}$	$R - \sqrt{R^2 - \dfrac{l^2}{100}}$
$\dfrac{2l}{10}$	$R - \sqrt{R^2 - \dfrac{l^2}{25}}$
$\dfrac{3l}{10}$	$R - \sqrt{R^2 - \dfrac{9l^2}{100}}$
$\dfrac{4l}{10}$	$R - \sqrt{R^2 - \dfrac{4l^2}{25}}$
$\dfrac{5l}{10}$	$R - \sqrt{R^2 - \dfrac{l^2}{4}}$
$\dfrac{6l}{10}$	$R - \sqrt{R^2 - \dfrac{9l^2}{25}}$
$\dfrac{7l}{10}$	$R - \sqrt{R^2 - \dfrac{49l^2}{100}}$
$\dfrac{8l}{10}$	$R - \sqrt{R^2 - \dfrac{16l^2}{25}}$
$\dfrac{9l}{10}$	$R - \sqrt{R^2 - \dfrac{81l^2}{100}}$
$\dfrac{10l}{10}$	$R - \sqrt{R^2 - l^2}$

TABLE 13-1 Value of x and y for Construction Centering for Thin-Shell Roof (See Figure 13-1)

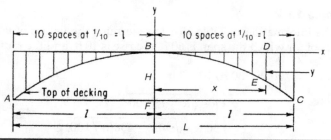

FIGURE 13-2   Elevations of points on a circular shell roof.

Elevations of Points on a Circular Arch

Line ABC in Figure 13-2 illustrates the top of the decking for a circular shell roof whose span AC is equal to L and whose rise BF is equal to H. The origin of the x and y axes is at B, the crown of the arch. Let l equal one-half the span, equal to $L/2$. Divide l into 10 equal parts, as illustrated. Let x equal the distance from B to any point D. The distance from D to point E on the decking is equal to y. The curve is symmetrical about the y axis. Table 13-2 gives the values of y for given values of x, expressed as fractions of l, for a curve whose span is 60 ft, whose rise is 10 ft, and whose radius is 50 ft.

Distance from Center of Arch	Value of x, ft	Value of y, ft
0	0	0.000
l/10	3	0.090
2l/10	6	0.361
3l/10	9	0.818
4l/10	12	1.461
5l/10	15	2.303
6l/10	18	3.352
7l/10	21	4.623
8l/10	24	6.137
9l/10	27	7.917
l	30	10.000

TABLE 13-2   Values of x and y for Determining the Elevations on the Circular Shell Roof in Example 13-2; for Span = 60 ft, Rise = 10 ft, and Radius = 50 ft (see Figure 13-2)

Example 13-2

Use the information in Table 13-1 to determine the values of y for a shell roof for which the span is 60 ft, the rise is 10 ft, and the radius is 50 ft.

For $x = l/10$, $x = 30$ ft/10 = 3 ft,

$$y = R - \sqrt{R^2 - l^2/100}$$
$$= 50\,\text{ft} - \sqrt{(50\text{ft})^2 - (30\text{ft})^2/100}$$
$$= 50\,\text{ft} - \sqrt{2,500 - 900/100}$$
$$= 50\,\text{ft} - \sqrt{2,491}$$
$$= 50\,\text{ft} - 49.9099$$
$$= 0.090\,\text{ft}$$

For $x = 2l/10$, $x = 6$ ft,

$$y = R - \sqrt{R^2 - l^2/25}$$
$$= 50\,\text{ft} - \sqrt{(50\text{ft})^2 - (30\text{ft})^2/25}$$
$$= 50\,\text{ft} - \sqrt{2,500 - 900/25}$$
$$= 50\,\text{ft} - \sqrt{2,491}$$
$$= 50\,\text{ft} - 49.639$$
$$= 0.361\,\text{ft}$$

Other values of y for increments of x are given in Table 13-2.

Forms for Circular Shell Roofs

The forms for circular shell roofs consist of decking plus the structural members that support the decking, called centering, as illustrated in Figure 13-3. The decking may be ⅝- or ¾-in. plywood supported by joists, usually of at least 2-in.-thick lumber. The joists are supported by lumber ribs, at least 2 in. thick that are sawed to provide the required curvature. The ribs may bear directly on shores, but they preferably bear on stringers that are supported by shores, as illustrated. It is extremely important that ribs be cut to permit the full vertical bearing of the ribs against the supporting stringers or shoring. A transverse row of shores is installed under each rib. The shores may be wood, patented, or metal scaffolding. The centering must include adequate braces, both horizontal and diagonal, in the transverse and longitudinal directions. Stringers should be securely fastened to shores with scabs and nails, as illustrated.

The ribs are long enough to provide laps of approximately 12 in. over the stringers. Also, the bottoms of the ribs are fabricated to permit full bearing on the tops of the stringers.

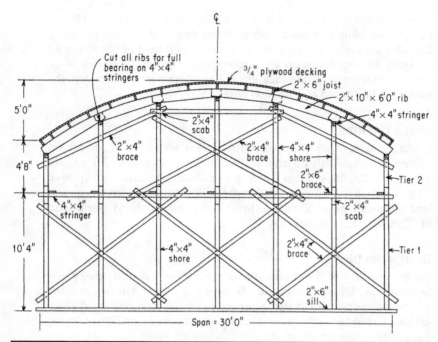

Figure 13-3 Formwork for a circular shell roof using lumber.

After the concrete has gained the desired strength, the decking can be lowered several inches below the roof slab. If the entire structure is to be lowered and moved longitudinally to a new location for reuse, the 2 × 6 sills should be installed longitudinally instead of in the position shown in order that the structure may slide on them. Also, it may be desirable to use larger sills, such as 2 × 12. If the structure is to be moved and reused, all braces and scabs should be left intact.

Design of Forms and Centering for a Circular Shell Roof

Use the information shown in Figure 13-3, together with that given in this section, to design the forms and centering for a circular shell roof. The span is 30 ft and the rise is 5 ft. The thickness of the concrete slab is 4 in. The live load will be 50 lb per sq ft. Use ¾-in. Class I Plyform decking and No. 2 grade Southern Pine lumber.

Space the Joists

The spacing of the joists will be governed by the allowable span length of the Plyform decking. The Plyform will be laid with the face grain across the joists. The dead load of 4-in.-thick concrete having a density of 150 lb per cu ft will be (4/12 ft)(150 lb per cu ft) = 50 lb per sq ft. Combining this dead load with the live load of 50 lb per sq ft, the total load will be 100 lb per sq ft. For a total load of 100 lb per sq ft, Table 4-13 indicates an allowable span length of 24 in. for ¾-in. Class I Plyform with the face grain across supports. Therefore, the joists will be spaced at 24 in. on centers.

Space the Ribs

Consider using 2 × 6 S4S joists. With joists spaced at 24 in., or 2 ft, on centers, the uniformly distributed load on the joists will be 2 ft × 100 lb per sq ft = 200 lb per lin ft. Table 4-17 indicates a maximum span of 73 in. for 2 × 6 S4S No. 2 grade Southern Pine lumber with an equivalent uniform load of 200 lb per lin ft. For constructability use a span of 72 in., or 6 ft, between ribs.

Design the Ribs

Set the spacing of the shores under the ribs at 5 ft 0 in., measured horizontally. The spans for the ribs near the crown will be 5 ft 0 in., whereas the spans near the outer edges of the roof will be slightly greater. Consider each rib member to be a simple beam 5 ft, or 60 in., long with a maximum of three concentrated loads from the joists, providing bending moment in the ribs. For the loading shown in Figure 13-4, the bending moment at the center of the rib is calculated as follows:

$$\text{Load } P \text{ from each joist} = (100 \text{ lb per sq ft })(2 \text{ ft})(6 \text{ ft})$$
$$= 1{,}200 \text{ lb}$$
$$\text{Total load on each rib} = 3(1{,}200 \text{ lb})$$
$$= 3{,}600 \text{ lb}$$
$$\text{Reaction at each end of rib} = (3{,}600 \text{ lb})/2$$
$$= 1{,}800 \text{ lb}$$

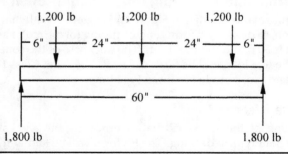

Figure 13-4   Loads on a rib member.

The maximum bending moment will occur at the center of the rib. Summing the moments at the center of a rib,

$$M = (1{,}800 \text{ lb})(30 \text{ in.}) - (1{,}200 \text{ lb})(24 \text{ in.})$$
$$= 25{,}200 \text{ in.-lb}$$

An alternative method of calculating the maximum moment is by using the equation for maximum moment of beam 5 in Table 5-2 as follows:

$$M = P[l/4 + a]$$
$$= 1{,}200[60/4 + 6]$$
$$= 25{,}200 \text{ in.-lb}$$

Consider 2×10 No. 2 grade Southern Pine with a load duration less than 7 days; Tables 4-2 and 4-4 give the allowable stresses:

Allowable bending stress, $F_b = C_D \times$ (reference bending stress)
$$= 1.25(1{,}050)$$
$$= 1{,}312 \text{ lb per sq in.}$$

Allowable shear stress, $F_v = C_D \times$ (reference shear stress)
$$= 1.25(175)$$
$$= 218 \text{ lb per sq in.}$$

The strength requirements for bending and shear will be considered, but deflection criteria will not be considered. Following are checks for bending and shear. For bending, the required section modulus can be calculated as follows:

From Eq. (5-11), $S = M/F_b$
$$= 25{,}200/1{,}312$$
$$= 19.2 \text{ in.}^3$$

The rib should have a section modulus of at least 19.2 in.^{3} An inspection of Table 4-1 indicates that a 2×10 S4S section has a section modulus of 21.39 in.^{3}, which is greater than 19.2. Therefore the 2×10 is adequate for bending.

The rib must also have adequate shear strength. The maximum shear force will occur at the support and is equal to 1,800 lb. The shear stress in the rib can be calculated as follows:

From Eq. (5-13), $f_v = 3V/2bd$
$$= 3(1{,}800 \text{ lb})/2(1.5 \text{ in.})(9.25 \text{ in.})$$
$$= 194.6 \text{ lb per sq in.}$$

Because the allowable shear stress F_v of 218 lb per sq in. is larger than the applied shear stress of 194.5 lb per sq in., the 2 × 10 S4S size rib is satisfactory in shear. However, due to required curvature the rib members must be sawn from 2 × 12 size lumber. Therefore the stresses must be checked for a 2 × 12, and the net depth of the member must be checked to ensure it is adequate.

The maximum depth that will be removed from the ends of a rib member can be determined by applying Eq. (13-3). For the required curve, R is approximately 24.5 ft and x is 2.5 ft, measured from the midpoint of a member. Let the depth to be removed equal y.

$$y = R - \sqrt{R^2 - x^2}$$
$$= 24.5 - \sqrt{(24.5)^2 - (2.5)^2}$$
$$= 0.128 \text{ ft, or } 1.54 \text{ in.}$$

The loss in depth at the ends of a 2 × 12 rib member will be 1.54 in., which will leave a net depth of approximately 11.25 in. − 1.54 in. = 9.7 in. Because the 9.7 in. is greater than the 9.25 in. for a 2 × 10, it is satisfactory. Because the 2 × 10 will be cut from a 2 × 12, the allowable bending stress for a 2 × 12 must be checked against the section modulus for a 2 × 10. The allowable bending stress from Tables 4-2 and 4-4 will be:

Allowable bending stress, $F_b = C_D \times$ (reference bending stress)
$$= 1.25(975)$$
$$= 1{,}218 \text{ lb per sq in.}$$

The actual bending stress can be calculated from Eq. (5-7) as follows:

Actual bending stress, $f_b = M/S$
$$= 25{,}200 \text{ in.-lb}/21.39 \text{ in}^3$$
$$= 1{,}178 \text{ lb per sq in.}$$

The allowable bending stress of 1,218 lb per sq in. is greater than the applied bending stress of 1,178 lb per sq in. Therefore cutting the curvature from 2 × 12 to make a 2 × 10 rib is satisfactory.

Determine the Load on the Shores

The circular shell roof formwork system in Figure 13-2 will be spaced at 5 ft in the horizontal direction. The spacing of shores in the 30-ft direction will be 6 ft on center. Therefore the area supported by each shore will be 5 ft × 6 ft = 30 sq ft. The load will be:

$$P = (100 \text{ lb per sq ft})(30 \text{ sq ft})$$
$$= 3{,}000 \text{ lb}$$

This is the vertical load on the interior shores. For the outer shores, the load will be larger because of the slope of the roof. However, shores with capacities of 4,000 lb will be satisfactory. Shores must be adequately braced.

Determine the Elevations of the Top of the Decking

The elevations of critical points on the top of the decking should be determined. These points are 0, 5, 10, and 15 ft, measured horizontally, from the centerline of the decking. Equation (13-3) gives the vertical distance from the crown, or centerline, down to these points. R is equal to 25 ft.

$$y = R - \sqrt{R^2 - x^2}$$

For $x = 0$, $y = 0$

For $x = 5$ ft, $y = 25 - \sqrt{625 - 25} = 0.506$ ft

For $x = 10$ ft, $y = 25 - \sqrt{625 - 100} = 2.087$ ft

For $x = 15$ ft, $y = 25 - \sqrt{625 - 225} = 5.000$ ft

Determine the Slope of the Decking at the Outer Edges

It is desirable to know the slope of the decking at the outer edges, where the slope is a maximum, to determine whether it will be necessary to use top forms for a portion of the concrete slab. Concrete with a low slump may be placed without a top form for slopes up to approximately 35°, but for steeper slopes a form may be necessary. Let a equal the angle between a tangent to the decking at the outer edge of the roof and a horizontal line. Reference to Figure 13-1 shows that

$$\sin a = l/R \qquad \text{(a)}$$

where $l = (30 \text{ ft})/2$

$= 15$ ft

and $R = 25$ ft

Substituting these values in Eq. (a) gives

$$\sin a = 15/25 = 0.600$$

$$\text{arcsin } a = 0.60$$

$$a = 36° \, 52'$$

It is possible that a concrete with a low slump can be placed without a top form.

Centering for Shell Roofs

Figure 13-3 showed wood shores used as centering to support the forms for a shell roof. Adjustable patented shores equipped with U-heads to receive the stringers have several advantages when compared

FIGURE 13-5   Formwork for a circular shell roof supported by tubular-steel-frame scaffolds. (*Source: Safway Steel Products*)

with wood shores, as discussed in Chapter 6. Using shores with screw-type adjustments facilitates setting the heights. Also, lowering the forms for removal can be accomplished easily and quickly. Patented shores should be also be adequately braced.

Figure 13-5 illustrates the use of tubular scaffolding as centering for a shell roof. Where these frames are interconnected with both horizontal and diagonal braces, they form a rigid system that is capable of supporting substantial loads. The top frames should be equipped with adjustable U-heads to support the stringers or ribs under the decking at the exact elevations required.

In the event that the forms are to be lowered and moved to a new location for reuse, the bottom legs of the forms can be equipped with casters onto which the frames may be lowered during the moving operation only. This will permit an entire assembly of forms and centering to be moved and reset for additional use within a relatively short time period.

Use of Trusses as Centering

Wood or steel bowstring trusses have been used as centering for forms for circular shell roofs. Each end of a truss is supported by a post or column whose length may be adjusted for exact height. Wood joists, extending from truss to truss, support the decking. An entire form assembly may be constructed and erected in such a way that it can be lowered and moved to a new location for reuse within a short time.

Figure 13-6 shows the seven trusses required for one section of forms in position for the installation of joists and decking. Figure 13-7 shows two sections of the forms lowered for moving.

Figure 13-6   Wood trusses installed to support formwork for circular shell roof. (*Source: Construction Methods and Equipment*)

Figure 13-7   Formwork for a circular shell roof lowered for moving to new location. (*Source: Construction Methods and Equipment*)

Decentering and Form Removal

Decentering and removal of formwork for thin-shell roof slabs is more important than removal of shoring and formwork for walls and floor slabs in the interior of a building. Upon removal of centering and forms, the roof slab may be subjected to the permanent live load of wind, whereas the permanent live load on a floor slab will not likely occur until the full 28-day strength of the concrete is attained.

Due to high deflections and high dead-load to live-load ratios that are common in thin-shell roof slabs, decentering and removal of formwork should not be permitted until appropriate tests have been performed to demonstrate that the specified concrete strength and modulus of elasticity have been achieved. It is possible that adequate compressive strength may be attained, but the modulus of elasticity may not be adequate. ACI 318 requires the use of field-cured beam test specimens if form removal time is based on development of a specified modulus of elasticity. The value of modulus of elasticity must be obtained from flexural testing of field-cured beam specimens, not as a proportion of the compressive strength from the beam test specimen.

Decentering methods should be planned to prevent any concentrated reaction on any part of the permanent structure. Decentering should begin at locations of maximum deflection and progress to locations of minimum deflection, with decentering of edge members proceeding simultaneously with the adjoining shell.

CHAPTER 14

Forms for Architectural Concrete

Forms for Architectural versus Structural Concrete

When concrete is used for structural purposes only and the appearance of the surface is not of primary importance, the essential requirements of the forms are strength, rigidity, and economy. This permits greater freedom in the choice of materials for forms, and the demands of quality of workmanship in fabricating and erecting forms are not as great as for forms used to produce architectural concrete.

Architectural concrete differs from structural concrete in that the appearance or color of the exposed surfaces of the forms may be more important than the strength of the members. The properties of fresh concrete are such that it may be cast or molded to produce any shape that forms can be made. A variety of colors may also be obtained by adding a color admixture to the concrete mix or by adding a surface coating after the concrete has hardened.

The surfaces may be extremely smooth or they may be quite rough, depending on the desired effects. The appearances of large and possibly drab areas can be improved greatly by the use of recessed or raised panels, rustications, ornaments, and other designs. Metal molds can produce concrete having extremely smooth surfaces. A variety of surface patterns can be obtained using form liners that are placed on the inside of plywood sheathing. A large variety of shapes is available from companies that supply formwork accessories. The use of milled wood forms and waste molds permits the forming of concrete with a great variety of intricate details.

Because the quality of the finished concrete is limited by the quality of the forms, it is necessary to exercise care in selecting the materials for the forms. Also, high-quality craftsmanship is required in building and stripping the forms of architectural concrete if the desired effects are to be achieved with a satisfactory degree of perfection. It may be desirable to provide 2- by 2-ft samples to the worker before starting construction to define the quality of workmanship that is required of the final product.

Concrete Coloring

Concrete that is mixed with normal cement has the traditional gray color. For architectural concrete, a wide variety of colors may be achieved by adding admixtures to the concrete mix or by adding a color hardener to the surface of the concrete after it is placed.

Colored, water-reducing admixtures are added to the concrete mix to produce a permanent nonfading uniform color. Admixtures are available in a natural range of earth-tone colors, from rich grays to warm reds and colored buffs. Custom-matched colors may also be obtained for individual designs.

A multicolored textured stone-like appearance for floors may be achieved by treating freshly placed concrete. Specially graded stones, selected for attractiveness and brilliance, are intermingled in a granulated colored cementitious material. When incorporated into the surface of freshly placed concrete and exposed by sandblasting, the result is a wear-resistant surface of the desired color. The appearance of granites or marbles may be achieved in a variety of colors, from pinks with undertones of black and green to earthen hues with creamy foregrounds.

Floor hardeners are dry-shake hardeners for freshly placed concrete flatwork. These hardeners are often used to form patterned or unique shapes for commercial floors, around pools, and for imprinted concrete paving. Such hardeners also provide the added benefit of reduced scaling from freeze/thaw cycles and salts.

Stained Concrete

Stained concrete provides an unlimited variety of coloring for concrete floor slabs. The desirability of acid stains is that the resulted are varied and often unpredictable because the chemistry of concrete and condition surrounding its placement has a large impact on the outcome. Patterns or textures can be applied before the cement sets to further enhance the final appearance of the concrete.

Acid stains are only one way to color hardened concrete, but it is the most permanent coloring method because it has a high resistance to abrasive wear and resistance to ultraviolet light damage. Stains consist primarily of metallic salt compounds, water, and muriatic acid, which react with concrete. Types II and V cements are best suitable for stain applications.

The surface area of the concrete is cleaned of grease or any other discoloration on the concrete. The stain may be applied by a brush, mop, roller, or sprayer to cover the concrete work area. Different results will be obtained; depending on the method the stain is applied.

For example, a sprayer will produce a diffused pattern, whereas an acid brush yields in a deeper, richer coat. A darker appearance can be obtained by applying additional coats of stains. A short-napped roller will ensure additional texture.

Acidic residue is removed after the stain has been allowed to dry and cure. The procedure for removing acidic residue is generally provided by the manufacturer of the concrete stain product. After the staining procedure is completed the concrete surface is sealed using a brush, roller, or sprayer.

Stamped Concrete

Stamped or patterned concrete involves pouring a slab concrete and then impressing both patterns and textures onto the concrete before it fully sets. Stamped concrete can produce textures that duplicate many different surfaces, such as cobblestones, brick, pavers, or wood. Also, pigment can be added to stamped concrete to further enhance the appearance.

The maximum size aggregate for most concrete is 1 to 1½ in. These size aggregates do not work well for stamped or patterned concrete. The maximum size of aggregate should not exceed ⅜ in. It is desirable to use round aggregate, sometime called pea gravel, for stamped concrete.

Stamped concrete should be no less than 4 in. thick. The concrete is placed, screeded, and bull-floated just like ordinary concrete. The concrete may be troweled before the stamping process. However, it should not be troweled more than once; otherwise the concrete may bleed water to the surface.

If it is desired to use coloring in the concrete, the color is added before the stamping process is started. Coloration can be achieved by ordering colored concrete or adding the color after the concrete has arrived at the job. Color added at the jobsite is introduced by shaking special dried pigments on top of the wet concrete. The pigments are then troweled into the concrete. The pigments penetrate ⅛ to ¼ in. into the concrete surface. The concrete can also be stained after it is placed.

Stamps are usually made of polyurethane material to resemble ordinary building products, such as flagstone, brick, or natural stone. The stamps are pushed into the concrete and then removed to leave the pattern. The freshly placed concrete must have sufficient hardness so the stamp will not sink, but must be soft enough for a clean stamp. A test of the stamp on a demonstration location should be performed before applying the stamp on the permanent surface.

Form Liners

Architectural concrete form liners may be used to create virtually unlimited textured or sculpted effects in concrete surfaces. The surfaces may simulate wood grains, rough brick, blocks, or an irregular texture, as illustrated in Figure 14-1. Form liners may be rigid or flexible.

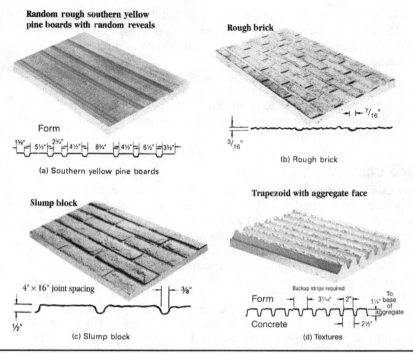

FIGURE 14-1   Illustrative patterns of form liners. (*Source: Greenstreak Plastic Products Company, Inc.*)

Sheets of rigid form liners are attached to the formwork or casting bed prior to placing the concrete. Screws or nails are recommended for mounting form liners. Following placement and normal curing time, the formwork and liner are stripped, leaving a textured concrete surface. The sheets generally are 4 ft wide and 10 ft long with square edges to allow the placement of adjacent sheets for large areas of concrete. The hard, void-free surface will not absorb moisture or cause discoloration. Sheets of rigid form liners are lightweight and easily stripped.

Rustication or reveal strips are recommended at the liner joints that do not blend with the pattern. A properly sized rustication will complement the pattern and enhance the overall appearance of the structure. For uniformity of color and texture, the concrete should be secured from a single supplier so that all of the ingredients will come from the same source.

A tremie should be used for placing the concrete to minimize aggregate segregation. The concrete must be properly vibrated to eliminate lift lines and minimize air voids. Pumping the concrete into the formwork from the bottom will generally reduce air voids in the surface of the concrete. This method will also raise form pressures significantly, which may damage the form liner.

A releasing agent should be used to minimize stripping forces and shorten form cleanup between pours. The release agent should be applied before each pour.

Sealing Form Liner Joints

All form liner joints and tie holes should be sealed to prevent localized water loss and subsequent discoloration of the concrete. Silicon sealant is recommended for cast-in-place jobs. Once cured, it is flexible, has good adhesion, and will not discolor or stick to the concrete. Sealant or putty on fastener heads will hide them in the finished concrete.

Heavy duct tape or double-sided foam tape stabilizes and seals the joint when applied to the formwork side of the form liner. Auto body putty is good for precast or other applications where the formwork is not flexed or moved.

Smooth-Surfaced Concrete

Exceptionally smooth concrete surfaces are generally more difficult to achieve than a textured surface. If a smooth surface is desired, patented metal forms, hardboards, or high-density overlay (HDO) Plyform sheathing are usually used. Even though the individual panels may provide smooth flat surfaces, the effect of joints appearing between panels may mar the surface of the concrete. Thus, special care must be exercised during erection of the panel forms to ensure smooth joints between the form panels.

Patented metal form panels are frequently used for continuous straight or curved concrete surfaces. They are manufactured with heavy-gauge metal to prevent waves in the surface. Because they are jig built in a factory with precision dimensions, the panel joints facilitate alignment and fit on the job. The panels are also generally large so that there will be a minimum of linear feet of joints.

The properties of plywood are discussed in Chapter 4. Plyform is the special product designed for use in forming concrete. It is exterior-type plywood limited to certain species of wood and grades of veneer to ensure high performance as a form material. It is especially useful for curved smooth surfaces. Table 4-19 provides the minimum bending radii for dry Plyform.

Hardboard

Hardboard, whose properties are discussed in Chapter 4, is frequently used to line forms for architectural concrete where smooth surfaces, entirely free from grain markings, are desired. The physical properties are shown in Table 4-20 and the minimum bending radii are listed in Table 4-21. Only tempered boards that have been treated to minimize absorption should be used with formwork.

The edges of adjacent sheets should be nailed to the same backing boards to prevent slight offsets that may accentuate the joints. Blue 3d shingle nails or other nails with thin, flat heads and thin shanks should be used to fasten the sheets to the backup lumber.

Joints between adjacent sheets may be filled with plaster, putty, or tape. A light sanding with No. 0 sandpaper will make the joint smooth and practically invisible. Holes for form ties should be drilled from the face side of the form, the smooth side of the hardboard, with a worm-center bit to avoid tearing the board. The hardboard surface should be oiled, and the material should also be thoroughly wet several hours before being used.

Wetting and Oiling Forms

Wood forms that will be in contact with concrete should be wetted thoroughly with water at least 12 hours before the concrete is placed. This tends to close the joints in sheathing and decking, prevents the absorption of water from the concrete, and ensures easier stripping.

Plywood and hardboard forms should be oiled or treated with a suitable bond-breaking agent before they are used. Excess oil must be avoided. Bond breakers or form oil should be applied before panels are erected because if the treatment is applied after the forms are set in place, the bonding of the concrete and reinforcing steel may become contaminated.

Nails for Forms

Most wood formwork is assembled with nails. The structural members should be fastened together with nails of sufficient sizes and quantities to ensure adequate strength. It is proper to use common nails for this purpose. However, box or thin-shank nails should be used to attach sheathing to studs, to attach plywood or hardboard linings, and to attach wood molds to sheathing. The thin shank permits easy stripping with minimal damage to the lumber and reduces the cost of labor required to strip the forms.

Form Ties

Various types of form ties are described in Chapter 9. For architectural concrete it is usually specified that form ties be withdrawn from the concrete. In the event that a portion is left in the concrete, no part of the tie shall be closer than 1½ to 2 in. from the surface of the concrete. Also, limitations may be specified for the sizes of the holes left in the concrete when the ties are removed. Figure 14-2 illustrates a coil tie with a loose (spreader) cone. The loose cones are used where architectural concrete finish is specified.

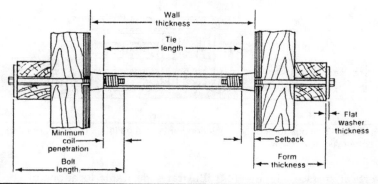

FIGURE 14-2 Coil tie with loose (spreader) cones for architectural concrete. (*Source: Dayton Superior Corporation*)

Construction Joints

All concrete structures, except possibly small and simple types, require construction joints, which serve several purposes. Concrete undergoes shrinkage as it sets and cures, which usually induces tension stresses. If these stresses are excessive, and if no provision is made for relieving them, it is probable that cracks will occur. Such cracks are unsightly, and they cannot be patched satisfactorily. Cracks are especially objectionable if they appear in the surfaces of architectural concrete.

Because forms for architectural concrete frequently contain molds and other ornament-producing features, excessive stresses or movements in the concrete resulting from shrinkage may damage or destroy some of the ornaments. These stresses and movements can be minimized by locating construction joints reasonably close together. For example, if concrete heads above windows are placed in one operation, it is probable that the shrinkage will cause cracks to appear at their tops. For this reason, it is good practice to provide horizontal construction joints directly under the window heads.

Another reason for using construction joints is that concrete for most structures is placed in stages, rather than continuously. Therefore, it is desirable to provide construction joints to control the appearance at each stage of concrete placement. When an architect or an engineer designs a concrete structure, he or she usually designates the locations of construction joints, giving consideration to the effects on the appearance of the structure and also to the quantity of concrete required between construction joints.

Because the surface appearance of architectural concrete is of significant importance, the location of construction joints should be selected carefully. Frequently, joints may be located so that they are inconspicuous. Horizontal joints may be located along the sills or heads of windows, or they may coincide with recesses in the concrete surface extending around the structure. Vertical construction joints may be located along the edges of pilasters or other intersections of

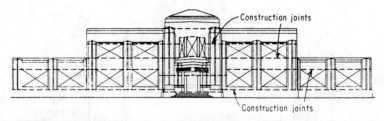

Figure 14-3   Locations of construction joints in a building. (*Source: Portland Cement Association*)

vertical surfaces. Figure 14-3 illustrates desirable locations of construction joints for a building.

Detailing Forms

Specifications for architectural concrete frequently require the contractor to submit, for approval by the architect or engineer, large-scale drawings showing in detail the formwork that he or she plans to use in casting the various members or units. Regardless of such requirements, fully dimensioned plans showing elevations, sections, and other information are desirable for use by the carpenters or mill workers who make and assemble the component parts of the formwork. These plans should show the exact dimensions and locations of all features that are to appear in or on the concrete. Figure 14-4 illustrates a set of detail plans for the formwork for the frieze around a concrete building.

Figure 14-5 illustrates a form detail showing a horizontal section through the wall of a portion of a concrete building, extending from the entrance at the left of Figure 14-5(a) to the corner at the right side of Figure 14-5(b). Exact sizes, shapes, and locations of all details must be shown on the final drawings.

Order of Erecting Forms for a Building

The order of erecting the forms for a building should be scheduled before materials are ordered and the formwork is fabricated. The materials that will be erected first should be ordered and fabricated first. The schedule selected will depend largely on the type of building to be constructed. There are three general plans for erecting formwork for structures using architectural concrete.

Plan 1: For one- and two-story buildings and the lower stories of tall buildings, which are usually more highly ornamented than the stories above, the following procedure may be used:

1. Erect the outside wall forms and bring them to alignment.
2. Erect the inside wall forms and the floor forms.
3. Bring all forms to the proper alignment, tighten the braces, and secure all form parts.

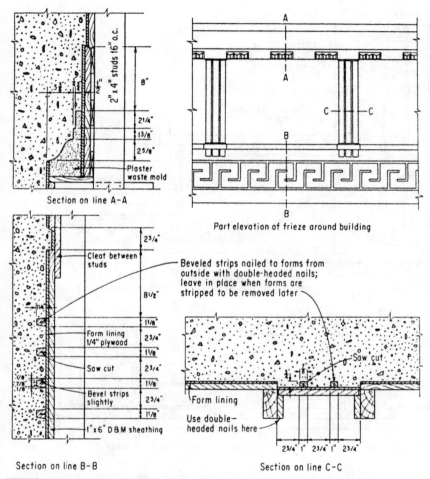

FIGURE 14-4   Details of forms to produce designs on architectural concrete. (*Source: Portland Cement Association*)

Plan 2: For buildings consisting principally of columns and spandrels, which are generally constructed with prefabricated panel forms that can be handled most conveniently from the deck, the following schedule may be used:

1. Erect the inside wall forms and the floor forms.
2. Erect the outside wall forms.
3. Bring all forms to the proper alignment, then brace and securely fasten all form parts.

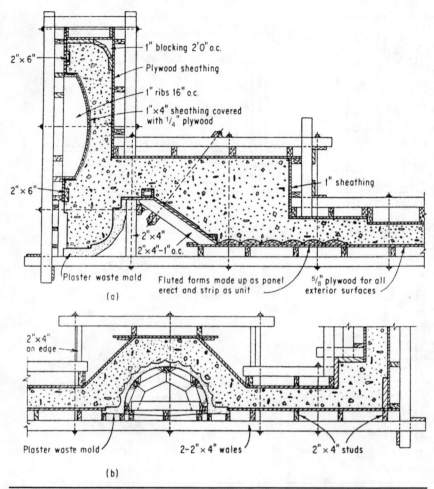

FIGURE 14-5 Details of forms to produce architectural concrete. (*Source: Portland Cement Association*)

Plan 3: For tall buildings requiring considerable ornamentation and necessitating the use of waste molds, milled wood molds, and other special forms, the following schedule may be used:

1. Erect the floor forms.

2. Erect the outside wall forms and bring them into approximate alignment.

3. Touch up the outside forms and fill any open joints in plaster or wood molds, if necessary.

4. Erect the inside wall forms.

5. Bring all forms to the proper alignment, then brace and securely fasten all form parts.

Order of Stripping Forms

At the time the details for making forms are prepared, the order and methods of stripping should be considered. Unless care is exercised in making and erecting forms, it may be discovered that some parts, panels, or members cannot be removed without wholly or partly destroying them. Because many panels are to be removed and reused, it is important to erect them in a manner that will permit their removal without damage. Parts or panels that are to be removed first should not be placed behind other parts that will be removed later. Such parts should be fastened together with double-headed nails, which can be removed easily.

Wood Molds

Wood molds are used as formwork to cast ornamental designs for cornices, pilasters, belt courses, water tables, and other members for architectural concrete. These molds are best adapted to ornaments requiring the use of standard moldings or moldings that can be made in a mill. A form unit consists of an assembly of several parts and pieces to produce the specified profile or design for the concrete, as illustrated in Figure 14-6.

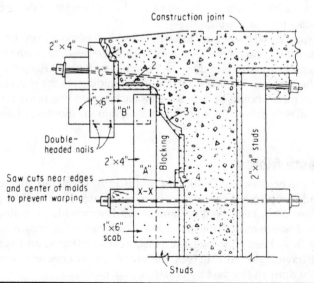

FIGURE 14-6 An assembly of wood molds. (*Source: Portland Cement Association*)

Much time can be saved in the erection and stripping of the forms for a detail involving many pieces of run moldings if brackets are made in the mill to a template to fit the general profile of the detail. The section in Figure 14-6 illustrates this point. The studs for the wall forms are cut off at the line X-X. The brackets, consisting of pieces A, B, and C, which have been assembled in the mill, are scabbed to the studs. The wales bearing on piece A and the lower half of the wale bearing on piece C are put in place to hold the brackets, which are spaced about 16 in. apart, in alignment. The cornice members are then applied. Pieces 1, 2, 3, and 4 are moldings, and all other pieces are ripped to size from stock lumber.

Even when wood is carefully oiled prior to its use with formwork, there is a tendency for it to swell when it is wet. For this reason, consideration should be given to the selection of the types, sizes, and arrangements of the component parts when detailing wood molds. Because wide and thick molds will swell more than narrow and thin molds, the former are more likely to bind and damage the ornaments in the concrete. Thus, it is desirable to use the narrowest and thinnest pieces from which the mold can be made with adequate strength. The danger of molding swelling and damaging the concrete details may be reduced by using longitudinal saw cuts along the back sides of the pieces, as specified in Figure 14-6.

Recesses in concrete, made by using narrow strips of wood molding, should always be beveled to give a width at the surface greater than that at the back. If this is not done, it will be necessary to split the strip with a wood chisel prior to removing it, which may damage the concrete. Even though the strip of molding is beveled, it is good practice to run a longitudinal saw cut along the full length of the back side of it to facilitate stripping.

Frequently it is necessary to leave individual pieces of molding in the concrete for several days after the molds are removed, to permit them to dry out and shrink so that they can be removed without danger of damaging the concrete. If this is to be done, the moldings or the parts that will be left in place should be nailed to the backing lumber, using relatively few small nails driven through the backing into the molding pieces.

Plaster Waste Molds

Where the ornamental details for architectural concrete are so intricate and complicated that it is impossible to use wood molds, plaster molds are used. Because these molds are destroyed in stripping, they are called waste molds. Plaster waste molds are not as common today as they have been in the past. Casting resins, plastics, and synthetic rubber have replaced much of the architectural concrete work that has been done in the past by plaster waste molds.

Waste molds are made of casting plaster containing jute fiber, and they are reinforced to prevent breakage during handling and erection.

A full-size model of the ornament is made of wood, plaster, or some other suitable material. Then, using this model as a pattern, the waste mold is cut, either as a single unit or in parts to be assembled later. If a single-piece mold will be too heavy for easy erection, the mold should be cast in two or more parts. The back of a mold should be shaped to permit it to fit accurately against the backing forms, sheathing, studs, wales, or blocks. Figure 14-7 illustrates form assemblies using waste molds and backup lumber.

Some molds are attached to the structural members of forms with nails that are driven through the plaster into the lumber. The nail heads are countersunk, and the holes are pointed with patching plaster. If the plaster is too thick to permit the use of nails to fasten the molds to the form members, holes may be drilled through the plaster to permit the use of wires for fastening purposes. The wires can be embedded in the plaster and covered with patching plaster to restore the original surface condition to the mold.

It is usually necessary to destroy a waste mold when it is stripped from the concrete, which may require some chipping with a cold

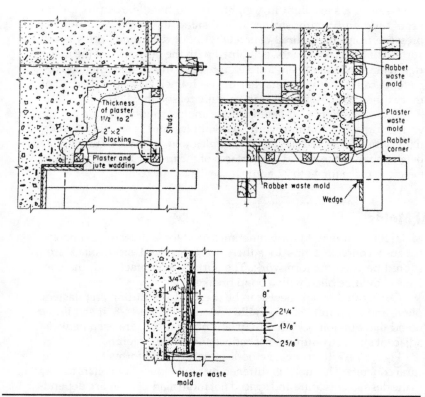

Figure 14-7   Plaster waste molds. (*Source: Portland Cement Association*)

chisel. It may be desirable to use a thin layer of colored plaster next to the concrete as a warning, to prevent cutting into the concrete during the chipping operation.

Because concrete will adhere to the untreated surface of plaster, it is necessary to paint the contact surface of a waste mold with shellac prior to placing the concrete. Any pointing of joints or patchwork should be sanded smooth and painted with shellac. Also, all contact surfaces of a mold should be greased with a light coat of grease, which may be thinned with kerosene if it is too thick. Any surplus must be wiped off.

Plastic Molds

Manufacturers of plastic products have perfected methods of making plastic molds for use with architectural concrete. The designs, which are attached to back-up sheathing as form liners, may vary in individual sizes from less than 1 ft square to approximately 4 by 7 ft. Several liner sheets may be installed adjacent to each other to produce larger designs.

These liners are light in weight, easy to handle, rustproof, dent proof, and sufficiently strong to withstand the weight of, and the pressure from, the poured concrete without distortion. The smooth surface of the plastic liners, combined with the flexibility of the material, will permit the liners to release readily from the hardened concrete. Because, with reasonable care, each liner may be reused 10 or more times, the cost per use can be quite reasonable.

The concrete placed against these molds should contain an air-entraining agent to improve its workability, and it should be internally vibrated to release all air bubbles and to assure full contact between the concrete and the surfaces of the molds. The liners should be left in place for 48 to 72 hours.

Metal Molds

Metal forms and molds are sometimes used for architectural concrete. Because concrete tends to adhere to galvanized steel, black iron should be used for formwork. The surface in contact with the concrete should be oiled with a bond breaker.

Corrugated metal sheets can be used to form fluting on pilasters, piers, and spandrel beams, as illustrated in Figure 14-8. It should be noted that butt joints, rather than laps of the metal, are used between adjacent sheets to eliminate or reduce the effects of joints.

Figure 14-9 illustrates methods of using metal forms for architectural concrete. The metal is stiffened by wood blocks or collars sawed to the desired curvature and spaced not more than 12 in. apart, depending on the gauge of metal used.

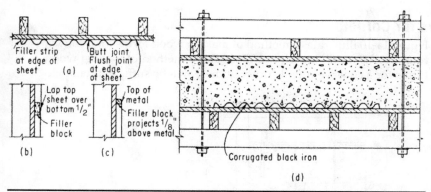

FIGURE 14-8 Fluting formed with corrugated iron. (*Source: Portland Cement Association*)

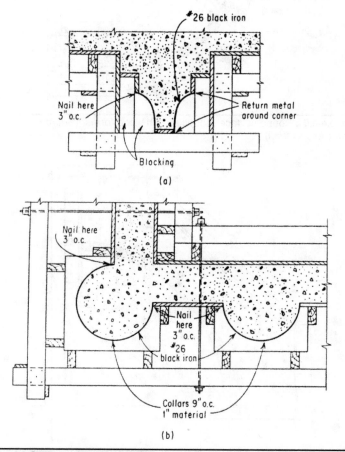

FIGURE 14-9 Round surfaces produced with metal forms. (*Source: Portland Cement Association*)

Forms for Corners

Figure 14-10 illustrates a method of making wood forms for a convex round corner and Figure 14-11 illustrates a method of making wood forms for a concave round corner. If there will be sufficient reuses to

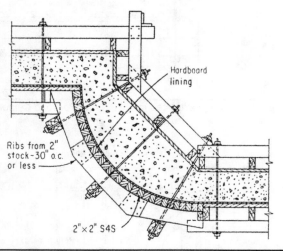

FIGURE 14-10   Round surface produced with wood forms. (*Source: Portland Cement Association*)

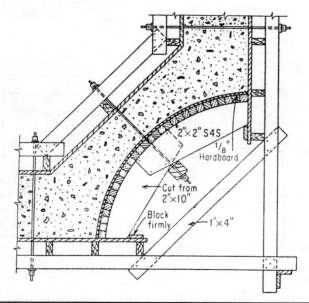

FIGURE 14-11   Round surface produced with wood forms. (*Source: Portland Cement Association*)

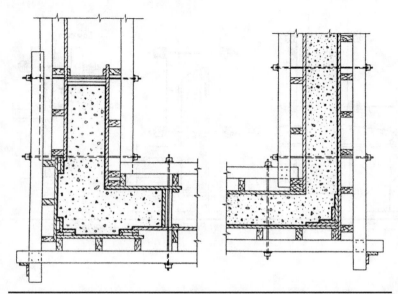

Figure 14-12 Forms for producing special corner designs. (*Source: Portland Cement Association*)

justify the possible higher initial cost, consideration might be given to using sheet-metal forms, with wood ribs sawed to the required curvature, spaced not more than 12 in. apart, and used to back up the metal to prevent distortion. Figure 14-12 illustrates two types of form-work for producing special corner designs in buildings or other concrete structures.

Forms for Parapets

The forms for several types of parapet walls are illustrated in Figure 14-13. On the exposed sides of the walls the forms are constructed to reproduce the specified details and ornaments. The forms for the back wall may be supported by 1- by 4-in. pieces of lumber, pointed on the lower ends, which rest on the slab decking. The supporting pieces are removed before the concrete slab hardens, and the holes are filled with concrete.

Forms for Roof Members

Figures 14-14 and 14-15 illustrate the pleasing aesthetics that can be achieved by using plywood in making forms for roof members.

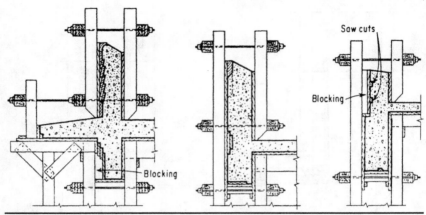

Figure 14-13   Forms for parapets. (*Source: Portland Cement Association*)

Figure 14-14   Plywood sheathing used to cast concrete roof member. (*Source: APA—The Engineered Wood Association*)

FIGURE 14-15 Concrete roof cast on plywood sheathing. (*Source: APA—The Engineered Wood Association*)

References

1. Gates & Sons, Inc., Denver, CO.
2. APA—The Engineered Wood Association, Tacoma, WA.
3. Greenstreak Plastic Products Company, Inc., St. Louis, MO.
4. "Forms for Architectural Concrete," Portland Cement Association, Skokie, IL.
5. "Architectural Concrete for Small Buildings," Portland Cement Association, Skokie, IL.

Slipforms

Introduction

The slipform method of concrete construction is used for forming both horizontal and vertical concrete structures. It is often used for forming highway pavements as a continuous operation. The slipform method has also been used for forming various types of vertical concrete structures, such as the 568-ft-high Reunion Tower in Dallas, Texas (Figure 15-1). Following are typical types of concrete structures that have been constructed with the slipform method:

1. Single-cell silos
2. Multi-cell silos
3. Buildings
4. Piers
5. Towers
6. Water reservoirs
7. Vertical shafts for tunnels and mines
8. Vertical shafts for missile launching bases
9. Chimneys

These are vertical structures, most of which have walls requiring slipforms with two surfaces to confine the concrete. However, some piers or sections of piers are solid concrete, requiring forms with external surfaces only. The concrete linings for vertical shafts are generally placed with the external surface against the natural earth or rock, requiring form linings for the inside surfaces only.

The Forms

The forms for the slipform method of construction consist of the following basic parts:

1. Sheathing
2. Wales or ribs

FIGURE 15-1
Reunion Tower
constructed by
slipform method.
(*Source: Sundt
Corporation*)

3. Yokes

4. Working platform or deck (one or more)

5. Suspended scaffolding

6. Lifting jacks

These essential parts of a slipform assembly are illustrated in Figure 15-2, which is a section through a wall under construction. This form is raised by hydraulic jacks furnished by Scanada International [1].

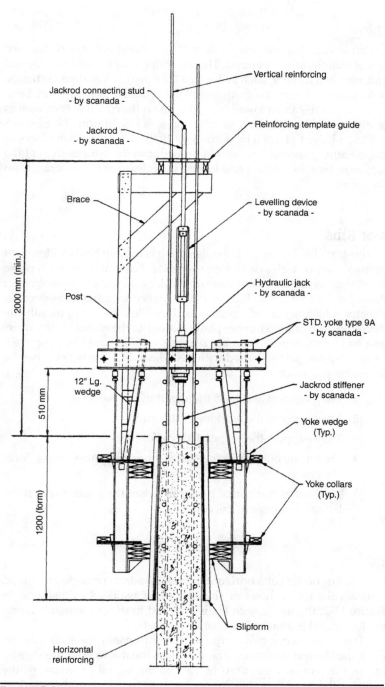

Vertical reinforcing

Jackrod connecting stud
- by scanada -

Reinforcing template guide

Jackrod
- by scanada -

Brace

Levelling device
- by scanada -

Post

Hydraulic jack
- by scanada -

2000 mm (min.)

STD. yoke type 9A
- by scanada -

12" Lg.
wedge

510 mm

Jackrod stiffener
- by scanada -

Yoke wedge
(Typ.)

Yoke collars
(Typ.)

1200 (form)

Slipform

Horizontal
reinforcing

FIGURE 15-2 Essential parts of a slipform assembly. (*Source: Scanada International, Inc.*)

Sheathing

For structures that include walls with inner and outer surfaces, two sets of sheathing are required. The sheathing may be made of dressed and matched lumber, such as 1 × 4 or 1 × 6 boards installed vertically, ¾-in. plywood with the grain installed vertically, or sheet steel. Steel has a longer life and a lower friction drag on the concrete than boards or plywood. The sheathing usually varies in height from 3 ft 6 in. to 5 ft 0 in.; a height of 4 ft 0 in. is commonly used. The opposite faces of the sheathing should be about ¼ in. wider at the bottom than at the top, to reduce the friction and to ease the release of the concrete from the forms.

Wales or Ribs

As illustrated in Figure 15-2, the sheathing is usually held in alignment by two rows of wales or ribs on each side. For structures with plane faces, such as buildings and piers, the wales are typically assemblies of three-ply 2 × 8 or 2 × 10 lumber planks nailed together. However, for structures having curved surfaces, such as silos, the wales usually are three-ply 2 × 6 or 2 × 8 lumber planks sawed to the required curvature. The end joints between lumber planks in a rib should be staggered. Steel wales are generally used with steel sheathing. The wales serve the following purposes:

1. They support and hold the sheathing in position.
2. They support the working platform.
3. They support the suspended scaffolding.
4. They transmit the lifting forces from the yokes to the form system.
5. They act as horizontal beams between the yokes to resist the lateral pressures of the concrete.

Yokes

Each yoke consists of a horizontal cross member connected to a jack, plus a yoke for each set of sheathing and wales. As illustrated in Figure 15-2, the top of each leg is attached to the cross member, and the lower end is attached to the bottom wale.

The yokes serve two purposes. They transmit the lifting forces from the jacks to the wales. The yoke legs must also hold the sheathing in the required positions and resist the lateral pressures of the concrete, because form ties cannot be used to maintain the required spacing between the form sheathings.

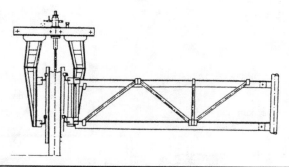

Figure 15-3   Structural members of a slipform assembly.

Working Platform

The working platform usually consists of 1-in.-thick boards or ¾-in.-thick plywood, supported by joists. The joists may be supported at the ends only by wales, or, for long spans, they may require intermediate supports by wood or steel trusses, steel joists, or steel beams, attached to the wales (Figure 15-3).

If the structure is to be finished with a concrete roof or cap, the working platform may be used as a form to support the concrete. For this purpose, pointed steel pins are driven through the sheathing into the concrete wall under the wales, and then the yokes are removed.

Suspended Scaffolding

The scaffolding suspended under the forms allows finishers to have access to the concrete surfaces, which usually require some finishing. The scaffolding should be assembled on the foundation and attached to the forms by means of wire rope slings before the placement of concrete has started. Safety railings are then added to the scaffolding after concrete placement has commenced and the form is raised to provide sufficient clearance. This procedure is normally performed without the need to temporarily stop concrete placement and form lifting (Figure 15-4).

Form Jacks

Jacks used to lift the forms are of three types: electric, hydraulic, and pneumatic. The jacks provide the forces required to pull the forms upward as the concrete is placed. Enough jacks must be used to lift the forms without excessive stresses on the jacks, yokes, and sections

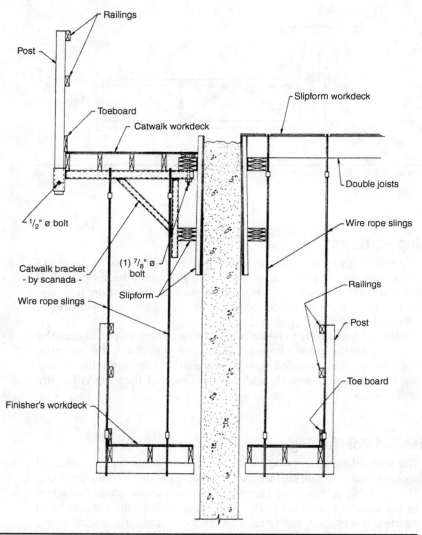

FIGURE 15-4 Details of working floor and scaffold assembly for slipforms. (*Source: Scanada International, Inc.*)

of the forms. If the jacks are overloaded, the upward movement of sections of the forms may not be uniform, which can cause distortions in the concrete structure. Usually, the jacks should be spaced from 6 to 8 ft apart along a wall.

A hydraulic jack is illustrated in Figure 15-2. A smooth steel jack rod, its lower end in the concrete, passes upward through the

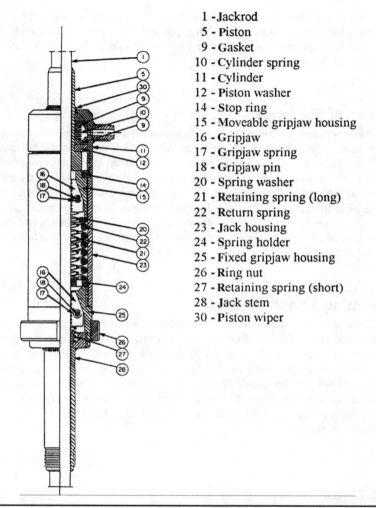

1 - Jackrod
5 - Piston
9 - Gasket
10 - Cylinder spring
11 - Cylinder
12 - Piston washer
14 - Stop ring
15 - Moveable gripjaw housing
16 - Gripjaw
17 - Gripjaw spring
18 - Gripjaw pin
20 - Spring washer
21 - Retaining spring (long)
22 - Return spring
23 - Jack housing
24 - Spring holder
25 - Fixed gripjaw housing
26 - Ring nut
27 - Retaining spring (short)
28 - Jack stem
30 - Piston wiper

FIGURE 15-5 Details of a hydraulic jack for lifting slipforms. (*Source: Scanada International, Inc.*)

hollow jack, which is attached to a yoke. When pressure is applied to a jack, one element of it grips the jack rod and another element moves upward, lifting the yoke with it (see Figure 15-5). When pressure is released temporarily, the jack resets itself automatically for another upward movement. Because all jacks are interconnected to a centrally located pump, the oil pressure on each jack will be the same, which should assure uniform upward movements. However, any given jack can be operated individually or taken out of operation, if this is desired, to bring sections of the forms to uniform elevations. One person can operate the electric motor that drives the pump, or the motor can be operated automatically to any set cycle.

Operation of Slipforms

After a set of slipforms is completely assembled on a concrete base, the forms are filled slowly with concrete. When the concrete in the bottom of the forms has gained sufficient rigidity, the upward movement of the forms is started and continued at a speed that is controlled by the rate at which the concrete sets. Lifting rates may vary from 2 or 3 in. per hr to in excess of 12 in. per hr, depending on the temperature and other properties of the concrete. An experienced person should be present at all time to establish the rate of movement. The reinforcing steel is placed as the forms move upward.

Constructing a Sandwich Wall

Where a wall is desired that has better insulating properties than solid concrete, it is possible to place insulating material between the inner and outer sections of the wall. This method is illustrated in Figure 15-6. As the forms move upward, boards of insulating materials are set into the guides above the concrete to provide a continuous insulating medium. Metal ties should be installed between the two sections of the wall at frequent intervals to hold the insulation boards together.

FIGURE 15-6
Method of
constructing a
sandwich wall
using slipforms.

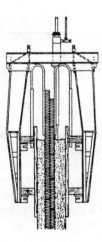

Silos and Mills

Concrete storage silos and feed processing buildings are the most common structures constructed with slipforms. Silos may be individual cells of various diameters, some exceeding 120 ft in height, or they may be interconnected multiple cells where all silos, sometimes 20 or more, are slipformed simultaneously.

Feed processing buildings are typically tall rectangular structures with multiple interior walls that divide the building into small square and rectangular bins. Figure 15-7 shows a completed multiple cell slipformed silo structure that is adjacent to a feedmill, which is in the process of being slipformed. These types of structures are usually cast monolithically in a continuous 24-hr operation, achieving average speeds of 12 in. per hr, or 24 ft per day.

FIGURE 15-7   Grain storage and mill building slipformed in South Sioux City, Nebraska. (*Source: Younglove Construction Company*)

Tall Bridge Piers

Figure 15-8 is a vertical section through a pier for a highway bridge across the Pecos River in Texas, which was constructed with slipforms. The bottom section has plan dimensions of 44 by 15 ft with

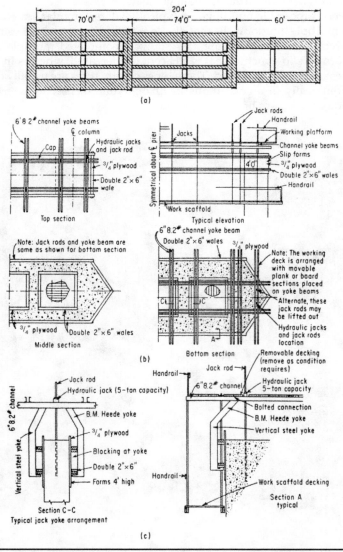

FIGURE **15-8** Bridge pier constructed with slipforms: (a) Section through piers. (b) Plan and elevation details of slipform construction. (c) Typical arrangement of jack yoke for slipform.

three 8- by 9-ft hollow vertical cells. The middle section has plan dimensions of 40 by 12 ft, with three 8- by 8-ft hollow vertical cells. The top section consists of two 8- by 7-ft solid rectangular columns constructed with slipforms. Two equal piers were constructed for the bridge.

Towers

The Reunion Tower in Dallas, Texas, is shown in Figure 15-9. Form-work for the tower consisted of three 14-ft-diameter elements around a central 27-ft-diameter core, which was slipformed as one unit. Three slipform decks, spaced 10 ft apart, were supported by horizontal and vertical trusses and by diagonal cable bracing to tie the three levels together. Crews working two shifts of about 9½ hours averaged more than 10 ft of progress per day. On the best day, the crew completed 18 ft of vertical height on the tower. The day shift slipformed and the night shift stocked material. The crane delivered steel and concrete to the top level, workers on the lower level received and placed the reinforcing steel, and concrete masons on the bottom level finished the tower legs. The tower was topped out in 68 working days.

FIGURE 15-9   Slipform assembly for 568-ft tower. (*Source: Sundt Corporation*)

Figure 15-10   CN Tower in Toronto, Canada: 1,815 ft total height, 1,476 ft slipform height.

Another example of the application of slipforming towers is the CN Tower in Toronto, Canada (Figure 15-10). At a total height of 1,815 ft, it is the world's tallest freestanding structure and holds the record as the tallest structure that has been built by the slipform method of construction, 1,476 ft of slipforming. The slipform was specially designed to be continuously adjustable during upward movement to achieve the graceful taper of the tower's triangular pedestal.

Concrete Buildings

Figure 15-11 illustrates the use of slipforms to construct the elevator and stair core of the ARCO II building in Denver, Colorado. This 45-story structure rose to a height of 544 ft in 50 working days. Unlike other slipformed structures, buildings are usually not slipformed on a continuous 24 hour per day schedule. Instead, the slipforms are

FIGURE 15-11   ARCO II Building in Denver, Colorado: 544-ft-high slipformed elevator shaft. (*Source: Sundt Corporation*)

Figure 15-12   Slipform core of 785 ft for Two Liberty Place in Philadelphia, Pennsylvania.

started and stopped each day and on weekends to avoid the high costs associated with continuous operations. Workers, equipment, and material delivery are planned to permit a production rate of one story height of walls per day. Figure 15-12 shows Two Liberty Place in Philadelphia, Pennsylvania. This structure is the world's tallest slipform core, a total height of 785 ft.

Linings for Shafts

Slipforms have been used to place the concrete linings for vertical shafts for tunnels and missile launching bases. The linings for the latter frequently must be constructed to tolerances of ⅛ in.

Figure 15-13 shows a section through the slipforms used to place concrete linings in vertical shafts located in White Sands, New Mexico, Victorville, California, and Helms Creek, California. These shafts ranged in diameters from 33 to 48 ft and in depths from 200 to over 400 ft. Special cable lifting jacks were used both to lower the slipforms

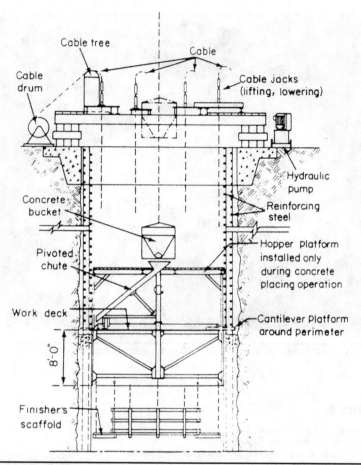

Cable tree

Cable

Cable
drum

Cable jacks
(lifting, lowering)

Hydraulic
pump

Concrete
bucket

Reinforcing
steel

Pivoted
chute

Hopper platform
installed only
during concrete
placing operation

Work deck

Cantilever Platform
around perimeter

8'-0"

Finisher's
scaffold

FIGURE 15-13   Schematic cross-section of slipform system for shaft work.
(*Source: Scanada International, Inc.*)

in order to permit preplacement of reinforcing steel and to lift the
slipforms for the placement of the concrete lining.

Slipforms for Special Structures

The slipform method of construction has been used successfully to
form special structures, such as the offshore platform shown in
Figure 15-14. In 1986, Norwegian contractors achieved a world
record by constructing the largest slipform project. The structure
has 30 cells, each with a diameter of 82 ft. The platform covers an
area of 182,920 sq ft and a linear edge length of 11,972 ft.

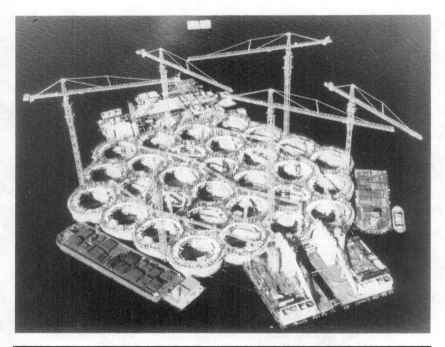

Figure 15-14 Slipformed oil platform with 30 cells, each 82 ft in diameter.

References

1. Scanada International, Bow, NH.
2. Sundt Corporation, Tucson, AZ.
3. Younglove Construction Company, Sioux City, IA.

CHAPTER 16

Forms for Concrete Bridge Decks

Wood Forms Suspended from Steel Beams

Figure 16-1 illustrates a method of suspending wood forms from steel beams that support a concrete bridge deck. Coil bolts attach into hanger frames resting on the tops of the beam flanges, and provide support to the double stringers that span between the steel beams. The single-span double stringers support the deck joists. The forms are stripped by removing the coil bolts and lowering the stringers. The decking and joists can be prefabricated in panels of convenient widths, then erected and removed intact for additional uses. The following example illustrates one method of designing the component parts of the formwork.

Example 16-1

Design the component parts of the formwork for Figure 16-1 for the following conditions:

Thickness of concrete slab, 8 in.

Center-to-center spacing of steel beams, 6 ft 6 in.

Clear span between steel beams, 5 ft 6 in.

Live load on forms, 75 lb per sq ft

Decking, ¾-in. Plyform Class I

Lumber will be No. 2 grade Southern Pine

Assume dry condition and load-duration less than 7 days

Permissible deflection, $l/360$

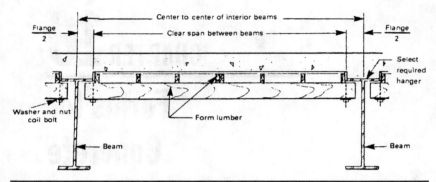

Figure 16-1   Forms for a concrete bridge slab. (*Source: Dayton Superior Corporation*)

Design of Decking to Resist Concrete Pressure and Live Loads

The total vertical pressure on the decking will be the dead load of the concrete plus the 75 lb per sq ft live load as follows:

$$\text{Concrete dead load } (150 \text{ lb per cu ft})(8/12 \text{ ft}) = 100 \text{ lb per sq ft}$$
$$\text{Live load on forms} = \underline{ 75 \text{ lb per sq ft}}$$
$$\text{Design pressure on decking, } w = 175 \text{ lb per sq ft}$$

The physical properties and allowable stresses for ¾-in.-thick Class I Plyform decking are shown in Tables 4-11 and 4-12. From Table 5-4, determine the spacing of the joists based on bending, shear, and deflection of the decking as follows:

For bending, from Eq. (5-41),

$$l_b = [120F_b S_e / w_b]^{1/2}$$
$$= [120(1,930)(0.455)/175]^{1/2}$$
$$= 24.5 \text{ in.}$$

For rolling shear, from Eq. (5-48),

$$l_s = 20F_s(Ib/Q)/w_s$$
$$= 20(72)(7.187)/175$$
$$= 59.1 \text{ in.}$$

Deflection criteria less than $l/360$, from Eq. (5-57a),

$$l_d = [1,743EI/360w_d]^{1/3}$$
$$= [1,743(1,650,000)(0.199)/360(175)]^{1/3}$$
$$= 20.9 \text{ in.}$$

Following is a summary for the decking:

For bending, the maximum span = 24.5 in.

For rolling shear, the maximum span = 59.1 in.

For deflection, the maximum span = 20.9 in.

For this example, deflection governs the spacing of the Plyform decking. For constructability, the joists will be spaced at 20 in. on centers.

Design of Joist to Support Decking

If the joists and the decking are to be prefabricated, 2×6 S4S joists may be desirable. However, if the joists are to be placed across stringers followed by placement of the decking on the joists, it may be desirable to use 4×4 or 4×6 S4S joists because the 4-in. width is more easily laid across the stringers and will stand in place without falling sideways. Also, the larger size joists will allow longer spans for the joist that will reduce the cost for the labor of erecting and removing the forms.

For this example assume that the joists and decking will be prefabricated in panels of convenient lengths. With joists spaced at 20 in. on centers, the uniform load on a joist can be calculated as follows:

$$w = (175 \text{ lb per sq ft})(20/12 \text{ ft})$$
$$= 292 \text{ lb per lin ft}$$

Consider 2×6 joists. The physical properties for S4S lumber are presented in Table 4-1. For dry condition and load-duration less than 7 days, the allowable stresses for No. 2 grade Southern Pine can be obtained from Tables 4-2 and 4-4 as follows:

Allowable bending stress, $F_b = C_D \times$ (reference bending stress)
$$= 1.25(1,250)$$
$$= 1,562 \text{ lb per sq in.}$$

Allowable shear stress, $F_v = C_D \times$ (reference shear stress)
$$= 1.25(175)$$
$$= 218 \text{ lb per sq in.}$$

Modulus of elasticity, $E = 1,600,000$ lb per sq in.

Determine the maximum permissible span length of 2×6 S4S joists (Table 5-3) as follows:
For bending, from Eq. (5-34),

$$l_b = [120F_bS/w]^{1/2}$$
$$= [120(1,562)(7.56)/292]^{1/2}$$
$$= 69.6 \text{ in.}$$

For shear, from Eq. (5-36),

$$l_v = 192F_vbd/15w + 2d$$
$$= 192(218)(1.5)(5.5)/15(292) + 2(5.5)$$
$$= 89.8 \text{ in.}$$

Deflection less than $l/360$, from Eq. (5-37a),

$$l_\Delta = [1,743EI/360w]^{1/3}$$
$$= [1,743(1,600,000)(20.8)/360(292)]^{1/3}$$
$$= 82.0 \text{ in.}$$

Below is a summary for the allowable span lengths of the joists:

For bending, the maximum span = 69.6 in.

For shear, the maximum span = 89.8 in.

For deflection, the maximum span = 82.0 in.

In this example, bending governs the allowable span length of the joists. For constructability, the double stringers must be spaced at 60 in., or 5 ft on centers.

Design of Stringers to Support Joists

The double stringers must support the joists. The loads applied to the stringers from the joists will be every 20 in. along the stringer. The magnitude of these loads can be calculated by multiplying the design pressure by the spacing and span length of the joists as follows:

Point load on stringers, $P = (175 \text{ lb per sq ft})(20/12 \text{ in.})(60/12 \text{ in.})$
$$= 1,458 \text{ lb}$$

Figure 16-2 shows the point loads from the joists on the stringers. The length of the stringers will be less than the distance between the steel beams because the stringers must fit between the steel beams. For this example, the length of the stringers is 66 in., or the 5 ft 6 in., the distance between the coil bolts of the hangers. The load on each hanger will be the 2,916 lb as shown in Figure 16-2.

The maximum bending moment will occur at the center of the stringer and can be calculated by summing the moments of all forces as follows:

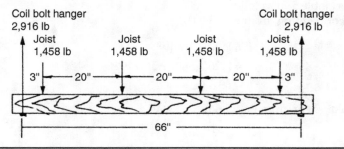

FIGURE 16-2 Concentrated loads from joists on stringer.

$$M = (2{,}916 \text{ lb})(33 \text{ in.}) - (1{,}458 \text{ lb})(30 \text{ in.}) - (1{,}458 \text{ lb})(10 \text{ in.})$$
$$= (96{,}228 \text{ in.-lb}) - (43{,}740 \text{ in. lb}) - (14{,}580 \text{ in.-lb})$$
$$= 37{,}908 \text{ in.-lb}$$

An alternate method of calculating the maximum bending moment is to use the equation for maximum moment in Beam 4 in Table 5-2, combined with the technique of superposition as follows:

$$M = Pa_1 + Pa_2$$
$$= 1{,}458 \text{ lb } (3\text{-in.}) + 1{,}458 \text{ lb } (23\text{-in.})$$
$$= 4{,}374 \text{ in.-lb} + 33{,}534 \text{ in.-lb}$$
$$= 37{,}908 \text{ in.-lb}$$

Consider using double 2×8 S4S lumber for stringers. From Table 4-1, the section modulus for a double 2×8 is double the value of a single 2×8 as follows:

Section modulus for double 2×8 stringers, $S = 2(13.14 \text{ in.}^3)$
$$= 26.28 \text{ in.}^3$$

The applied bending stress, from Eq. (5-7),

$$f_b = M/S$$
$$= (37{,}908 \text{ in.-lb})/26.28 \text{ in.}^3$$
$$= 1{,}442 \text{ lb per sq in.}$$

The allowable bending stresses for 2×8 No. 2 grade Southern Pine for can be determined from Tables 4-2 and 4-4 as follows:

Allowable bending stress, $F_b = C_D \times$ (reference bending stress)
$$= 1.25(1{,}200)$$
$$= 1{,}500 \text{ lb per sq in.}$$

The allowable bending stress of $F_b = 1{,}500$ lb per sq in. is greater than the applied bending stress of $f_b = 1{,}442$ lb per sq in. Therefore, the double 2×8 stringer is adequate for bending.

The maximum shear force at the end support of the stringer is 2,916 lb, which will occur at the support hanger. Since the stringer is a double member, the shear area of bd will be two times the area of a single 2×8, or $2 \times 10.88 \text{ in.}^2 = 21.76 \text{ in.}^2$, reference Table (4-1). The applied shear stress can be calculated as follows:

The applied shear stress, from Eq. (5-13) can be calculated as follows:

$$f_v = 3V/2bd$$
$$= 3(2{,}916)/2(21.76)$$
$$= 201.0 \text{ lb per sq in.}$$

The allowable shear stress from Tables 4-2 and 4-4 can be calculated as follows:

Allowable shear stress, $F_v = C_D \times$ (reference shear stress)
$$= 1.25(175)$$
$$= 218 \text{ lb per sq in.}$$

The allowable shear stress $F_v = 218$ lb per sq in. is greater than the calculated applied shear stress $f_v = 201.0$. Therefore, the double 2×8 stringer is satisfactory for shear.

The deflection due to the four 1,458 lb concentrated loads can be calculated by using the deflection equation for Beam 4 in Table 5-2, combined with the technique of superposition. From Table 4-1, the moment of inertia of the double 2×8 stringer can be calculated as $2(47.63 \text{ in.}^4)$, or 95.26 in.^{4} From Table 4-2, the modulus of elasticity for a 2×8 No. 2 grade Southern Pine is 1,600,000 lb per sq in. The deflection can be calculated using the technique of superposition as follows:

$$\Delta = \frac{Pa_1[3(l^2)/4 - (a_1)^2]}{6EI} + \frac{Pa_2[3(l^2)/4 - (a_2)^2]}{6EI}$$

$$= \frac{1,458(3)[3(66)^2/4 - (3)^2]}{6(1,600,000)(95.26)} + \frac{1,458(23)[3(66)^2/4 - (23)^2]}{6(1,600,000)(95.26)}$$

$$= 0.016 \text{ in.} + 0.10 \text{ in.}$$

$$= 0.12 \text{ in.}$$

The permissible deflection is $l/360$, or $66/360 = 0.18$ in. Thus, the calculated deflection of 0.12 in. is less than the permissible deflection of 0.18 in. Therefore, the double 2×8 stringer is satisfactory for deflection.

An approximate deflection can be calculated by transforming the four concentrated loads into an equivalent uniformly distributed load having the same total value as the sum of the concentrated loads. Using this method, the deflection can be approximated as follows:

The equivalent uniformly distributed load on stringers is

$$w = 4(1,458 \text{ lb})/(66/12 \text{ ft})$$
$$= 1,060 \text{ lb per lin ft}$$

Because the stringer is a single-span member, the deflection for a single-span beam can be calculated using Eq. (5-26) as follows:

The deflection, from Eq. (5-26):
$$\Delta = 5wl^4/4,608EI$$
$$= 5(1,060)(66)^4/(4,608)(1,600,000)(95.26)$$
$$= 0.14 \text{ in.}$$

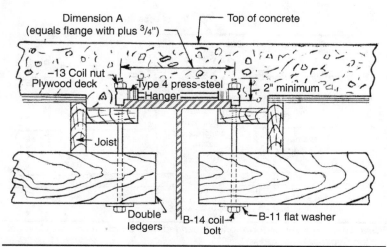

Dimension A
(equals flange with plus 3/4")
Top of concrete
–13 Coil nut
Plywood deck
Type 4 press-steel
Hanger
2" minimum
Joist
Double
ledgers
B-14 coil
bolt
B-11 flat washer

Figure 16-3   Coil rod hanger for bridge decks suspended from steel beams.
(*Source: Dayton Superior Corporation*)

For this particular condition, the calculated approximate deflection of 0.14 in. using an equivalent uniformly distributed load is slightly higher than the calculated 0.12 in. deflection using the four concentrated loads.

Hanger to Support Stringer

The coil bolt hanger that rests on the top flange of the steel beam must have adequate strength to hold the 2,916-lb load from the ends of the double stringers. The safe working load of the coil rod hanger assemblies generally ranges from 2,000 to 6,000 lb. For this example, a 3,000-lb hanger will be required.

Figure 16-3 shows the details of a coil bolt hanger that is available for forming bridge decks. The hanger rests on the top flange of the steel beam and is to be used with full bearing under the end sections. There should be a careful check of the exact beam size before ordering hangers because steel beams are often oversize and concrete beams are often undersize.

Coil bolt hangers provide adjustment quickly and easily from the top of the formwork. These hangers consist of specially formed stamped steel end sections electrically welded to each end of a connecting strut. Both adjustable coil bolt assemblies and fixed-length coil bolt assemblies are available.

Wood Forms for Deck Slab with Haunches

Coil rod hangers are available for hanging formwork from structural steel beams with haunches of any height. These hangers are designed to break back inside the concrete. Breakback is accomplished by placing

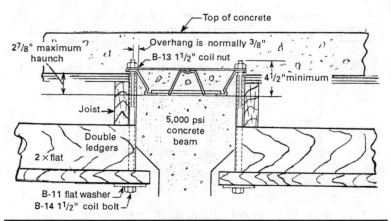

Figure 16-4   Hanger for formwork suspended from concrete beams. (*Source: Dayton Superior Corporation*)

a length of pipe over the exposed end of the hanger and working the end back and forth, similar to snap ties for wall forms.

Wood Forms for Deck Slab Suspended from Concrete Beams

Figure 16-4 illustrates a hanger frame designed for a precast prestressed concrete beam. Provision is also made for forming a haunch above the beam. The top strap is high-tensile steel and the hanger is reinforced with bracing chairs for adequate strength on wider beams. The design of the formwork components may be performed as illustrated in Example 16-1.

Forms for Overhanging Deck Constructed on Exterior Bridge Beams

Figure 16-5 illustrates a method of constructing an overhanging deck on steel beams. The formwork hanger has two types of end sections. A 90° end section is used to support the formwork on the interior side of the exterior beam, whereas a 45° end section is used to suspend bridge overhang brackets on the exterior side. This same method may also be used for forming an overhang from precast prestressed bridge beams by using a formwork hanger specially made for concrete beams.

Figure 16-6 illustrates several methods of overhang formwork when the bridge design will not allow the use of conventional bridge overhang brackets. In Figure 16-6(a), a wide overhang is formed on a

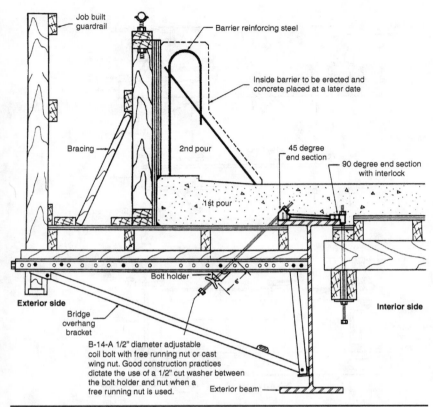

Job built
guardrail

Barrier reinforcing steel

Inside barrier to be erected and
concrete placed at a later date

Bracing

2nd pour

45 degree
end section

90 degree end section
with interlock

1st pour

Exterior side

Interior side

Bolt holder

Bridge
overhang
bracket

B-14-A 1/2" diameter adjustable
coil bolt with free running nut or cast
wing nut. Good construction practices
dictate the use of a 1/2" cut washer between
the bolt holder and nut when a
free running nut is used.

Exterior beam

FIGURE 16-5 Overhang formwork for a bridge slab. (*Source: Dayton Superior Corporation*)

shallow steel beam. Formwork for a short overhang is illustrated in Figure 16-6(b) for a shallow steel beam and in Figure 16-6(c) for a shallow concrete beam.

Deck Forms Supported by Steel Joists

If bridge piers are close enough together, and if there is enough reuse to justify the initial cost, open-web steel joists may be used to support the forms for a concrete deck, as illustrated in Figure 16-7. At the time a concrete cap is cast, anchors should be set for bolts to support steel angles. The ends of the joists are supported by these angles. If necessary, internal steel-plate fillets may be welded across the legs of the angle to provide increased stiffness. Nailer planks should be attached to the tops of the joists. Variable-depth wood joists may be used to provide camber or to provide for variations in the thickness of the concrete slab between piers.

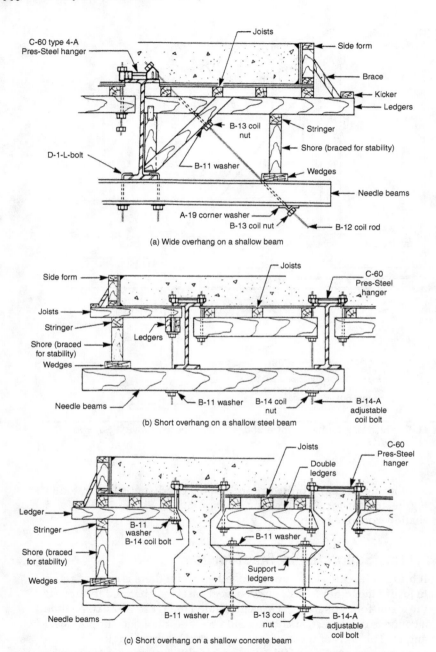

FIGURE 16-6 Methods of overhang formwork for bridge slabs. (*Source: Dayton Superior Corporation*)

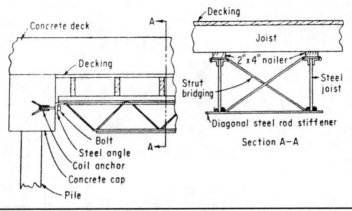

FIGURE 16-7   Deck forms supported by steel joists.

As illustrated in Figure 16-7, horizontal and diagonal bridging must be installed between adjacent joists to prevent transverse movements of the bottom chords of the joists, following the recommendations of the Steel Joist Institute. Horizontal bridging consists of two continuous horizontal steel members, one attached to the top chord and the other to the bottom chord. Diagonal bridging consists of cross-bracing. The number of rows of bridging depends on the joist designation and span length in accordance with standards of the Steel Joist Institute.

Because the joists will be reused several times on a given project, the bottom chords and possibly the top chords of several rows of joists should be interconnected by welding ¾-in. steel rods diagonally across and between adjacent joists. This will permit the interconnected joists to be lowered onto a truck-mounted traveler as a single assembly and moved to a new location. The diagonal rods will prevent racking during the handling and moving of the joists.

Also, if ground conditions under a bridge permit the operation of a truck equipped with a variable-height traveler, it is possible that an assembly of several interconnected steel joists, wood joists, and decking can be lowered intact onto the traveler and moved to a new location, which can produce considerable economy in formwork. If this procedure is to be followed, the wood joists should be cut to lengths that conform to the width of an assembly unit unless the forms for an entire bay will be moved as a unit.

Table 16-1 provides allowable total safe loads for K-series steel joists with allowable tensile stress of 30,000 lb per sq in. Dead loads, including joists, must be deducted to determine the load-carrying capacity of joists. The values in the tables are pounds per linear foot along the joist. The component parts of the formwork should be designed as illustrated in the following example.

Joist Type	20K4	20K5	20K6	20K7	22K5	22K6	22K7	24K5	24K6	24K7	24K8
Depth in.	20	20	20	20	22	22	22	24	24	24	24
Weight lb/ft	7.6	8.2	8.9	9.3	8.8	9.2	9.7	9.3	9.7	10.1	11.5
Clear Span, ft											
20	550 (550)	550 (550)	550 (550)	550 (550)	— —	— —	— —	— —	— —	— —	— —
21	550 (520)	550 (520)	550 (520)	550 (520)	— —	— —	— —	— —	— —	— —	— —
22	514 (461)	550 (490)	550 (490)	550 (490)	550 (548)	550 (548)	550 (548)	— —	— —	— —	— —
23	469 (402)	529 (451)	550 (468)	550 (518)	550 (518)	550 (518)	550 (518)	— —	— —	— —	— —
24	430 (353)	485 (396)	528 (430)	550 (448)	536 (483)	550 (495)	550 (495)	550 (495)	550 (544)	550 (544)	550 (544)
25	396 (312)	446 (350)	486 (380)	541 (421)	493 (427)	537 (464)	550 (474)	540 (511)	550 (520)	550 (520)	550 (520)
26	366 (277)	412 (310)	449 (337)	500 (373)	455 (379)	496 (411)	550 (454)	499 (453)	543 (493)	550 (499)	550 (499)

27	339 (247)	382 (277)	416 (301)	463 (333)	422 (337)	459 (367)	512 (406)	503 (404)	550 (439)	550 (479)	550 (479)
28	315 (221)	355 (248)	386 (269)	430 (298)	392 (302)	427 (328)	475 (364)	429 (362)	467 (393)	521 (436)	550 (456)
29	393 (199)	330 (223)	360 (242)	401 (268)	365 (272)	398 (295)	443 (327)	400 (325)	435 (354)	485 (396)	536 (429)
30	274 (179)	308 (201)	336 (218)	374 (242)	341 (245)	371 (266)	413 (295)	373 (293)	406 (319)	453 (353)	500 (387)

Notes:

1. Loads are based on allowable tensile stress of 30,000 lb per sq in., for K-series joists adopted by the Steel Joist Institute.

2. Top figures give total safe uniformly distributed load-carrying capacities, in lb per lin ft. Dead loads, including joists, must be deducted to determine the load-carrying capacity of joists.

3. Numbers in parentheses represent linear load per foot of joist which will produce a deflection of $l/360$ of span. For a deflection of $l/240$, multiply the number in parentheses by $360/240$, or 1.5. For a deflection of $l/270$, multiply the number in parentheses by $360/270$, or 1.333.

TABLE 16-1 Allowable Total Safe Load on K-Series Steel Joists, in Pounds per Linear Foot

Example 16-2

Determine the size, required strength, and safe spacing for steel joists for a concrete bridge deck, given the following conditions:

Distance face to face between bridge piers, 25 ft 6 in.

Length of span for steel joists, 25 ft 0 in.

Unit weight of concrete, 150 lb per cu ft

Thickness of concrete slab, 14 in.

Assumed live load on slab, 50 lb per sq ft

Proposed decking, ⅞-in. Class I Plyform

Proposed spacing of wood joists, 24 in.

Deflection criterion, $l/270$, not to exceed ¼ in.

The uniform pressure on the decking will be

$$\text{Concrete (150 lb per cu ft)(14/12 ft)} = 175 \text{ lb per sq ft}$$
$$\text{Live load} = \underline{50 \text{ lb per sq ft}}$$
$$\text{Total pressure} = 225 \text{ lb per sq ft}$$

The size of the wood joists selected will depend on the weight from the concrete and the strength and safe spacing of the steel joist, one or more of which should be assumed initially. For this example, the proposed ⅞-in. decking is proposed for placement on wood joists spaced at 24 in. on centers.

Table 16-1 shows that, for a span of 25 ft 0 in., the allowable safe total load on the listed steel joists, including the weight from the concrete, the forms, and the joists, varies from 396 to 550 lb per lin ft, or from 312 to 520 lb per lin ft for deflection limited to $l/360$ of the span. Consider a 20K4 joist, for which the safe load is 312 lb per lin ft for a deflection of $l/360$. Adjusting for the $l/270$ deflection criterion for this example, the allowable load will be (360/270)(312), or 416 lb per lin ft. However, this value exceeds the maximum permissible safe load of 396 lb per lin ft for this steel joist (see Table 16-1). Thus, the safe load is limited to 396 lb per lin ft. Because this steel joist weighs 7.6 lb per lin ft, the net safe load on a joist will be 396 lb per lin ft less the 7.6 lb per lin ft dead load of the joist, or 388.4 lb per lin ft. Assuming that the decking, wood joists, and bridging for the steel joists weigh 6 lb per sq ft, the total weight on the joist will be 231 lb per sq ft. For this condition, the maximum safe spacing of the joist will be 388.4 lb per lin ft divided by the 231 lb per sq ft, or 1.68 ft, which is 20.2 in. Therefore, the spacing of the steel joists must not exceed 20.2 in. For constructability, choose a 20-in. center-to-center spacing of the steel joists.

A check must be made to determine the adequacy of the wood formwork above the steel joists, the wood joists that will rest on the steel joists, and the decking that will rest on the wood joists.

The wood joists will be placed in a direction perpendicular to the steel joists. Thus, the wood joists that rest on top of the steel joists must have adequate strength and rigidity to span the 20-in. distance between the supporting steel joists. Assume a 5 lb per sq ft dead load to include the weight of the Plyform decking and wood joists. Then the design load on the wood joists will be 225 + 5 = 230 lb per sq ft. Since the proposed spacing of the wood joists is 24 in., the uniformly distributed load on the wood joists will be (230 lb per sq ft) (24/12 ft) = 460 lb per lin ft. Table 4-17 indicates a 2 × 4 S4S wood joist, which is continuous over four or more supports and has a maximum span length of 32 in. for an equivalent uniform load of 600 lb per lin ft. Because the design load of 460 lb per lin ft is less than the 600 lb per lin ft, and the 32 in. is greater than 22 in., the 24-in. spacing of the wood joists on centers with a 20-in. span length will be satisfactory.

Now the decking must be checked for adequate strength and rigidity to sustain the 225 lb per sq ft pressure between the supporting wood joists. The ⅞-in. Class I Plyform decking will rest on the wood joists, which are proposed at a 24-in. spacing on centers. Therefore, the decking must have adequate strength and rigidity to sustain the 225 lb per sq ft pressure over the 24-in. span length, the spacing of the supporting wood joists.

From Table 4-11, the physical properties of ⅞-in. Class I Plyform for stress applied parallel to grain are

Section modulus, $S_e = 0.584$ in.^{3}

Moment of inertia, $I = 0.296$ in.^{4}

Rolling shear constant $(Ib/Q) = 8.555$ in.^{2}

From Table 4-12, the allowable stresses for Class I Plyform are

Allowable bending stress, $F_b = 1,930$ lb per sq in.

Allowable shear stress, $F_s = 72$ lb per sq in.

Modulus of elasticity, $E = 1,650,000$ lb per sq in.

The Plyform decking will be continuous over multiple wood joists. Therefore, the equations for stress and deflection for span conditions of three or more spans will apply (see equations in Table 5-4). Determine the spacing of wood joists based on bending, shear, and deflection of the Plyform decking joists as follows:

For bending in the ⅞-in. Plyform decking, from Eq. (5-41),

$$l_b = [120F_b S_e / w_b]^{1/2}$$
$$= [120(1,930)(0.584)/225]^{1/2}$$
$$= 24.5 \text{ in.}$$

For shear in the ⅞-in. Plyform decking, Eq. (5-48),

$$l_s = 20F_s(Ib/Q)/w_s$$
$$= (20)(72)(8.555)/(225)$$
$$= 54.7 \text{ in.}$$

For a deflection less than $l/270$, substituting 270 for 360 in Eq. (5-57a) of Table 5-4:

$$l_d = [1,743EI/270w_s]^{1/3}$$
$$= [1,743(1,650,000)(0.296)/270(225)]^{1/3}$$
$$= 24.1 \text{ in.}$$

For deflection not to exceed ¼ in., substituting ¼ in. for ⅟₁₆ in., Eq. (5-57b) of Table 5-4:

$$l_d = [1,743EI/4w_s]^{1/4}$$
$$= [1,743(1,650,000)(0.296)/4(225)]^{1/4}$$
$$= 31.1 \text{ in.}$$

Below is a summary of the allowable span lengths of the ⅞-in. Class I Plyform decking:

For bending, the maximum allowable span length = 24.5 in.
For shear, the maximum allowable span length = 54.7 in.
For deflection, the maximum allowable span length = 24.1 in.

For this example, the spacing of the wood joists must not exceed the maximum allowable span length of the Plyform decking, which is 24.1 in. Thus, the proposed 24-in. spacing of the wood joists with ⅞-in. Class I Plyform decking is satisfactory.

Deck Forms Supported by Tubular Steel Scaffolding

Where ground conditions under a bridge deck permit the use of tubular steel scaffolding, this method of supporting the formwork can be satisfactory and economical, as illustrated in Figure 16-8.

FIGURE 16-8 Deck forms supported by tubular steel scaffolding. (*Source: Patent Construction Systems*)

Adjustable Steel Forms for Bridge Decks

Commercial steel forms which are adjustable are available for forming bridge decks. As illustrated in Figure 16-9, this forming system may be used for steel girders or prestressed concrete bridge girders. This system consists of two distinctively different components: the overhang system and the interior, or girder-to-girder, system.

The interior form system rests on the bottom flange of a concrete or steel girder, as shown in Figure 16-10. Two sets of adjustable pipe struts support a pair of slider assemblies that carry two interior form soffit panels and are typically 12 ft long. A T-pan is a flat plate that makes up the variable dimension between the two interior form soffit panels. Interior form units are set in place with a crane and are stripped with special handling and access equipment.

The overhang form system illustrated in Figure 16-11 has clamps on the bottom flange of a concrete or steel girder. Structural steel framing and walkboards create access for working the pipe struts

FIGURE 16-9 Adjustable bridge deck and overhang forming system. (*Source: Symons Corporation*)

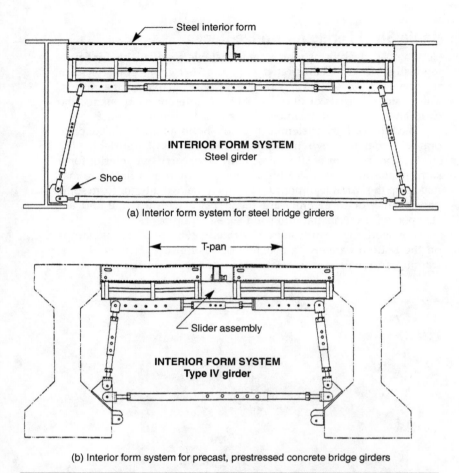

(a) Interior form system for steel bridge girders

(b) Interior form system for precast, prestressed concrete bridge girders

FIGURE 16-10 Adjustable interior form system for bridge decks. (*Source: Symons Corporation*)

and shoes. An all-steel soffit panel, which is typically 12 ft long, is supported by two sets of pipe struts and shoes. The soffit panel is designed to allow the slab edge form to be bolted down from the top, and is adjustable for various width overhangs. The slab edge form adjusts for superelevation with the adjustable-height steel chamfer. A crane-handled C-hood is used to set and strip the overhang form units.

The bridge deck forming system has an all-steel adjustable and reusable form that eliminates most of the costly cutting and fitting of plywood forms between steel or concrete bridge girders, which reduces scrap and scrap removal costs. It is adjustable for various girder depths

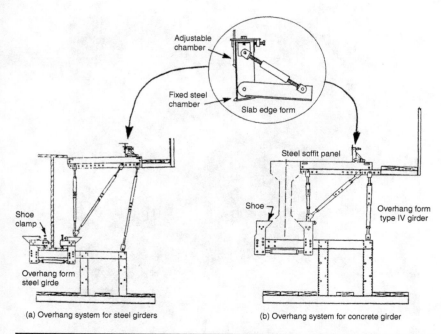

Figure 16-11 Overhang form system for bridge decks. (*Source: Symons Corporation*)

and types. Simple and quick clamping attachment shoes allow fast setting and stripping in typical 12-ft modules with a small crew.

Overhang forms strip and set complete with slab edge forms, hand rail posts, and lower personnel access scaffolding by crane as one unit. The slab edge form is adjustable so that the form face will be plumb even on superelevated curves.

All-Steel Forms for Bridge Structures

Modular all-steel forms are available for forming bridges, as illustrated in Figure 16-12. This forming system, manufactured by EFCO, has the ability to be self-supporting and self-spanning. The steel ribs on the form panels serve two purposes: as web stiffeners and as beams to transfer the horizontal pressures of the liquid concrete from the form face plate to the form panel top and bottom flanges.

This form system is available in modular lengths that can be bolted together easily, which provides flexibility for changing job conditions. The self-supporting ability of this forming system enables the contractor to pour aerial concrete without shoring between spans.

FIGURE 16-12   All-steel forming system for bridge structures. (*Source: EFCO*)

References

1. Dayton Superior Corporation, Dayton, OH.
2. EFCO Corporation, Des Moines, IA.
3. Patent Construction Systems, Paramus, NJ.
4. Symons Corporation, Elk Grove Village, IL.

CHAPTER 17

Flying Deck Forms

Introduction

The term "flying deck form" is used in this book to designate a system of components that are assembled into units, called decks, for forming concrete slabs in multistory buildings. The same set of flying deck forms is used repeatedly to form multiple floor slabs in a building. After the concrete that has been placed in a slab is sufficiently cured, the flying deck form for the slab is removed (without disassembly of the parts), moved (flown) horizontally outward, away from the building, and then moved up and back inward to the building to a new location, and used again to form another concrete slab. The term is derived from the process of moving (flying) the form outward, away from the building, as it is moved upward to the next floor level in the structure.

Each unit, or flying deck form, consists of various structural components, such as trusses, stringers, joists, and plywood decking. The unit is rigidly assembled to be used and reused in molding the concrete slabs of a building. The forms may be used to support concrete beams, girders, slabs, and other parts of a structure.

Figures 17-1 through 17-3 illustrate how the forms may be used in constructing high-rise buildings. However, the forms may also be used in constructing structures other than high-rise buildings, provided there will be enough reuses to justify the initial cost of fabricating the forms.

Advantages of Flying Forms

If a high-rise building is specifically designed for the use of flying deck forms for constructing its concrete slabs and beams, the cost of the formwork can be substantially less than the cost of forming by conventional methods.

When conventional methods are used for forming a cast-in-place concrete structure, it is common practice to fabricate and assemble the forms for the columns, beams, girders, slabs, and shear walls in place. This requires considerable labor. After the concrete has been

FIGURE 17-1   Forms moving out to be flown to higher floor level. (*Source: Form-Eze Systems, Inc.*)

FIGURE 17-2   Landing a flying deck form in a new position. (*Source: Form-Eze Systems, Inc.*)

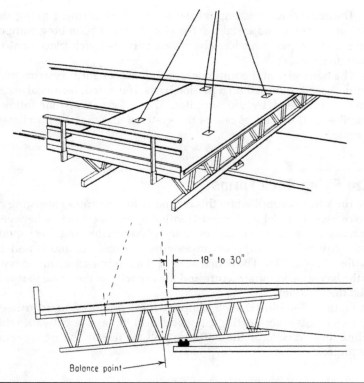

18" to 30"

Balance point

FIGURE 17-3   Flying deck form being removed for lifting to a higher location.
(*Source: Form-Eze Systems, Inc.*)

placed and cured sufficiently, the forms are removed piece by piece, or in relatively small sections, and moved to another location in the building where they are reassembled for additional uses. Following this procedure results in considerable cost for labor. In addition, it delays the completion of the project.

If flying deck forms are used, the forms are preassembled into sections or modules at ground level, using labor-saving equipment and methods. They are then moved (flown) into position for concreting. Each deck module is located in its proper position, and its height is adjusted to the correct elevation. Filler strips are then placed between the deck modules. Reinforcing steel is placed, electrical and plumbing components are installed, and any other services that are required are performed. Then the concrete is placed.

When the concrete has attained sufficient strength, the flying deck forms are lowered from beneath the newly placed concrete slab, moved to the edge of the supporting concrete slab, and hoisted to a higher level for reuse. The most recently placed concrete slab should be reshored immediately before any loads are placed on it.

The removing, hoisting, relocating the deck forms, placing the reinforcing steel, and installing the electrical and plumbing components may be performed easily, which permits quick placement of concrete for a slab.

The hardware and equipment for flying deck form systems may be purchased or rented on a per-job basis. If desired, technical assistance is provided by the supplier. The information that follows describes and illustrates one of the systems that may be purchased or rented.

Form-Eze Flying Deck Forms

The deck forms supplied by this company are modules consisting of two or more parallel steel trusses (with diagonal steel braces between the trusses) that support steel joists and plywood decking. Deck modules frequently are supplied in lengths in excess of 100 ft and in widths of up to 30 ft. The actual sizes of the modules are limited only by the bay sizes in the structure and the capacity of the crane that will hoist the modules.

Figure 17-4 illustrates a simple arrangement of three trusses with steel joists supporting plywood decking and a catwalk with railing located outside the building slab for use by the workers on the project.

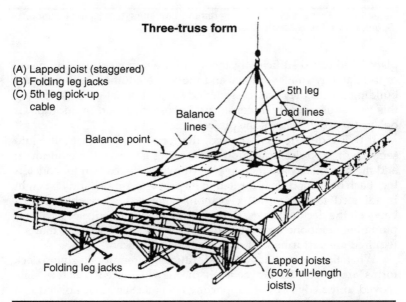

Three-truss form

(A) Lapped joist (staggered)
(B) Folding leg jacks
(C) 5th leg pick-up
    cable

5th leg

Load lines

Balance
lines

Balance point

Folding leg jacks

Lapped joists
(50% full-length
joists)

FIGURE 17-4 Representative members of a flying assembly. (*Source: Form-Eze Systems, Inc.*)

The sequence of operations for removing and relocating flying deck forms is shown in Figure 17-5. Tilting form guides are placed under the trusses to assist in removing the forms. The decks are flown onto dollies, which may be used to move forms across a slab for final manual positioning.

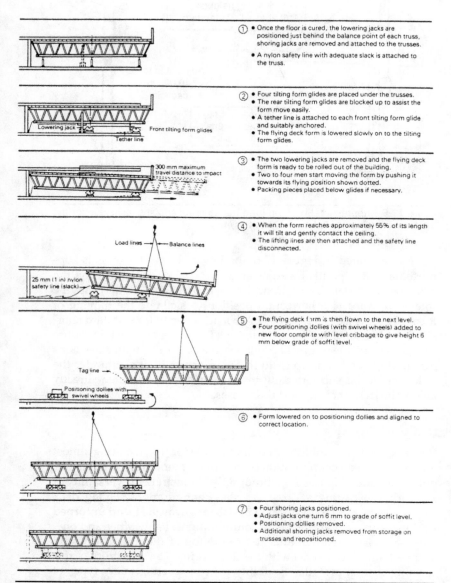

① • Once the floor is cured, the lowering jacks are positioned just behind the balance point of each truss, shoring jacks are removed and attached to the trusses.
   • A nylon safety line with adequate slack is attached to the truss.

② • Four tilting form glides are placed under the trusses.
   • The rear tilting form glides are blocked up to assist the form move easily.
   • A tether line is attached to each front tilting form glide and suitably anchored.
   • The flying deck form is lowered slowly on to the tilting form glides.

③ • The two lowering jacks are removed and the flying deck form is ready to be rolled out of the building.
   • Two to four men start moving the form by pushing it towards its flying position shown dotted.
   • Packing pieces placed below glides if necessary.

④ • When the form reaches approximately 55% of its length it will tilt and gently contact the ceiling.
   • The lifting lines are then attached and the safety line disconnected.

⑤ • The flying deck form is then flown to the next level.
   • Four positioning dollies (with swivel wheels) added to new floor complete with level cribbage to give height 6 mm below grade of soffit level.

⑥ • Form lowered on to positioning dollies and aligned to correct location.

⑦ • Four shoring jacks positioned.
   • Adjust jacks one turn 6 mm to grade of soffit level.
   • Positioning dollies removed.
   • Additional shoring jacks removed from storage on trusses and repositioned.

FIGURE 17-5   Sequence of operations for removing and relocating flying deck forms. (*Source: Form-Eze Systems, Inc.*)

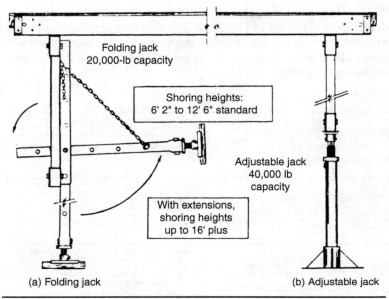

Folding jack
20,000-lb capacity

Shoring heights:
6' 2" to 12' 6" standard

Adjustable jack
40,000 lb
capacity

With extensions,
shoring heights
up to 16' plus

(a) Folding jack

(b) Adjustable jack

FIGURE 17-6   Representative flying form trusses with cribbing jacks supporting the trusses. (*Source: Form-Eze Systems, Inc.*)

As illustrated in Figure 17-6, the lower chord of a truss is supported at panel points by adjustable screw jacks, which rest on the concrete floor slab. The folding jack legs in Figure 17-6(a) may be rotated upward to a horizontal position to facilitate the moving of forms. Folding jacks have a 20,000-lb load capacity for standard shoring heights from 6 ft 2 in. to 12 ft 6 in. The adjustable jack in Figure 17-6(b) is available with a 40,000-lb load capacity and with extensions for shoring heights of up to 16 ft. The concrete floor slab under the jacks should be supported by reshores, which are located directly below the jacks that support the trusses.

Versatility of Forms

It is possible to assemble form units suitable for use in casting almost any concrete slab configuration by combining modules, components, and accessories. Figures 17-7 through 17-12 illustrate the versatility of the forms in casting a variety of slab configurations. Flying deck forms may be used to form beam in slab, one-way joist, and upturned or down-turned spandrel beam configurations. Pan waffles and one-way pan joists are often attached to the top of the deck forms and stripped and flown intact with the deck forms. The versatility of this forming system provides flexibility for constructing many types of concrete structures. Figure 17-7 illustrates the use of the deck forms to cast various shapes of beam and slab designs.

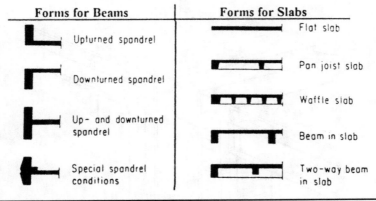

Forms for Beams	Forms for Slabs
Upturned spandrel	Flat slab
Downturned spandrel	Pan joist slab
Up- and downturned spandrel	Waffle slab
Special spandrel conditions	Beam in slab
	Two-way beam in slab

FIGURE 17-7 Representative concrete sections constructed with flying forms. (*Source: Form-Eze Systems, Inc.*)

Figure 17-8 illustrates a deck form that permits the use of metal, plywood, or fiberglass pans and pan joist systems for concrete floor slabs. The pans remain on the flying deck forms and are flown intact with the deck forms to the next higher floor level. A large variety of types and sizes of pans is available, as presented in Chapter 12.

Figure 17-9 illustrates one method of joining two modules with the framing installed to cast a concrete beam with the concrete slab. This forming system is used when casting a large beam or girder.

Figure 17-10 illustrates a form arrangement that can be used to cast a concrete spandrel beam.

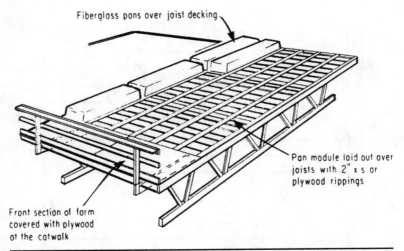

Fiberglass pans over joist decking

Pan module laid out over joists with 2" x s or plywood rippings

Front section of form covered with plywood at the catwalk

FIGURE 17-8 Representative members of a flying form assembly for a pan-joist concrete slab. (*Source: Form-Eze Systems, Inc.*)

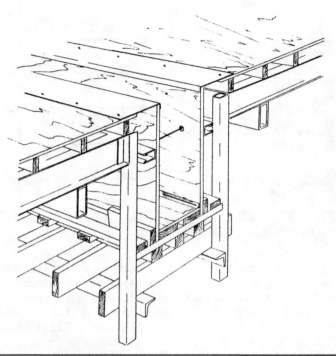

Figure 17-9   Representative details for casting a concrete beam in a slab using flying forms. (*Source: Form-Eze Systems, Inc.*)

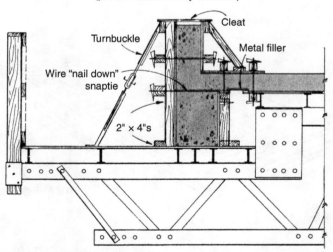

Up & downturned spandrel
(poured monolithically with slab)

Figure 17-10   Details of forming a concrete slab with a spandrel beam. (*Source: Form-Eze Systems, Inc.*)

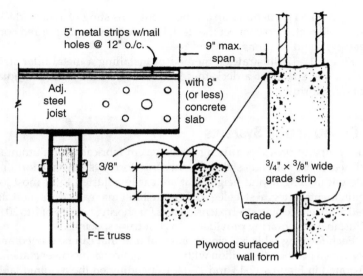

FIGURE 17-11 Representative details for casting a concrete slab over shear walls and columns; 14-gauge galvanized metal filler strip (deck to wall). (*Source: Form-Eze Systems, Inc.*)

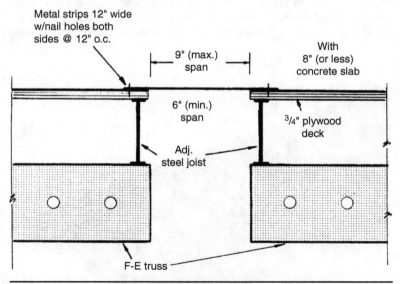

FIGURE 17-12 Representative method of installing filler strips between the flying forms; 14-gauge galvanized metal filler strip (deck to deck). (*Source: Form-Eze Systems, Inc.*)

Figure 17-11 illustrates an arrangement consisting of a metal deck-to-wall filler strip to connect the deck form to previously poured concrete walls and columns.

Figure 17-12 illustrates a method of installing a metal filler strip to close the space for a deck-to-deck connection between two adjacent deck forms.

Patent Construction Systems

This company provides a flying form system fabricated of aluminum and steel that is fully adjustable. Patent's Interform system has a forming table width, length, and height that are easily adjusted to fit most job requirements. Integral steel leg sockets are built into each truss module, fitted with telescoping steel extension legs for shoring heights of 7 to 20 ft. Adjustable steel bracing provides 6- to 20-ft truss spacings.

Each forming table is a single, integral unit that can be moved and flown from position to position with an average-capacity tower crane, as illustrated in Figures 17-13 and 17-14. Depending on the required table width, two or three trusses are used. Each truss unit is held in a fixed rigid position by adjustable diagonal steel bracing. Table trusses are then spanned laterally with extruded aluminum joists. The aluminum joists provide the basic surface on which the deck form is mounted.

The modular units are available in lengths of 2½, 5, 10, 20, and 30 ft, as shown in Figure 7-15(a). The trusses and joists are manufactured with lightweight aluminum. Individual truss modules may be joined together with splice sets to quickly assemble a specific table length.

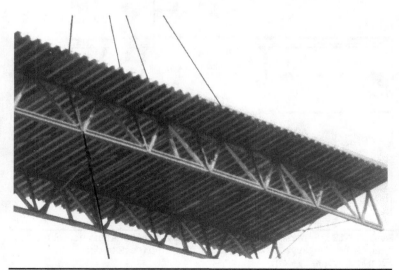

FIGURE 17-13   Lifting a flying form table unit to a new location. (*Source: Patent Construction Systems*)

FIGURE 17-14   Landing a flying form table at a new location. (*Source: Patent Construction Systems*)

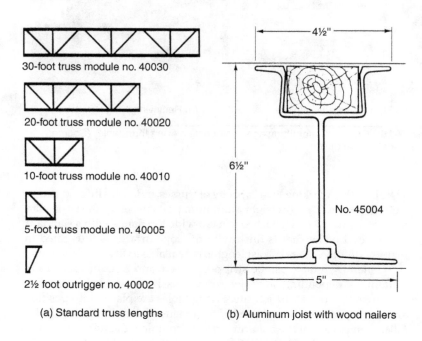

30-foot truss module no. 40030

20-foot truss module no. 40020

10-foot truss module no. 40010

5-foot truss module no. 40005

2½ foot outrigger no. 40002

(a) Standard truss lengths

4½"

6½"

5"

No. 45004

(b) Aluminum joist with wood nailers

FIGURE 17-15   Truss module lengths and aluminum joists. (*Source: Patent Construction Systems*)

(a) Lifting a flying form table

(b) Landing a flying form table

(c) Flying form table in place

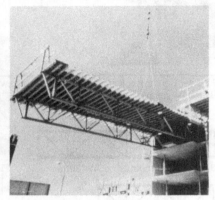

(d) Removing flying form table

Figure 17-16   Operations for lifting and positioning a flying form table. (*Source: Patent Construction Systems*)

The allowable working load capacity of trusses, individually or joined, is 2,000 lb per lin ft. Lightweight aluminum joists make up the basic platform of the flying table. Each joist has a wide top flange with a standard 2 × 3 wood nailer that is flush with the top surface, as illustrated in Figure 7-15(b). A range of joist lengths is available to fit a particular job.

Generally, a table can be stripped, flown, and reset in about 15 to 20 minutes, depending on the job conditions. Preassembled truss modules are delivered to the job site. Flying tables are placed into position and set to grade by adjustment of the extension legs and screw jacks. Fillers are placed in the column strips to complete the deck form. The concrete is poured, and once the slab has attained adequate strength, the table is stripped and lowered onto rollout units. The table is slowly rolled out and the crane harness is attached. The table is flown to its new position and reset to grade. Figure 17-16 shows the lifting and positioning operations of a flying table.

References

1. Form-Eze Systems, Inc., Albuquerque, NM.
2. Patent Construction Systems, Paramus, NJ.
3. Symons Corporation, Elk Grove Village, IL.

Dimensional Tolerances for Concrete Structures

The plans or drawings for concrete structures frequently specify the permissible variations from lines, grades, and dimensions that the contractor is expected to observe. The tolerances should be realistic, considering the nature of the structure. Tolerances that are more rigid than justified will increase the cost of a structure unnecessarily.

The following tolerances are excerpts from the American Concrete Institute publication ACI 117-90, "Standard Tolerances for Concrete Construction." The ACI standard also contains tolerances related to materials, precast concrete, and masonry. The following is reprinted with permission from the American Concrete Institute.

Tolerances of Cast-in-Place Concrete for Buildings

1. Vertical alignment:
   (a) For height 100 ft or less:
   Lines, surfaces, and arrises .... 1 in.
   Outside corner of exposed corner columns and control
       joint grooves in concrete exposed to view .... ½ in.
   (b) For heights greater than 100 ft:
   Lines, surfaces, and arrises, $\frac{1}{1000}$ times the
       height but not more than .... 6 in.
   Outside corner of exposed corner columns and
       control joint grooves in concrete, $\frac{1}{1000}$ $\frac{1}{600}$ times
       the height but not more than .... 3 in.

2. Lateral alignment:
   (a) Members .... 1 in.
   (b) In slabs, centerline location of openings 12 in. or
       smaller and edge location of larger openings .... ½ in.
   (c) Sawcuts, joints, and weakened plane embedments
       in slabs .... ¾ in.

3. Level alignment:
  (a) Top of slabs:
      Elevation of slabs-on-grade .... ¾ in.
      Elevation of top surfaces of formed slabs before
        removal of supporting shores................................... ¾ in.
  (b) Elevation of formed surfaces before removal of
      shores.. ¾ in.
  (c) Lintels, sills, parapets, horizontal grooves, and
      other lines exposed to view...................................... ½ in.

4. Cross-sectional dimensions:
  Members, such as columns, beams, piers, walls (thickness
  only), and slabs (thickness only):
  (a) 12 in. dimension or less............................... +⅜ in. to −¼ in.
  (b) More than 12 in. dimension but not
      over 3 ft .... +½ in. to −⅜ in.
  (c) Over 3 ft dimension.. +1 in. to −¾ in.

5. Relative alignment:
  (a) Stairs:
      Difference in height between adjacent risers.............. ⅛ in.
      Difference in width between adjacent treads .... ¼ in.
  (b) Grooves:
      Specified width 2 in. or less.. ⅛ in.
      Specified width more than 2 in. but not more
        than 12 in .... ¼ in.
  (c) Formed surfaces may slope with respect to the speci-
      fied plane at a rate not to exceed the following amounts
      in 10 ft:
      Vertical alignment of outside corner of exposed
        corner columns and control joint grooves in
        concrete exposed to view .... ¼ in.
      All other conditions .... ⅜ in.
  (d) The offset between adjacent pieces of formwork
      facing material shall not exceed:
      For surfaces prominently exposed to public
        view where appearance is of special
        importance.. ⅛ in.
      Coarse-textured concrete-formed surfaces
        intended to receive plaster, stucco, or
        wainscoting .... ¼ in.
      General standard for permanently exposed
        surfaces where other finishes are not
        specified .... ½ in.
      Minimum quality surface where roughness is
        not objectionable, usually applied where
        surfaces will be concealed.. 1 in.

Erection Tolerances

1. Alignment of exposed wall panels:
   (a) Width of joints between exposed wall panels .... ¼ in.
   (b) Taper (difference in width) of joint between
       adjacent exposed wall panels, the greater of ¹⁄₄₀ in.
       per linear foot of joint or... ¹⁄₁₆ in.
   (c) Alignment of joints at adjoining corners..................... ¼ in.

2. Offset of top surfaces of adjacent elements in erected position:
   (a) With topping slab .... ¾ in.
   (b) Floor elements without topping slab........................... ¼ in.
   (c) Roof elements without topping slab............................ ¾ in.
   (d) Guideway elements to be used as riding surface ...... ¹⁄₁₆ in.

Tolerances for Concrete Footings

1. Lateral alignment:
   As cast to the center of gravity as specified: 0.02 times
       width of footing in direction of misplacement
       but not more than .... 2 in.

2. Level alignment:
   (a) Top of footings supporting masonry .... ½ in.
   (b) Top of other footings .... +½ in. to –2 in.

3. Cross-sectional dimensions:
   (a) Horizontal dimensions of formed
       members... +2 in. to –½ in.
   (b) Horizontal dimension of unformed members cast
       against soil:
       2 ft or less.. +3 in. to –½ in.
       Greater than 2 ft, but less than 6 ft.............. +6 in. to –½ in.
       Over 6 ft .... +12 in. to –½ in.

4. Relative alignment:
   Footing side and top surfaces may slope with respect
       to the specified plane at a rate not to exceed the
       following amount in 10 ft.. 1 in.

Cast-in-Place, Vertically Slipformed Building Elements

1. Vertical alignment, translation and rotation from a fixed point
   at the base of the structure:
   (a) For heights 100 ft or less .... 2 in.
   (b) For heights greater than 100 ft, ¹⁄₆₀₀ times the height but
       not more than .... 8 in.

2. Lateral alignment between adjacent elements..................... 2 in.

3. Cross-sectional dimensions of walls................ + ¾ in. to –⅜ in.

4. Relative alignment:
Formed surfaces may slope with respect to the
specified plane at a rate not to exceed the
following amount in 10 ft... ¼ in.

Tolerances of Cast-in-Place Bridges

1. Vertical alignment:
   (a) Exposed surfaces .... ¾ in.
   (b) Concealed surfaces... 1½ in.

2. Lateral alignment:
   (a) Centerline alignment...................................... 1 in.

3. Level alignment:
   (a) Profile grade .... 1 in.
   (b) Top of other concrete surfaces and horizontal grooves:
       Exposed.. ¼ in.
       Concealed .... 1½ in.
   (c) Mainline pavements in longitudinal direction,
       the gap between a 10-ft unleveled straightedge
       resting on highspots shall not exceed.......................... ⅛ in.
   (d) Mainline pavements in transverse direction, the
       gap below a 10-ft unleveled straightedge resting
       on highspots shall not exceed... ¼ in.
   (e) Ramps, sidewalks, and intersections, in any direction,
       the gap below a 10-ft unleveled straightedge
       resting on highspots shall not exceed.......................... ¼ in.

4. Cross-sectional dimensions:
   (a) Bridge slabs vertical dimension
       (thickness)....................................... ¼ in. to – ⅛ in.
   (b) Members such as columns, beams, piers,
       walls, and other (slabs thickness only)...... +½ in. to –¼ in.
   (c) Openings through concrete members .... ½ in.

5. Relative alignment:
   (a) Location of openings through concrete
       members.. ½ in.
   (b) Formed surfaces may slope with respect to the
       specified plane at a rate not to exceed the following
       amounts in 10 ft:
       Watertight joints .... ⅛ in.
       Other exposed surfaces .... ½ in.
       Concealed surfaces... 1 in.
   (c) Unformed exposed surfaces, other than pavements and
       sidewalks, may slope with respect to the specified plane
       at a rate not to exceed the following:
       In 10 ft .... ¼ in.
       In 20 ft .... ⅜ in.

Tolerances of Canal Lining

1. Lateral alignment:
   - (a) Alignment of tangents .... 2 in.
   - (b) Alignment of curves... 4 in.
   - (c) Width of section at any height: 0.0025 times
     specified width W plus 1 in.......................... $0.0025W + 1$ in.

2. Level alignment:
   - (a) Profile grade .... 1 in.
   - (b) Surface of invert.. ¼ in.
   - (c) Surface of side slope... ½ in.
   - (c) Height of lining: 0.005 times established
     height H plus 1 in. ....$0.005H + 1$ in.

APPENDIX B

Guidelines for Safety Requirements for Shoring Concrete Formwork

Reprinted with permission from the Scaffolding, Shoring, and Forming Institute, Inc., Cleveland, Ohio.

TABLE OF CONTENTS

471

INTRODUCTION

The "Recommended Safety Guidelines for Shoring Concrete Formwork" was developed to fill the need for a single, authoritative source of information on shoring as used in the construction industry, and to provide recommendations for the safe and proper use of this type of equipment.

The following safety guidelines were designed to promote safety and the safe use of shoring in the construction industry. This document does not purport to be all inclusive or to supplant or replace other additional safety and precautionary measures to cover usual or unusual conditions. Local, State, or Federal statutes or regulations shall supersede these guidelines if there is a conflict, and it is the responsibility of each employee to comply.

These safety guidelines set forth common sense procedures for safety, utilizing shoring equipment. However, equipment and systems differ and accordingly, reference must always be made to the instructions and procedures of the supplier of the equipment. Since field conditions vary and are beyond the control of the Institute, safe and proper use of the equipment and its maintenance is the responsibility of the user, and not the Institute.

This document was prepared by the Shoring Engineering Committee of the Scaffolding, Shoring & Forming Institute, Inc., 1230 Keith Building, Cleveland, Ohio, 44115.

SECTION 1 — DEFINITIONS

1.1 **Adjustable Beams** — Beams whose length can be varied within predetermined limits.

1.2 **Adjustment Screw or Screw Jack*** — A load carrying device composed of a threaded screw and an adjusting handle used for vertical adjustment of shoring and formwork.

1.3 **Allowable Load** — The ultimate load divided by the factor of safety.

1.4 **Base Plate** — A device used between post, leg, or screw and foundation to distribute the vertical load.

1.5 **Beam** — A horizontal structural load bearing member.

1.6 **Bracket** — A device used on scaffolds to extend the width of the working platform. Also refers to a member that projects from a wall or other structure and designed to support a vertical load.

1.7 **Clear Span** — The distance between facing edges of supports.

1.8 **Component** — An item used in a shoring system.

1.9 **Coupler or Clamp*** — A device used for fastening component parts together.

1.10 **Coupling Pin** — An insert device used to align lifts or tiers vertically.

1.11 **Cross Bracing** — A system of members which connect frames or panels of shoring laterally to make a tower or continuous structure.

1.12 **Dead Load (Shoring)** — The actual weight of forms, stringers, joists, reinforcing rods, and concrete to be placed.

1.13 **Deck Form** — A temporary support system consisting of primary load gathering members, joists, and decking material, all fastened together to form a rigid unit.

1.14 **Design Stress (Timber)** — Allowable stress for stress-grade lumber conforming to the recommended unit stress indicated in the tables in the "Wood Structure Design Data Book" by National Forest Products Association (formerly National Lumber Manufacturers Association), Washington, D.C.

1.15 **Extension Device** — Any component, other than an adjustment screw, used to obtain vertical adjustment of shoring towers.

1.16 **Factor of Safety or Safety Factor*** — The ratio of ultimate load to the imposed load, or, the ratio of the beam bending load at yield to the load imposed on the beam.

1.17 **Failure** — Load refusal, breakage, or separation of component parts. Load refusal is the point at which the ultimate strength is exceeded.

1.18 **Flying Deck Forms** — A formwork system for floor slabs that is moved in large sections by mechanical equipment (crane, forklift, etc.)

1.19 **Formwork** — The total system of support for freshly placed or partially cured concrete including the mold or sheathing that contacts the concrete as well as all supporting members, hardware, and bracing.

1.20 **Foundation** — The solid ground or constructed base upon which the shoring or reshoring is supported, including all underlying ground strata.

1.21 **Frame or Panel*** — The principal prefabricated, welded structural unit in a tower.

1.22 **Horizontal Shoring Beam** — Adjustable or fixed length members, either solid or fabricated type, which spans between supporting members and carry a load.

1.23 **Joists** — Horizontal members which directly support sheathing.

1.24 **Lifts or Tiers*** — The number of frames erected in a vertical direction.

1.25 **Live Load (Shoring)** — The total weight of workers, equipment, buggies, vibrators and other loads that may exist and move about due to the method of placement, leveling and screeding of the concrete pour.

---

* These terms, and others so marked, may be used synonymously.

SECTION 1 — DEFINITIONS (Continued)

1.26 **Locking Device** — A device used to secure components.

1.27 **Post Shore or Pole Shore*** — Individual vertical member used to support load.

1.28 **Reshoring** — The construction operation in which the original shoring is adjusted or replaced to support partially cured concrete and other imposed loads.

1.29 **Roller Assembly** — The mechanism of the rolling shore bracket that allows movement of the deck form across it.

1.30 **Roller Beam** — The primary load supporting member of the deck form supported by the rolling shore bracket.

1.31 **Rolling Shore Bracket** — A prefabricated welded or cast structural unit affixed to a wall or column and capable of supporting the loads transmitted from a deck form, and also the rolling load transmitted by the movement of the deck form.

1.32 **Roll-Out** — The act of moving the deck form across the roller assemblies to its position for next use or for crane pickup.

1.33 **Safe Leg Load** — The ultimate leg load divided by the appropriate safety factor.

1.34 **Sheathing** — The material forming the contact face of forms; also called lagging or sheeting.

1.35 **Shore Heads** — Flat or formed members placed and centered on vertical shoring members.

1.36 **Shoring or Falsework*** — The elements used, excluding the formwork, to support fresh concrete and/or structural members during construction.

1.37 **Shoring Layout** — An engineering drawing prepared prior to erection showing arrangements of equipment for proper shoring based on manufacturer's recommended use and loading criteria.

1.38 **Sill or Mud Sill*** — A member (usually wood) designed to distribute the vertical shoring loads to the ground or slab below.

1.39 **Spacing** — The horizontal center-to-center distance between support members.

1.40 **Stringers or Ledgers*** — Horizontal structural members which directly support joists.

1.41 **System Scaffold Shoring** — An assembly used as a load-carrying structure, consisting of fabricated columns, braces, ties, adjustable bases, and other components which connect to the uprights to form the structure.

1.42 **Towers** — A composite structure of frames, braces, and accessories.

1.43 **Tube and Coupler Shoring** — An assembly used as a load-carrying structure, consisting of tube or pipe which serves as posts, braces and ties, a base supporting the posts, and special couplers which serve to connect the uprights and join the various members.

1.44 **Ultimate Load** — The minimum load which may be placed on a structure causing failure.

---

* These terms, and others so marked, may be used synonymously.

SECTION 2 — GENERAL REQUIREMENTS FOR SHORING

Design

2.1   Shoring installations constructed in accordance with these recommended safety requirements shall require a shoring layout.

2.2   The shoring layout shall include details accounting for unusual conditions such as heavy beams, sloping areas, ramps and cantilevered slabs, as well as plan and elevation views, and applicable construction notes.

2.3   A copy of the shoring layout shall be available and used on the job site at all times.

2.4   The shoring layout shall be prepared or approved by a person qualified to analyze the loadings and stresses which are induced during the construction process.

2.5   The minimum total design load for any formwork and shoring used in a slab and beam structure shall be not less than 100 lbs. per square foot for the combined live and dead load regardless of slab thickness; however, the minimum allowance for live load shall be not less than 20 lbs. per square foot.

2.6   When motorized carts or buggies are used, the design load, as described in Section 2.5, shall be increased 25 lbs. per square foot.

2.7   Allowable loads shall be based on a safety factor consistent with the type of shoring used and as set forth in Sections 3, 4, 5, 6, 7, and 8.

2.8   The design stresses for form lumber and timbers shall commensurate with the grade, conditions, and species of lumber used, in accordance with the current edition of the "National Design Specification for Stress-Grade Lumber and Its Fastenings". (National Forest Products Association.)

2.9   Design stresses may be increased for short term loading conditions as provided in the current edition of "The Wood Structure Design Data Book". (National Forest Products Association.)

2.10   The design stresses used for form lumber and timber shall be shown on all drawings, specifications and shoring layouts.

Installation

2.11   The sills for shoring shall be sound, rigid and capable of carrying the maximum intended load without settlement or displacement. The load should be applied to the sill in a manner which will avoid overturning of the tower or the sill.

2.12   When shoring from soil, an engineer or other qualified person shall determine that the soil is adequate to support the loads which are to be placed on it.

2.13   Weather conditions can reduce the load-carrying capacity of the soil below the sill. Under these conditions, the sill design must be evaluated and adjusted to compensate for these conditions.

2.14   When shoring from fill, or when excessive earth disturbance has occurred, an engineer or other qualified person shall supervise the compaction and reworking of the disturbed area, and determine that it is capable of carrying the loads which are to be imposed on it.

2.15   Suitable sills shall be used on a pan or grid dome floor, or any other floor system involving voids where vertical shoring equipment could concentrate an excessive load on a thin concrete section.

2.16   Formwork, together with shoring equipment, shall be adequately designed, erected, braced and maintained so that it will safely support all vertical and lateral loads that might be applied, until such loads can be supported by the concrete structure.

2.17   Construction requirements for forming shall be in accordance with the provisions of the current issue of "Recommended Practice for Concrete Formwork", published by the American Concrete Institute.

2.18   When temporary storage of reinforcing rods, material, or equipment on top of formwork becomes necessary, special consideration shall be given to these areas and they shall be strengthened to meet these loads.

SECTION 2 — GENERAL REQUIREMENTS FOR SHORING (Continued)

Use

2.19 Prior to erection, all shoring equipment shall be inspected to verify that it conforms to the type of equipment specified on the shoring layout.

2.20 Damaged equipment shall not be used for shoring.See individual Sections for details.

2.21 Erected shoring equipment shall be inspected by the contractor who is responsible for placement of concrete immediately prior to pour, during pour, and after pour, until concrete is set.

2.22 If any deviation is necessary because of field conditions, the person who prepared the shoring layout shall be consulted for his approval of the actual field setup before concrete is placed, and the shoring layout shall be revised to indicate any approved changes.

2.23 The shoring setup shall be checked by the contractor who erects the equipment to determine that all details of the layout have been met.

2.24 The contractor who erects the shoring equipment shall ensure that the lateral stability bracing specified in the shoring layout is installed as the erection progresses.

2.25 The completed shoring setup shall have all specified bracing installed.

2.26 All vertical shoring equipment shall be plumb in both directions, unless otherwise specified in the layout. The maximum allowable deviation from the vertical centerline of the leg is 1/8 inch in 3 feet, but this maximum deviation in the completed structure shall not exceed the radius of vertical member. If this tolerance is exceeded, the shoring equipment shall not be used until readjusted.

2.27 Any erected shoring equipment that is damaged or weakened shall be immediately removed and replaced by adequate shoring.

2.28 Prior to pouring of concrete, the method of placement should be evaluated to insure that the additional load, i.e., vibrators, pump hoses, etc., will not adversely affect the shoring structure.

Dismantling

2.29 Loaded shoring equipment shall not be released or removed until the supported concrete is sufficiently cured.

2.30 Release and removal of loads from shoring equipment shall be sequenced so that the equipment which is still in place is not overloaded.

2.31 Slabs or beams which are to be reshored should be allowed to take their actual permanent deflection before reshoring equipment is installed.

2.32 While the reshoring is underway, no construction loads shall be permitted on the partially cured concrete.

2.33 The allowable load on the supporting slab shall not be exceeded when reshoring.

2.34 The reshoring shall be thoroughly checked by the engineer of record to determine that it is properly placed and that it has the allowable load capacity to support the areas that are being reshored.

2.35 Do not use stability bracing as a work platform.

2.36 Bracing members shall not be used as a support for a work platform.

2.37 Independent work platforms should be used to support workers and materials.

2.38 When stripping the equipment, lower the components gently, do not allow them to fall onto work platform.

2.39 Begin dismantling at the top. Do not throw shoring or forming equipment or other material to the ground.

2.40 Do not strip shoring by hammering or pulling at the base of vertical equipment and allowing it and other supported equipment to fall.

2.41 Stock-pile dismantled equipment in an orderly manner.

2.42 The work area should be kept clear of personnel not involved in the dismantling process.

SECTION 3 — TUBULAR WELDED FRAME SHORING

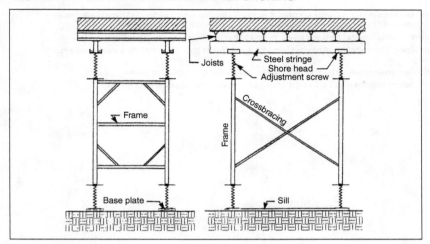

Tubular Frame Shoring

3.0   See Section 2, "General Requirements for Shoring".

3.1   Metal tubular frames used for shoring shall have allowable loads based on tests conducted according to a standard test procedure for the Compression Testing of Scaffolds and Shores, as established by the Scaffolding, Shoring & Forming Institute, Inc.

3.2   Design of shoring layouts using tubular welded frames shall be based on allowable loads which were obtained using these test procedures, and at least a 2.5 to 1 safety factor.

3.3   All metal frame shoring equipment shall be inspected before erection by the contractor who erects the equipment.

3.4   Metal frame shoring equipment and accessories shall not be used if excessively rusted, bent, dented, rewelded beyond the original factory weld locations, having broken welds, or having other defects.

3.5   All components shall be in good working order and in a condition similar to that of original manufacture.

3.6   When checking the erected shoring frames with the shoring layout, the spacing between towers and cross brace spacing shall not exceed that shown on the layout, and all locking devices shall be in the closed position.

3.7   Devices to which the external lateral stability bracing are attached shall be securely fastened to the legs of the shoring frames.

3.8   Base plates, shore heads, extension devices, or adjustment screws shall be used in top and bottom of each leg of every shoring tower.

3.9   All base plates, shore heads, extension devices or adjustment screws shall be in firm contact with the footing sill, and/or the form material, and shall be snug against the legs of the frame.

3.10 There shall be no gaps between the lower end of one frame and the upper end of the other frame.

3.11 Any component which cannot be brought into proper alignment or contact with the component into or onto which it is intended to fit, shall be removed and replaced.

SECTION 3 — TUBULAR WELDED FRAME SHORING (Continued)

3.12 When two or more tiers of frames are used, they shall be cross braced. Towers shall have lateral bracing in accordance with manufacturers' recommendations and as shown on the shoring layout.

3.13 Eccentric loads on shore heads and similar members shall be avoided.

3.14 Special precautions shall be taken when formwork is at angles or sloping, or when the surface shored from is sloping.

3.15 Adjustment screws shall not be adjusted to raise formwork during concrete placement.

3.16 Work platform brackets shall not be used for carrying shoring loads. (See Section 1.6)

SECTION 4 — TUBE AND COUPLER SHORING

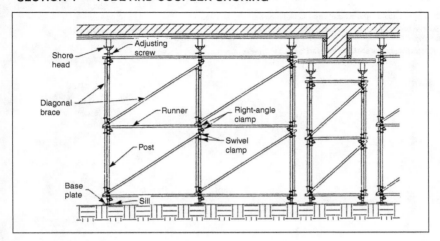

Tube & Coupler Shoring

4.0   See Section 2, "General Requirements for Shoring".

4.1   Tube and coupler systems towers used for shoring shall have allowable loads based on tests conducted according to a standard test procedure for the Compression Testing of Scaffolds and Shores, such as the one established by the Scaffolding, Shoring & Forming Institute, Inc., or its equivalent.

4.2   Design of tube and coupler systems towers in the shoring layouts shall be based on allowable loads which were obtained using these test procedures, and at least a 2.5 to 1 safety factor.

4.3   All tube and coupler systems components shall be inspected by the contractor who erects the equipment before being used.

4.4   Components structures shall not be used if excessively rusted, bent, dented, or having other defects.

4.5   All components shall be in good working order and in a condition similar to that of original manufacture. Couplers (clamps) shall not be used if deformed, broken, having defective or missing threads on bolts, or other defects.

4.6   The material used for the couplers (clamps) shall be of a structural type such as drop-forged steel, malleable iron, or structural grade aluminum. Gray cast iron shall not be used.

4.7   When checking the erected shoring towers with the shoring layout, the spacing between posts shall not exceed that shown on the layout, and all locking devices shall be in their closed position.

4.8   Base plates, shore heads, extension devices, or adjustment screws shall be used in top and bottom of each post.

4.9   All base plates, shore heads, extension devices, or adjustment screws shall be in firm contact with the footing sill and/or the form material, and shall be snug against the posts.

4.10 Any component which cannot be brought into proper alignment or contact with the component into or onto which it is intended to fit, shall be removed and replaced.

SECTION 4 — TUBE AND COUPLER SHORING (Continued)

4.11 Eccentric loads on shore heads and similar members shall be avoided.

4.12 Special precautions shall be taken when formwork is at angles or sloping, or when the surface shored from is sloping.

4.13 Adjustment screws shall not be used to raise formwork during concrete placement.

SECTION 5 — SYSTEM SCAFFOLD SHORING

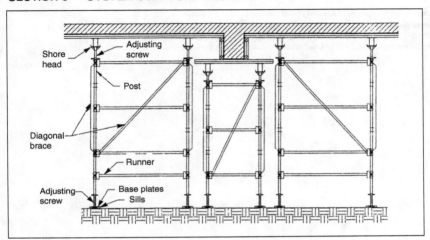

Systems Shoring

5.0 See Section 2, "General Requirements for Shoring".

5.1 System scaffold towers used for shoring shall have allowable loads based on tests conducted according to a standard test procedure for the Compression Testing of Scaffolds and Shores, such as the one established by the Scaffolding, Shoring & Forming Institute, Inc., or its equivalent.

5.2 Design of system scaffold towers in the shoring layouts shall be based on allowable loads which were obtained using these test procedures, and at least a 2.5 to 1 safety factor.

5.3 All system components shall be inspected by the contractor who erects the equipment before being used.

5.4 Components of shoring structures shall not be used if excessively rusted, bent, dented, or having other defects.

5.5 All components shall be in good working order and in a condition similar to that of original manufacture. Couplers (clamps) shall not be used if deformed, broken, having defective or missing threads on bolts, or other defects.

5.6 When checking the erected shoring towers with the shoring layout, the spacing between the posts.shall not exceed that shown on the layout, and all locking devices shall be in the closed position.

5.7 Devices to which the external lateral stability bracings are attached shall be securely fastened to the posts.

5.8 Devices to which the external lateral stability bracing are attached shall be firmly attached to the post.

5.9 Base plates, shore heads, extension devices, or adjustment screws shall be used in top and bottom of each post.

5.10 All base plates, shore heads, extension devices, or adjustment screws shall be in firm contact with the footing sill and/or the form material, and shall be snug against the posts.

5.11 Any component which cannot be brought into proper alignment or contact with the component into or onto which it is intended to fit, shall be removed and replaced.

SECTION 5 — SYSTEM SCAFFOLD SHORING (Continued)

5.12 Eccentric loads on shore heads and similar members shall be avoided.

5.13 Special precautions shall be taken when formwork is at angles or sloping, or when the surface shored from is sloping.

5.14 Adjustment screws shall not be used to raise formwork during concrete placement.

SECTION 6 — SINGLE POST SHORES

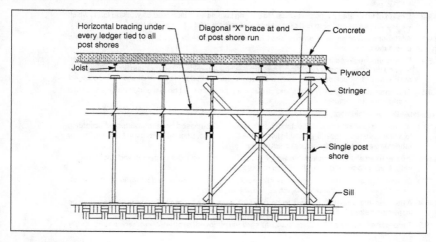

Single Post Shore

6.0 See Section 2, "General Requirements for Shoring".

6.1 When checking erected single post shores with the shoring layout, the spacing between shores in either direction shall not exceed that shown on the layout, and all clamps, screws, pins and all other components should be in the closed or engaged position.

6.2 For stability, single post shores shall have adequate bracing provided in both the longitudinal and transverse directions, and adequate diagonal bracing shall be provided.

6.3 Devices to which the external lateral stability bracing are attached shall be securely fastened to the single post shores.

6.4 All base plates or shore heads of single post shores shall be in firm contact with the footing sill and form material.

6.5 Eccentric loads on shore heads shall be prohibited, unless the post shore is designed to accommodate such loads.

6.6 Special precautions shall be taken when formwork is at angles, or sloping, relative to the post shore stringers (bearers), etc.

6.7 When post shores rest on a sloping surface, they shall be braced to compensate for the lateral forces involved.

6.8 Adjustment of single post shores to raise formwork shall not be made after concrete is in place.

6.9 Fabricated single post shores.

 6.9.1 All working load ratings for fabricated single post shores shall be based on tests conducted according to a standard test procedure for fabricated single post shores, established by the Scaffolding, Shoring & Forming Institute, Inc., or its equivalent.

SECTION 6 — SINGLE POST SHORES (Continued)

6.9.2   Design of fabricated single post shores in shoring layouts shall be based on working loads which were obtained using these test procedures and at least a 3 to 1 safety factor.

6.9.3   All fabricated single post shores shall be inspected before being used by the contractor who erects the equipment.

6.9.4   Fabricated single post shores shall not be used if excessively rusted, bent, dented, rewelded beyond the original factory weld locations or have broken welds. If they contain timber, they shall not be used if timber is split, cut, has sections removed, is rotted or otherwise structurally damaged.

6.9.5   All clamps, screws, pins, threads and all other components shall be in a condition similar to that of original manufacturer.

6.10 Adjustable timber single post shores.

6.10.1   All working load ratings for adjustable post shores shall be based on tests conducted according to a standard test procedure established by the Scaffolding, Shoring & Forming Institute, Inc. for fabricated single post shores, or its equivalent.

6.10.2   Timber used shall have allowable working load for each size, grade, species and shoring height, as recommended by the clamp manufacturer.

6.10.3   Design of adjustable timber single posts in shoring layouts shall be. based on working loads which were obtained using the.test procedure and at least a 3 to 1 safety factor.

6.10.4   All timber and adjusting devices to be used for adjustable timber single post shores shall be inspected before erection by the contractor who erects the equipment.

6.10.5   Timber shall not be used if it is split, cut, has sections removed, is rotted, or is otherwise structurally damaged.

6.10.6   Adjusting devices shall not be used if excessively rusted, bent, dented, rewelded beyond the original factory weld locations, or have broken welds.

6.10.7   Hardwood wedges shall be used to obtain final adjustment and firm contact with footing sills and form material.

6.10.8   All nails used to secure bracing on adjustable timber single post shores shall be driven home and bent over if possible.

6.10.9   When placing an adjustable clamp, it shall be aligned parallel to the stringers.

6.11 Timber single post shores.

6.11.1   All safety factors and allowable working loads for timber used as single post shores shall be as recommended in tables for wooden columns in the current edition of the "Wood Structural Design Data Book", prepared by the National Forest Products Association, Washington, D.C.

6.11.2   Design of timber single posts in shoring layouts shall be based on working loads which were obtained by using the tables referred to in Paragraph 6.11.1.

6.11.3   All timber to be used for single post shoring shall be inspected before erection by the contractor who erects the equipment.

6.11.4   Timber shall not be used if it is split, cut, has sections removed, is rotted, or is otherwise structurally damaged.

6.11.5   The ends of supporting timbers shall be cut square to provide proper bearing.

6.11.6   Hardwood wedges shall be used to obtain final adjustment and firm contact with footing sills and form material.

6.11.7   All nails used to secure bracing on timber single post shores shall be driven home and bent over if possible.

SECTION 7 — HORIZONTAL SHORING BEAMS

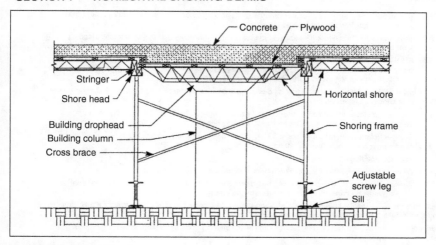

Horizontal Shore at Drophead

7.0   See Section 2, "General Requirements for Shoring".

7.1   Published horizontal shoring beams' allowable loads shall be based on tests conducted by the manufacturer to determine load necessary to:

    (1)   Eliminate camber on adjustable horizontal members with built-in camber.

    (2)   Deflect the member a maximum of 1/360th of the span, or 1/4 inch maximum.

Sufficient tests shall be conducted to be able to create a table for all spans and loads to which the member may be subjected.

In addition, horizontal shoring beams shall be tested to their ultimate load which shall be at least twice the allowable load.

7.2   The design of horizontal shoring components in shoring layouts shall be based on allowable loads which were obtained using these test procedures and taken into account the continuity factors in multispan cases.

7.3   All horizontal shoring beams shall be inspected before using by the contractor who erects the equipment.

7.4   Erected horizontal shoring beams shall be inspected to be certain that the span, spacing, types of shoring beams, and size, height and spacing of vertical shoring supports are in accordance with the shoring layout.

7.5   Adequate support shall be provided and maintained to properly distribute shoring loads. When supporting horizontal beams on:

    7.5.1   Masonry walls. Walls shall have adequate strength. Brace walls as necessary.

    7.5.2   Ledgers supported by walls using bolts or other means. The ledgers shall be properly designed and installed per recommendation of supplier or job architect/engineer. Actual anchor detail and design is the responsibility of the engineer of record or architect.

SECTION 7 — HORIZONTAL SHORING BEAMS (Continued)

7.5.3　Formwork. The formwork designer has the responsibility to design the formwork to carry the additional loads imposed by the shoring beams.

7.5.4　Structural steel framework. The ability of the steel framework to support this construction loading shall be checked and approved by the responsible project architect/engineer.

7.5.5　Steel hangers. Their bearing ends shall be fully engaged on the hangers. The hangers shall be designed to conform to the bearing end, and to safely support the shoring loads imposed. (Hanger manufacturers' recommendations shall be followed.)

7.6　When installing horizontal shoring beams or designing a shoring system using horizontal shoring beams, precautions should be taken when:

7.6.1　Secure cantilevered horizontal shoring beams to stringers whenever end-loading presents a danger of tipping or up-ending the horizontal shoring beams.

7.6.2　Stringer height/width ratio exceeds 2.5 to 1. Under no circumstances shall horizontal shoring beams bear on a single "two by" stringer.

7.6.3　Eccentric loading conditions exist.

7.6.4　When stringer consists of multiple members, (i.e., double 2 x 6, 2 x 8, etc.).

7.6.5　Varying elevation of horizontal shoring beam prong to match prefabricated joist (any additional blocking must be considered in the base-to-height ratio).

7.7　Bearing ends of horizontal shoring beams shall be properly supported, and locking devices, if any, properly engaged before placing any load on beams.

7.8　Horizontal shoring beams with bearing prongs shall not be supported other than at the bearing prongs, unless recommended by supplier.

7.9　When a horizontal shoring beam is supported by a second horizontal beam, there shall be full bearing between beams as determined by the horizontal shoring beam supplier.

7.10　Do not nail adjustable horizontal beam bearing prongs to ledger.

7.11　Horizontal shoring beams with wooden nailing strips shall have the deck material nailed to the wooden nailer strip at the edge intersection of each piece of deck material and metal shoring beam as the deck material is installed.

7.12　Adjustable horizontal shoring beams shall not be used as part of a reshoring system.

7.13　Adjustable horizontal shoring beams shall not be used as a stringer for other horizontal shoring beams, unless recommended and approved by the supplier/manufacturer.

7.14　Horizontal shores at an angle to a stringer or beamside must have full bearing support under the bearing prong. The male bearing prong should be used at the beamside to facilitate full bearing.

SECTION 8 — ROLLING SHORE BRACKETS (WALL OR COLUMN MOUNTED DECK FORMS

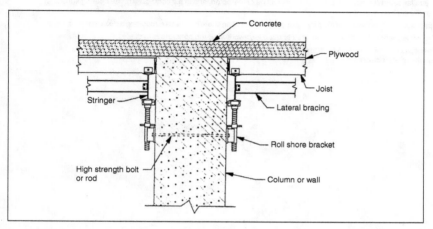

Roll Shore System

8.0 See Section 2, "General Requirements for Shoring".

8.1 Rolling shore brackets shall have allowable loads based on tests conducted by the manufacturer and witnessed by an independent testing organization.

8.2 Design of shoring layouts using rolling shore brackets shall be based on allowable loads which were obtained using the manufacturers' tests, and at least a 2.5 to 1 safety factor.

8.3 The shoring is to be approved by the engineer of record, relative to:

    a. Method used to anchor shoring bracket to column or wall.

    b. Eccentric load, on column or wall, caused by shoring bracket load.

8.4 All rolling shore brackets shall be inspected before erection by the contractor who erects the equipment.

8.5 Rolling shore brackets shall not be used if any component is excessively rusted, bent, rewelded beyond the original factory weld locations, has broken welds, or other defects.

8.6 All components shall be in good working order and in a condition similar to that of original manufacture.

8.7 When checking the erected rolling shore brackets with the shoring layout, the spacing between brackets and the size of the deck form being supported shall not exceed that shown on the layout.

8.8 The rolling shore brackets shall be plumb and level and tightened to the mounting surface with the manufacturers' designated size and grade of bolts/rods or inserts prior to concrete placement.

8.9 Check to see that the rolling shore bracket is not extended out from its vertical support face in excess of the manufacturers' recommendations. Means must be provided to prevent lateral movement of the deck form during roll-out.

SECTION 8 — ROLLING SHORE BRACKETS (Continued)

8.10 The roller beam (stringer) of the deck form shall be centered on, and in full contact with, the bearing surface of the rolling shore bracket.

8.11 Special precautions shall be taken by the user when the rolling shore bracket is supporting a deck form on a slope.

8.12 The adjustment screw shall not be used to raise formwork after the concrete is in place.

8.13 Do not place combined deck form and concrete loading on the roller assembly unless its design permits. Check with the manufacturer.

SECTION 9 — FLYING DECK FORMS

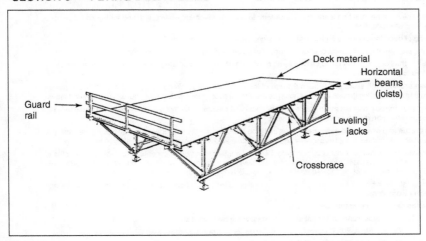

Flying Deck Form

9.0   See Section 2, "General Requirements for Shoring".

9.1   Flying deck forms shall have allowable loads as determined by detailed structural analysis. The results of the structural analysis shall be verified by tests conducted under, or equal to, field conditions, witnessed by an independent authority.

9.2   Design of shoring layouts using flying deck form systems shall be based on the above results and using factors of safety as follows: Trusses and proprietary or component supporting members - 2.5 to 1; all other members as per this shoring standard.

9.3   All flying deck components shall be inspected before use to insure that the erected form conforms to the shoring layout, that correct components are being used, and that no damaged components are being used.

9.4   Follow manufacturers' instructions, load ratings, and job site layout drawings for placement on each project.

9.5   No components shall be used if they are heavily rusted, bent, dented, torn, or otherwise defective.

9.6   All components are to be in good working order and in a condition similar to that of original manufacture.

9.7   All screw jacks and connections shall be checked for snugness.

9.8   Special precautions shall be taken by the user when the flying deck form is angling, sloping, or cantilevered.

9.9   Adjustment of flying deck forms to raise formwork shall not be made during concrete placement.

9.10 A method of positive control must be provided for the rolling-out operation by tie-back, braking system or other devices. The form must never be allowed to have uncontrolled horizontal movement.

9.11 Positive connection between the form and the crane must be provided by slings, eyes, or other devices recommended by the manufacturer.

9.12 Follow manufacturers' recommendation for roll-out, flying, and landing of the flying deck form.

SECTION 9 — FLYING DECK FORMS

9.13 A tag line must be attached to the flying deck form to control movement while flying, and for helping to position the form when landing.

9.14 Flying of deck forms during high winds is hazardous and not recommended.

9.15 Crane pick points shall not be changed without consulting with the form manufacturer.

9.16 Worker shall not be permitted to be on a flying deck form when it is completely attached to a crane and is being moved to its new location.

9.17 When shoring high slabs that require the use of long extension legs, diagonal bracing in both lateral planes resulting in braced leg groups of four (4) must be installed in accordance with manufacturer's recommendations.

9.18 Filler and hinged panels result in concentrated loads on the edge of the flying deck system and must be taken into consideration during movement and flying.

9.19 It is unsafe and unlawful to allow employees on flying deck forms without proper guardrailing. If guardrailing is impractical, employees must be properly protected by other equivalent means, such as safety belts and lanyards.

9.20 Use extreme caution against tipping hazards; small, light tables may be susceptible to tipping by weights of personnel and materials. Such tables must be restrained from tipping by rear tie-downs or front support shores and no personnel must be allowed on the deck until such restraints are installed.

9.21 Horizontal shoring beams shall be spaced over and clipped to the top chord of the flying form in accordance with supplier's drawings.

9.22 Plywood decks on flying forms shall have:

  (1) Face grain running at right angles to supporting horizontal shoring beams.

  (2) Conform to A.P.A. Plyform Class 1, B-B Exterior Grade in accordance with PSI-66 in "as new" condition, or be of equal or greater strength.

  (3) Edge-Nailed to the nailer strip on the horizontal shoring beam at supplier's recommended spacing.

HORIZONTAL SHORING BEAM SAFETY RULES

As Recommended by the

SCAFFOLDING, SHORING AND FORMING INSTITUTE, INC.

Following are some common sense rules designed to promote safety in the use of horizontal shoring beams. These rules are illustrative and suggestive only, and are intended to deal only with some of the many practices and conditions encountered in the use of horizontal shoring beams. The rules do not purport to be all-inclusive or to supplant or replace other additional safety and precautionary measures to cover usual or unusual conditions. They are not intended to conflict with, or supersede, any state, local, or federal statute or regulation; reference to such specific provisions should be made by the user. (See Rule II.)

    I. POST THESE SHORING SAFETY RULES in a conspicuous place and be sure that all persons who erect, use, or dismantle horizontal shoring beams are aware of them.

    II. FOLLOW ALL STATE, LOCAL AND FEDERAL CODES, ORDINANCES AND REGULATIONS pertaining to Shoring.

    III. INSPECT ALL EQUIPMENT BEFORE USING. Never use any equipment that is damaged.

    IV. A SHORING LAYOUT—shall be available and used on the jobsite at all times.

    V. INSPECT ERECTED SHORING AND FORMING FOR CONFORMITY WITH LAYOUT AND SAFETY PRACTICES: a. Immediately prior to pour. b. During pour. c. After pour, until concrete is set.

    VI. CONSULT YOUR SHORING EQUIPMENT SUPPLIER WHEN IN DOUBT. Shoring is his business, NEVER TAKE CHANCES.

    VII. CONSULT STEEL FRAME SHORING SAFETY RULES, SINGLE POST SHORE SAFETY RULES, VERTICAL SHORING SAFETY CODE, AND FRAME SHORING ERECTION PROCEDURE, developed by the Scaffolding and Shoring Institute.

A. USE MANUFACTURERS' RECOMMENDED SAFE WORKING LOADS AND PROCEDURES FOR:

  1. Span, spacing, and types of shoring beams.

  2. Types, sizes, heights, and spacing of vertical shoring supports.

B. USE LUMBER EQUIVALENT TO THE STRESS, species, grade and size used on the layout. Use only lumber that is in good condition. Do not splice between supports.

C. DO NOT MAKE UNAUTHORIZED CHANGES OR SUBSTITUTION OF EQUIPMENT; always consult your supplier prior to making changes necessitated by jobsite conditions.

D. PROVIDE AND MAINTAIN ADEQUATE SUPPORT TO properly distribute shoring loads. When supporting horizontal shoring beams on:

  1. Masonry walls, insure that masonry units have adequate strength. Brace walls as necessary.

  2. Ledgers supported by walls using bolts, or other means, they should be properly designed and installed per recommendation of supplier or job architect/engineer.

  3. Formwork, such formwork should be designed for the additional loads imposed by the shoring beams.

  4. Structural Steel Framework, the ability of the steel to support this construction loading should be checked and approved by the responsible project architect/engineer.

  5. When supporting horizontal beams on steel hangers, be sure that the bearing ends fully engage on the hangers. The hangers shall be designed to conform to the bearing end and shall have a rated strength to safely support the shoring loads imposed. (Follow hanger manufacturers' recommendations.)

  6. Vertical Shoring (see II and VII above).

E. SPECIAL CONSIDERATION MUST BE GIVEN TO THE INSTALLATION OF HORIZONTAL SHORING:

  1. When sloped or supported by sloping ledgers (stringers).

  2. When ledger (stringer) height/width ratio exceeds 2½-1. Under no circumstances shall horizontal shoring beams bear on a single "two by" ledger (stringer).

  3. When eccentric loading conditions exist.

  4. When ledger (stringer) consists of multiple members. (i.e., double 2x6, 2x8, etc.)

F. ASSURE THAT BEARING ENDS OF SHORING BEAMS ARE PROPERLY SUPPORTED and that locking devices are properly engaged before placing any load on beams.

G. IF MOTORIZED CONCRETE PLACEMENT EQUIPMENT IS TO BE USED, be sure that lateral and other forces have been considered and adequate precautions taken to assure stability.

H. HORIZONTAL SHORING BEAMS SHOULD NOT be supported other than at the bearing prongs unless recommended by supplier. Under no circumstances support or cantilever truss member of horizontal shore beam.

I. DO NOT NAIL BEAM BEARING PRONGS TO LEDGER.

J. PLAN CONCRETE POURING METHODS AND SEQUENCES TO insure against unbalanced loading of the shoring equipment. Take all necessary precautions to avoid uplift of shoring components and formwork.

K. AVOID SHOCK OR IMPACT LOADS FOR which the shoring was not designed.

L. DO NOT PLACE ADDITIONAL, TEMPORARY LOADS (such as rebar bundles) on erected formwork or poured slabs, without checking the capacity of the shoring and/or structure to safely support such additional loads.

M. DO NOT RELEASE ANY PART OF THE FORMWORK OR SHORING until proper authority has been obtained. Particular consideration must be given to reshoring procedures.

N. RESHORING PROCEDURES SHOULD ALWAYS be approved by the responsible project architect/engineer.

OSHA Regulations for Formwork and Shoring

The pages in this appendix are reproduced from the United States Occupational Safety and Health Act (OSHA), 2010. This material contains OSHA regulations (standards) for construction as published in the Federal Register and online at www.osha.gov.

OSHA Regulations (Standards)

Part Number 1926: Safety and Health Regulations for Construction

Standard Number 1926 Subpart Q: Concrete and Masonry Construction

Source: www.osha.gov

1926.700   Scope, Application, and Definitions Applicable to This Subpart

(a) *Scope and application.* This subpart sets forth requirements to protect all construction employees from the hazards associated with concrete and masonry construction operations performed in workplaces covered under 29 CFR Part 1926. In addition to the requirements in Subpart Q, other relevant provisions in Parts 1910 and 1926 apply to concrete and masonry construction operations.

(b) *Definitions* applicable to this subpart. In addition to the definitions set forth in 1926.32, the following definitions apply to this subpart.

    (1) "Bull float" means a tool used to spread out and smooth concrete.

(2) "Formwork" means the total system of support for freshly placed or partially cured concrete, including the mold or sheeting (form) that is in contact with the concrete as well as all supporting members including shores, reshores, hardware, braces, and related hardware.

(3) "Lift slab" means a method of concrete construction in which floor and roof slabs are cast on or at ground level and, using jacks, lifted into position.

(4) "Limited access zone" means an area alongside a masonry wall, which is under construction, and which is clearly demarcated to limit access by employees.

(5) "Precast concrete" means concrete members (such as walls, panels, slabs, columns, and beams) which have been formed, cast, and cured prior to final placement in a structure.

(6) "Reshoring" means the construction operation in which shoring equipment (also called reshores or reshoring equipment) is placed, as the original forms and shores are removed, in order to support partially concrete and construction loads.

(7) "Shore" means a supporting member that resists a compressive force imposed by a load.

(8) "Vertical slip forms" means forms which are jacked vertically during the placement of concrete.

(9) "Jacking operation" means the task of lifting a slab (or group of slabs) vertically from one location to another [e.g., from the casting location to a temporary (parked) location, or to its final location, or to its final location in the structure], during the construction of a building/structure where the lift-slab process is being used.

1926.701　General Requirements

(a) *Construction loads*.

No construction loads shall be placed on a concrete structure or portion of a concrete structure unless the employer determines, based on information received from a person who is qualified in structural design, that the structure or portion of the structure is capable of supporting the loads.

(b) *Reinforcing steel*.

All protruding reinforcing steel, onto which employees could fall, shall be guarded to eliminate the hazard of impalement.

(c) *Post-tensioning operations*.

(1) No employee (except those essential to the post-tensioning operations) shall be permitted to be behind the jack during tensioning operations.

    (2) Signs and barriers shall be erected to limit employee access to post-tensioning area during tensioning operations.

(d) *Riding concrete buckets.*
No employee shall be permitted to ride concrete buckets.

(e) *Working under loads.*
    (1) No employee shall be permitted to work under concrete buckets while buckets are being elevated or lowered into position.
    (2) To the extent practical, elevated concrete buckets shall be routed so that no employee, or the fewest number of employees, are exposed to the hazards associated with falling concrete buckets.

(f) *Personal protective equipment.*
No employee shall be permitted to apply a cement, sand, and water mixture through a pneumatic hose unless the employee is wearing protective head and face equipment.

Standard Number 1926.702 Requirements for Equipment and Tools

(a) *Bulk cement storage.*
    (1) Bulk storage bins, containers, and silos shall be equipped with the following:
        (i) Conical or tapered bottoms; and
        (ii) Mechanical or pneumatic means of starting the flow of material
    (2) No employee shall be permitted to enter storage facilities unless the ejection system has been shut down, locked out, and tagged to indicate that the ejection system is not to be operated.

(b) *Concrete mixers.*
Concrete mixers with one cubic yard (.8 m^3) or larger loading skips shall be equipped with the following:
    (1) A mechanical device to clear the skip of materials; and
    (2) Guardrails installed on each side of the skip.

(c) *Power concrete trowels.*
Powered and rotating type concrete troweling machines that are manually guided shall be equipped with a control switch that will automatically shut off the power whenever the hands of the operator are removed from the equipment handles.

(d) *Concrete buggies.*
Concrete buggy handles shall not extend beyond the wheels on either side of the buggy.

(e) *Concrete pumping systems.*
  (1) Concrete pumping systems using discharge pipes shall be provided with pipe supports designed for 100 percent overload.
  (2) Compressed air hoses used on concrete pumping system shall be provided with positive fail-safe joint connectors to prevent separation of sections when pressurized.

(f) *Concrete buckets.*
  (1) Concrete buckets equipped with hydraulic pneumatic gates shall have positive safety latches or similar safety devices installed to prevent premature or accidental dumping.
  (2) Concrete buckets shall be designed to prevent concrete from hanging up on top and the sides.

(g) *Tremies.*
  Sections of tremies and similar concrete conveyances shall be secured with wire rope (or equivalent materials) in addition to the regular couplings or connections.

(h) *Bull floats.*
  Bull float handles, used where they might contact energized electrical conductors, shall be constructed of nonconductive material or insulated with a nonconductive sheath whose electrical and mechanical characteristics provide the equivalent protection of a handle constructed of nonconductive material.

(i) *Masonry saws.*
  (1) Masonry saws shall be guarded with a semicircular enclosure over the blade.
  (2) A method for retaining blade fragments shall be incorporated in the design of the semicircular enclosure.

(j) *Lockout/Tagout procedures.*
  (1) No employee shall be permitted to perform maintenance or repair activity on equipment (such as compressors, mixers, screens, or pumps used for concrete and masonry construction activities) where the inadvertent operation of the equipment could occur and cause injury, unless all potentially hazardous energy sources have been locked out and tagged.
  (2) Tags shall read Do Not Start or similar language to indicate that the equipment is not to be operated.

1926.703 Requirements for Cast-in-Place Concrete

(a) *General requirements for formwork.*
  (1) Formwork shall be designed, fabricated, erected, supported, braced, and maintained so that it will be capable

of supporting without failure all vertical and lateral loads that may reasonably be anticipated to be applied to the formwork. Formwork which is designed, fabricated, erected, supported, braced, and maintained in conformance with the Appendix to this section will be deemed to meet the requirements of this paragraph.

(2) Drawings or plans, including all revisions, for the jack layout, formwork (including shoring equipment), working decks, and scaffolds shall be available at the jobsite.

(b) *Shoring and reshoring.*

(1) All shoring equipment (including equipment used in reshoring operations) shall be inspected prior to erection to determine that the equipment meets the requirements specified in the formwork drawings.

(2) Shoring equipment found to be damaged such that its strength is reduced to less than that required by 1926.703(a)(1) shall not be used for shoring.

(3) Erected shoring equipment shall be inspected immediately prior, during, and immediately after concrete placement.

(4) Shoring equipment that is found to be damaged or weakened after erection, such that its strength is reduced to less than that required by 1926.703(a)(1), shall be immediately reinforced.

(5) The sills for shoring shall be sound, rigid, and capable of carrying the maximum intended load.

(6) All base plates, shore heads, extension devices, and adjustment screws shall be in firm contact, and secured when necessary, with the foundation and the form.

(7) Eccentric loads on shore heads and similar members shall be prohibited unless these members have been designed for such loading.

(8) Whenever single post shores are used one on top of another (tiered) the employer shall comply with the following specific requirements in addition to the general requirements for formwork:

(i) The design of the shoring shall be prepared by a qualified designer and the erected shoring shall be inspected by an engineer qualified in structural design.

(ii) The single post shores shall be vertically aligned.

(iii) The single post shores shall be spliced to prevent misalignment.

(iv) The single post shores shall be adequately braced in two mutually perpendicular directions at the splice level. Each tier shall also be diagonally braced in the same two directions.

   (9) Adjustment of single post shores to raise formwork shall not be made after the placement of concrete.

  (10) Reshoring shall be erected, as the original forms and shores are removed, whenever the concrete is required to support loads in excess of its capacity.

(c) *Vertical slip forms.*

   (1) The steel rods or pipes on which jacks climb or by which the forms are lifted shall be:

     (i) Specifically designed for that purpose; and

    (ii) Adequately braced where not encased in concrete.

   (2) Forms shall be designed to prevent excessive distortion of the structure during the jacking operation.

   (3) All vertical slip forms shall be provided with scaffolds or work platforms where employees are required to work or pass.

   (4) Jacks and vertical supports shall be positioned in such a manner that the loads do not exceed the rated capacity of the jacks.

   (5) The jacks or other lifting devices shall be provided with mechanical dogs or other automatic holding devices to support the slip forms whenever failure of the power supply or lifting mechanism occurs.

   (6) The form structure shall be maintained within all design tolerances specified for plumbness during the jacking operation.

   (7) The predetermined safe rate of lift shall not be exceeded.

(d) *Reinforcing steel.*

   (1) Reinforcing steel for walls, piers, columns, and similar vertical structures shall be adequately supported to prevent overturning and to prevent collapse.

   (2) Employers shall take measures to prevent unrolled wire mesh from recoiling. Such measures may include, but are not limited to, securing each end of the roll or turning over the roll.

(e) *Removal of formwork.*

   (1) Forms and shores (except those used for slabs on grade and slip forms) shall not be removed until the employer determines that the concrete has gained sufficient strength to support its weight and superimposed loads. Such determination shall be based on compliance with one of the following:

     (i) The plans and specifications stipulate conditions for removal of forms and shores, and such conditions have been followed, or

(ii) The concrete has been properly tested with an appropriate ASTM standard test method designed to indicate the concrete compressive strength, and the test results indicate that the concrete has gained sufficient strength to support its weight and superimposed loads.

(2) Reshoring shall not be removed until the concrete being supported has attained adequate strength to support its weight and all loads in place upon it.

Appendix to 1926.703(a)(1)   General Requirements for Formwork

(This Appendix is non-mandatory.)

This appendix serves as a non-mandatory guideline to assist employers in complying with the formwork requirements in 1926.703(a)(1). Formwork which has been designed, fabricated, erected, braced, supported, and maintained in accordance with Sections 6 and 7 of the American National Standard for Construction and Demolition Operations Concrete and Masonry Work, ANSI A10.9-1983, shall be deemed to be in compliance with the provision of 1926.703(a)(1).

1926.704   Requirements for Precast Concrete

(a) Precast concrete wall units, structural framing, and tilt-up wall panels shall be adequately supported to prevent overturning and to prevent collapse until permanent connections are completed.

(b) Lifting inserts which are embedded or otherwise attached to tilt-up precast concrete members shall be capable of supporting at least two times the maximum intended load applied or transmitted to them.

(c) Lifting inserts which are embedded or otherwise attached to precast concrete members, other than the tilt-up members, shall be capable of supporting at least four times the maximum intended load applied or transmitted to them.

(d) Lifting hardware shall be capable of supporting at least five times the maximum intended load applied or transmitted to the lifting hardware.

(e) No employee shall be permitted under precast concrete members lifted or tilted into position except those employees required for the erection of those members.

1926.705   Requirements for Lift-Slab Operations

(a) Lift-slab operations shall be designed and planned by a registered professional engineer who has experience in lift-slab construction. Such plans and designs shall be implemented by the employer and shall include detailed instructions and sketches indicating the prescribed method of erection. These plans and designs shall also include provisions for ensuring lateral stability of the building/structure during construction.

(b) Jacks/lifting units shall be marked to indicate their rated capacity as established by the manufacturer.

(c) Jacks/lifting units shall not be loaded beyond their rated capacity as established by the manufacturer.

(d) Jacking equipment shall be capable of supporting at least two and one-half times the load being lifted during jacking operations and the equipment shall not be overloaded. For the purpose of this provision, jacking equipment includes any load bearing component which is used to carry out the lifting operation(s). Such equipment includes, but is not limited, to the following: threaded rods, lifting attachments, lifting nuts, hook-up collars, T-caps, shearheads, columns, and footings.

(e) Jacks/lifting units shall be designed and installed so that they will neither lift nor continue to lift when they are loaded in excess of their rated capacity.

(f) Jacks/lifting units shall have a safety device installed which will cause the jacks/lifting units to support the load in any position in the event any jacklifting unit malfunctions or loses its lifting ability.

(g) Jacking operations shall be synchronized in such a manner to ensure even and uniform lifting of the slab. During lifting, all points at which the slab is supported shall be kept within ½ inch of that needed to maintain the slab in a level position.

(h) If leveling is automatically controlled, a device shall be installed that will stop the operation when the ½ inch tolerance set forth in paragraph (g) of this section is exceeded or where there is a malfunction in the jacking (lifting) system.

(i) If leveling is maintained by manual controls, such controls shall be located in a central location and attended by a competent person whole lifting is in progress. In addition to meeting the definition in 1926.32(f), the competent person must be experienced in the lifting operation and with the lifting equipment being used.

(j) The maximum number of annually controlled jacks/lifting units on one slab shall be limited to a number that will permit the operator to maintain the slab level within specified tolerances of paragraph (g) of this section, but in no case shall that number exceed 14.

(k)(1) No employee, except those essential to the jacking operation, shall be permitted in the building/structure while any jacking operation is taking place unless the building/structure has been reinforced sufficiently to ensure its integrity during erection. The phrase "reinforced sufficiently to ensure its integrity" used in this paragraph means that a registered professional engineer, independent of the engineer who designed and planned the lifting operation, has determined from the plans that if there is a loss of support at any jack location, that loss will be confined to that location and the structure as a whole will remain stable.

(k)(2) Under no circumstances, shall any employee who is not essential to the jacking operation be permitted immediately beneath a slab while it is being lifted.

(k)3 For the purpose of paragraph (k) of this section, a jacking operation begins when a slab or group of slabs is lifted and ends when such slabs are secured (with either temporary connections or permanent connections.

(k)(4) Employers who comply with appendix A to 1926.705 shall be considered to be in compliance with the provisions of paragraphs (k)(1) through (k)(3) of this section.

(l) When making temporary connections to support slabs, wedges shall be secured by tack welding, or an equivalent method of securing the wedges to prevent them from falling out of position. Lifting rods may not be released until the wedges at that column have been secured.

(m) All welding on temporary and permanent connections shall be performed by a certified welder, familiar with the welding requirements specified in the plans and specifications for the lift-slab operation.

(n) Load transfer from jacks/lifting units to building columns shall not be executed until the welds on the column shear plates (weld blocks) are cooled to air temperature.

(o) Jacks/lifting units shall be positively secured to building columns so that they do not become dislodged or dislocated.

(p) Equipment shall be designed and installed so that the lifting rods cannot slip out of position or the employer shall institute other measures, such as the use of locking or blocking devices, which will provide positive connection between the lifting rods and attachments and will prevent components from disengaging during lifting operations.

Appendix to 1926.705   Lift-Slab Operations

(This Appendix is non-mandatory.)

In paragraph 1926.705(k), OSHA requires employees to be removed from the building/structure during jacking operations unless an independent registered professional engineer, other than the engineer who designed and planned the lifting operation, has determined that the building/structure has been sufficiently reinforced to insure the integrity of the building/structure. One method to comply with this provision is for the employer to ensure that continuous bottom steel is provided in every slab and in both directions through every wall or column head area. (Column head area means the distance between lines that are one and one-half times the thickness of the slab or drop panel. These lines are located outside opposite faces of the outer edges of the shearhead sections. See Figure C-1.)

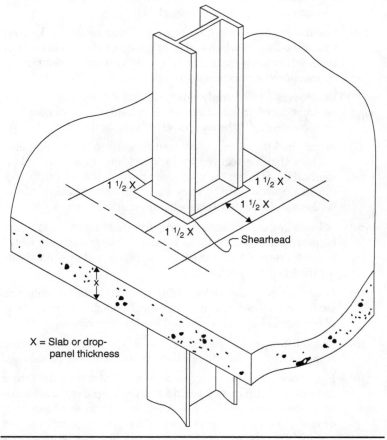

FIGURE C-1   Column head area.

The amount of bottom steel shall be established by assuming loss of support at a given lifting jack and then determining the steel necessary to carry, by catenary action over the span between surrounding supports, the slab service dead load plus any service dead and live loads likely to be acting on the slab during jacking. In addition, the surrounding supports must be capable of resisting any additional load transferred to them as a result of the loss of support at the lifting jack considered.

Appendix A: References to Subpart Q of Part 1926

(This Appendix is non-mandatory.)

The following non-mandatory references provide information which can be helpful in understanding and complying with the requirements contained in Subpart Q.

- Accident Prevention Manual for Industrial Operations; 8th Edition.
- National Safety Council.
- Building Code Requirements for Reinforced Concrete (ACI 318-83).
- Formwork for Concrete (ACI SP-4).
- Recommended Practice for Concrete Formwork (ACI 347-78).
- Safety Requirements for Concrete and Masonry Work (ANSI A10.9-1983).
- Standard Test Method for Compressive Strength of Cylindrical Concrete Specimens (ASTM C39-86).
- Standard Test Method for Making and Curing Concrete Test Specimens in the Field (ASTM C31-85).
- Standard Test Method for Penetration Resistance of Hardened Concrete (ASTM C803-82).
- Standard Test Method for Compressive Strength of Concrete Cylinder Cast-In-Place in Cylindrical Molds (ASTM C873-85).
- Standard Method for Developing Early Age Compressive Test Values and Projecting Later Age Strengths (ASTM C918-80).
- Recommended Practice for Inspection and Testing Agencies for Concrete Steel and Bituminous Materials as Used in Construction (ASTM E329-77).
- Method of Making and Curing Concrete Test Specimens in the Laboratory (ASTM C192-88).
- Methods of Obtaining and Testing Drilled Cores and Sawed Beams of Concrete (ASTM C42-87).

- Methods of Securing, Preparing and Testing Specimens from Hardened Lightweight Insulating Concrete for Compressive Strength (ASTM C513-86).

- Test Method for Comprehensive Strength of Lightweight Insulating Concrete (ASTM C495-86).

- Method of Making, Accelerating Curing, and Testing of Concrete Compression Test Specimens (ASTM C684-81).

- Test Method for Compressive Strength of Concrete Using Portions of Beams Broken in Flexure (ASTM C116-68,1980).

Conversion of Units of Measure between U.S. Customary System and Metric System

Multiply	by	To obtain
cm	0.3937	in.
cm^3	0.061024	$in.^3$
ft	0.3048	m
ft^3	0.028317	m^3
ft-lb	1.35582	J
in.	2.5401	cm
$in.^3$	16.387	cm^3
J	0.73756	ft-lb
kg	2.20462	lb
km	3280.8	ft
km/hr	0.62137	mi/hr
kPa	0.14504	$lb/in.^2$
lb	4.4482	N
lb/ft^2	6.9444×10^{-3}	$lb/in.^2$
$lb/in.^2$	6894.8	Pa
m	3.28083	ft
m^3	35.3147	ft^3
mi/hr	1.6093	km/hr
N	0.22481	lb
Pa	1.4504×10^{-4}	$lb/in.^2$

Directory of Organizations and Companies Related to Formwork for Concrete

Organizations

American Concrete Institute (ACI)
Box 19150, Redford Station
Detroit, Michigan 48219
Ph (313)532-2600 Fax (313)538-0655
ww.aci-int.org

American Forest & Paper Association (AF&PA)
National Design Standard (NDS)
1111 19th Street, N.W., Suite 800
Washington, D.C. 20036
Ph (202)463-4713 Fax (202)463-2791
www.afandpa.org

APA—The Engineered Wood Association (APA)
7011 S. 19th Street, P.O. Box 11700
Tacoma, Washington 98466-5333
Ph (253)565-6600 Fax (253)565-7265
www.apawood.org

American Institute of Timber Construction (AITC)
012 S. Revere Parkway, Suite 140
Englewood, CO 80112
www.aitcglulam.org

Canadian Wood Council (CWC)
99 Bank Street, Suite 400
Ottawa, Canada K1P 6B9
Ph (613)247-7077   Fax (613)247-7856
www.cwc.ca

Insulating Concrete Forms Association (ICFA)
1298 Cronson Boulevard, Suite 201
Crofton, Maryland 21114
Ph (888)854-4232   Fax (410)4518343
www.form.org

Portland Cement Association (PCA)
5420 Old Orchard Road
Skokie, Illinois 60017
Ph (847)966-6200   Fax (847)966-8389
www.cement.org

Scaffolding, Shoring, and Forming Institute (SSFI)
1300 Sumner Avenue
Cleveland, Ohio 44115
Ph (216)241-7333   Fax (216)241-0105
www.ssfi.org

Southern Pine Inspection Bureau (SPIB)
4709 Scenic Highway
Pensacola, Florida 32504
Ph (850)434-2611   Fax (850)433-5594
www.spib.org

Steel Joist Institute (SJI)
196 Stonebridge Drive, Unit 1
Myrtle Beach, South Carolina 29588
Ph (843)293-1995   Fax (843)293-7500
www.steeljoist.org

West Coast Lumber Inspection Bureau (WCLB)
6980 S.W. Varns
Tigard, Oregon 97223
Ph (503)639-0651   Fax (503)684-8928
www.wclib.org

Western Wood Products Association (WWPA)
522 Southwest 5th Avenue, Suite 500
Portland, Oregon 97204-2122
Ph (503)224-3930   Fax (503)224-3934
www.wwpa.org

Companies

Ceco Concrete Construction Company
10110 N. Ambassador Drive, Suite 400
Kansas City, Missouri 64153
Ph (816)459-7000   Fax (816)459-7135
www.cecoconcrete.com

Dayton Superior Corporation
7777 Washington Village Drive, Suite 130
Dayton, Ohio 45459
Ph (888)977-9600   Fax (937)428-9560
www.daytonsuperior.com

Deslauriers, Inc.
1245 Barnsdale Road
LaGrange Park, Illinois 60526
Ph (800)743-4106   Fax (877)743-4107
www.deslinc.com

Doka Formwork
214 Gates Road
Little Ferry, New Jersey 07643
Ph (201)329-7839
www.dokausa.com

EFCO Economy Forms Corporation
1800 N.E. Broadway Avenue
Des Moines, Iowa 50313-2644
Ph (515)266-1141   Fax (515)313-4424
www.efcoforms.com

Ellis Construction Specialties, Ltd.
12407 Holmboe Avenue
Oklahoma City, Oklahoma 73114-8128
Ph (800)654-9261   Fax (405)848-4474
www.ellisforms.com

Form-Eze Systems, Inc.
6201 Industrial Way Southeast, P.O. Box 3646
Albuquerque, New Mexico 87190
Ph (505)877-8100   Fax (505)877-8102

Gates & Sons, Inc.
90 South Fox Street
Denver, Colorado 80223
Ph (800)333-8382   Fax (303)744-6192
www.gatesconcreteforms.com

Greenstreak Plastic Products Company, Inc.
3400 Tree Court Industrial Boulevard
St. Louis, Missouri 63122-6614
Ph (800)325)9504 Fax (800)551-5145
www.greanstreak.com

Molded Fiber Glass Concrete Forms (MFG)
1018 West Sycamore Street
Independence, Kansas 67301
Ph (800)225-5632 Fax (620)331-8647
www.mfcgp.com

Patent Construction Systems
One Mack Centre Drive
Paramus, New Jersey 07652
Ph (800)969-5600 Fax (201)261-5600
www.pcshd.com

PERI Formwork Systems
7135 Dorsey Run Road
Elkridge, Maryland 21075
Ph (410)712-7225
www.peri-use.com

H. H. Robertson Company
450 19th Street
Ambridge, Pennsylvania 15003
Ph (442)299-8074
www.hhrobertson.com

Safway Steel Products
N19 W24200 Riverwood Drive
Waukesha, Wisconsin 55188
Ph (800)558-4772
www.safway.com

Scanada International, Inc.
8 Robinson Road
Bow, New Hampshire 03304
Ph (603)229-0014
www.scanda.com

Sonoco Products Company
P. O. Box 160
Hartsville, South Carolina 29550
Ph (800)337-2692
www.sonoco.com

Symons Corporation
2400 Arthur Avenue
Elk Grove Village, Illinois 60007
Ph (800)800-7966
www.symons.com

Titan Formwork
7855 S. River Parkway, Suite 105
Tempe, Arizona 85284
Ph (480)305-1900
www.titanformwork.com

Waco Formworks
800 Westchester Avenue, Suite 641N
Rye Brook, New York 10573
Ph (914)872-4000
www.wacoforms.com

Index

9 780071 639170